Processes and Systems

in

Industrial Chemistry

PRENTICE-HALL INTERNATIONAL SERIES
IN THE PHYSICAL AND CHEMICAL ENGINEERING SCIENCES

AMUNDSON *Mathematical Methods in Chemical Engineering: Matrices and Their Application*
ARIS *Elementary Chemical Reactor Analysis*
ARIS *Introduction to the Analysis of Chemical Reactors*
ARIS *Vectors, Tensors, and the Basic Equations of Fluid Mechanics*
BOUDART *Kinetics of Chemical Processes*
CROWE, ET AL. *Chemical Plant Simulation*
FREDRICKSON *Principles and Applications of Rheology*
HAPPEL AND BRENNER *Low Reynolds Number Hydrodynamics: With Special Applications to Particulate Media*
HIMMELBLAU *Basic Principles and Calculations in Chemical Engineering, 2nd Edition*
HOLLAND *Multicomponent Distillation*
HOLLAND *Unsteady State Processes with Applications in Multicomponent Distillation*
KOPPEL *Introduction to Control Theory with Applications to Process Control*
LEVICH *Physicochemical Hydrodynamics*
MEISSNER *Processes and Systems in Industrial Chemistry*
PETERSEN *Chemical Reaction Analysis*
PRAUSNITZ *Molecular Thermodynamics of Fluid-Phase Equilibria*
PRAUSNITZ AND CHUEH *Computer Calculations for High-Pressure Vapor-Liquid Equilibria*
PRAUSNITZ, ECKERT, ORYE, O'CONNELL *Computer Calculations for Multicomponent Vapor-Liquid Equilibria*
WHITAKER *Introduction to Fluid Mechanics*
WILDE *Optimum Seeking Methods*

PRENTICE-HALL, INC.
PRENTICE-HALL INTERNATIONAL, INC., UNITED KINGDOM AND EIRE
PRENTICE-HALL OF CANADA, LTD., CANADA

Processes and Systems in Industrial Chemistry

H. P. MEISSNER

Professor of Chemical Engineering
Massachusetts Institute of Technology

PRENTICE-HALL, INC.

Englewood Cliffs, N.J.

PRENTICE-HALL INTERNATIONAL, INC., *London*
PRENTICE-HALL OF AUSTRALIA, PTY. LTD., *Sydney*
PRENTICE-HALL OF CANADA, LTD., *Toronto*
PRENTICE-HALL OF INDIA PRIVATE LIMITED, *New Delhi*
PRENTICE-HALL OF JAPAN, INC., *Tokyo*

Current printing (last digit):
10 9 8 7 6 5 4 3 2 1

Library of Congress Catalog Card Number: 79-130374

13-723320-5

Printed in the United States of America

Preface

Industrial chemistry deals with the methods and reactions for making the products of the process industries on a large scale. Study of this enormous field is generally undertaken in either of three ways, each having advantages and disadvantages. In one approach, which might be called encyclopedic, coverage of the entire field is attempted. This approach provides the broad background which is so necessary for effective action but runs the danger of being superficial. It is difficult to present a meaningful survey, because of both the enormous number of processes involved and the fact that each is complex, being the result of much ingenuity and invention often reflecting decades of technical evolution. In a second approach, a few processes are selected for study in depth. This method insures rigor but may become parochial, since opportunites for developing and illustrating principles of general applicability are limited. In the third approach, reactions are grouped into the so-called unit processes such as oxidations, hydrations, nitrations, etc. The patterns uncovered are again valid and helpful, but the focus is on the reactor rather than on the overall flowsheet. Often, important reactions are not readily amenable to such classification.

None of the three approaches described is to be disparaged, especially since professional activity in the field benefits from each. The student, however, continues to have difficulties in bringing order out of what he calls the chaos of industrial chemistry. To help reduce these difficulties, I have made an attempt in this book to indicate that processes might also be classified by similarities in (1) types of equilibria, (2) types of energy management problems, (3) types of rate problems (including catalysis), and finally (4) types of flow sheet patterns and materials handling problems. The object here, of course, is only incidentally to classify and describe. Of primary importance is the development of a capacity to do effective and original thinking in the field, and such professional competence requires more than the absorption of a vast mass of factual material. Active work is essential on case-type problems in which are presented a little of the hurly-burly of the industrial world with its conflicting requirements, its superabundance of data or lack of data, and its need for judgment and invention. Some of the problems presented therefore call for analytical thinking; others require an imaginative approach to process thinking and to the solutions of process difficulties. These problems represent the heart of this book to which the text material is mainly supplementary.

As in industry, the mathematical preparation required to deal with these problems is not formidable, extending no further than calculus. The reader should have a good background in physical chemistry and thermodynamics. A certain amount of review material is included, particularly in kinetics and electrochemistry, since these fields are usually not intensively discussed in courses in physical chemistry. Again, a passing familiarity with chemical plant equipment such as heat exchangers, stills, and the like is assumed, but no more familiarity than a college senior or graduate student might have is needed.

I have made an attempt in this book to provide enough background to prepare the student for the great differences between an industrial plant and a university laboratory. That is, the student is accustomed to the simplicity and versatility of beakers, Erlenmeyer flasks, glass filters, etc., and is always surprised by the high degree of specialization of industrial equipment and processes. The care with which each piece of industrial equipment is designed for its task reflects the pressure to attain lowest possible production costs, which must be at least comparable with those of competitors' plants. In consequence, the industrial chemist is vitally concerned with the material efficiency of his plant, which will consume thousands and even millions of dollars worth of raw material during its lifetime, and also require large expenditures for supplies of various types for labor, power, water, etc. Design of an efficient plant obviously requires a far-reaching knowledge of the basic chemistry, physical chemistry of the process including fluid flow behavior, mass and heat transfer rates, reaction rates, and whatever else may be directly relevant to the specific process under study. The cost of the research and development effort required to obtain such knowledge must be balanced against the potential profits in prospect and, especially for larger installations, the cost of such studies are easily justified.

The difficulties of integration of a process, so that all its parts are kept in proper balance, are again novel to the student. For example, in continuous production of a material like ethylene by propane cracking, the heat addition to the reactor must be in balance with the feed, which contains both fresh and recycled raw material. These operations must all be properly matched to product cooling, to the stills for separating the product from the unconverted raw materials, etc. This is quite unlike a laboratory synthesis, where each stage is carried out in sequence and at leisure, receiving the full attention of the investigator. The preparation of a properly integrated flowsheet may be viewed as the chemical counterpart of the "systems engineering" function of the electrical and the mechanical industries.

The student often expects to find industrial chemistry to be somehow different than other kinds of chemistry. This is not the case, save that attention is focused on reactions which produce the products of industrial interest. Of course it often proves necessary to develop new reactions because of economic pressures to use new raw materials or to improve product characteristics. For example, the Solvay process for making soda ash was developed in a competition fostered by the French Government in the latter part of the eighteenth century for a process starting with limestone and common salt. Again, Haber's synthesis of ammonia from the elements by high pressure catalysis is an example of unorthodox operations which had not been studied previously. It was the increasing demand for ammonia and the inadequacy of the existing supply from coal distillation which provided the incentive for this pioneering effort. Similarly, the industrial chemist is always under pressure to find cheaper raw materials. To find a process for making ethylene glycol from ethylene which does not consume chlorine, the industrial chemist developed the ethylene oxide route, a direct oxidation which now competes successfully with the older ethylene chlorhydrin route. Often he develops a process which is basically not suited to small scale operation and so would not be used in the laboratory. The production of acetylene from hydrocarbons in a high temperature arc, followed by quench and recovery by scrubbing, is an obvious example.

The foregoing indicates that the industrial chemist often deals with reactions which are normally not of concern to the laboratory chemist. Many of the reactions of the so-called petrochemicals industry would not have been considered possible to the classic organic chemist of several decades ago. Examples here are the direct oxidation of cyclohexane to adipic acid, the chlorination of hydrocarbons with an oxygen–chlorine mixture to avoid the production of by-product hydrogen chloride, alkylation reactions such as the direct addition of ethylene to benzene to form ethyl benzene, and aromatizing reactions whereby normal paraffins are simultaneously dehydrogenated and converted to aromatics. The list can be extended to considerable length.

Often these reactions require a catalyst, especially in organic operations where the need for giving direction to reactions is particularly great. There is less need, or at any rate less use for, catalysis in inorganic industries, if the classic application of catalysis in SO_2 oxidation, ammonia synthesis, and ammonia oxidation to the nitrogen oxides is excluded. Again, the field of process metallurgy has so far found little application for catalysis. This should not be interpreted as an indication of backwardness, since process metallurgy deals with reactions at high temperature or with ionic reactions in aqueous solution where rates are usually great. Catalysis, nevertheless, is industrial chemistry's most important frontier and accounts for the space devoted here to this subject.

A further general responsibility of the industrial chemist is the modification of materials to fit changing specifications and use requirements. Problems are especially challenging in the fields of polymeric materials, namely plastics, textile fibers, paper, etc., and also in the production of composites such as laminates, paints, lubricants, etc. However, no field seems free from the problems of product modification, since use specifications are constantly changing. These areas are left to consideration of texts more specialized than this one.

An industrial chemist, therefore, is concerned with the methods and reactions for making chemicals on a large scale. His academic training is usually either in chemistry or chemical engineering, but his success in the field is dependent not so much on background as on his resourcefulness, ingenuity, and capacity to improvise and invent. The field is remarkably dynamic, changing constantly, and derives much of its fascination from the continuing challenges presented by its new problems and opportunities.

H. P. MEISSNER

Contents

Nomenclature

a	Stoichiometric coefficient in reactions (1-1), (2-2), and the like; also Activity, as defined in Eq. (2-1). Dimensionless.
A	Reaction partner in reactions (1-1), (2-2), and the like; also Area, usually in sq ft.
b	Stoichiometric coefficient in reactions (1-1), (2-2), and the like. Dimensionless.
B	Reaction partner in reactions (1-2) and (2-2); a constant in Eq. (8-20).
c	Heat capacity in cal/(g) (°C) or Btu/(lb) (°F); stoichiometric coefficient in reaction (1-1), etc.; c_A is concentration of any component A in units such as lb moles/cu ft or g moles/liter.
C	Reaction partner in reactions (1-1) and (2-2).
C	Number of Phase Rule components in Eq. (2-6).
D	Reaction partner in reactions (1-1) and (2-2); diffusion coefficient in Eq. (8-22) in units such as cm²/sec.
E	Activation energy, cal/g mole; cell voltage, with E° the standard state voltage.
f	Fugacity; also fraction conversion as in example 5-4.
$\mathcal{F}$	Faraday number, namely 96,454 amp sec/g equivalent, 23,010 cal/(V) (g equivalent), or 12,163 amp hr/(lb equivalent).
F	Feed rate to system, lb moles/sec.
ΔF°	Standard free energy change, usually in cal/g mole, or kcal/g mole, as in Eq. (1-10).
G	Constant, Eq. (8-25).
ΔH	Enthalpy change, usually in kcal/g mole, as in Eq. (1-10); ΔH° is the standard enthalpy change of a chemical reaction.
k	Reaction rate constant defined in Eqs. (6-4) and (7-9); also thermal conductivity as in Eq. (8-9), expressed in Btu/(hr) (sq ft) (°F)/ft.

k'	The effective coefficient of mass transfer through a porous solid, as defined in Eq. (8-3).
K	Equilibrium constant; K_a is the equilibrium constant expressed in terms of activities as in Eq. (1-2) and (2-3); K_p is expressed in terms of partial pressures, atm as in Eq. (1-6) and (2-4); K_B^A is the selectivity of an ion exchange resin as in Eq. (3-9).
L	Length, ft.
m	Stoichiometric coefficient for Eqs. (1-1) and (2-2).
M	Molecular weight; M_C referring to the core material of a solid particle as in Eq. (8-1); also reaction partner in reactions (1-1) and (2-2). MC_p is molal heat capacity.
n	Stoichiometric coefficient for reactions (1-1) and (2-2); also number of moles of a component present in a system, or flowing into a system, or formed within a system; $\dot{n}$ is moles reacting on a surface per unit time as in Eq. (8-20).
N	Reaction partner as in Eqs. (1-1) and (2-2); also particle number in Eq. (8-27a).
$\boldsymbol{N}$	Number of species in a system, as in the Phase Rule Eq. (2-6).
p	Partial pressure in atm; p_{ref} is partial pressure in the reference state; $p°$ is vapor pressure of a pure material; p_c in Eq. (8-8) is the partial pressure of a component at the surface of a particle core; p_∞ is the partial pressure in the gas phase a great distance from the particle surface.
$\boldsymbol{P}$	Number of phases in an equilibrium system, as in Eq. (2-6).
q	Heat flow rate, in Btu/hr. In Eq. (7-16), q_{chem} is heat generated "by an exothermal chemical reaction"; $q_{sensible}$ is the sensible heat which must be supplied; q_{cool} is cooling rate as in Eq. (7-17).
r	Rate of reaction, in lb moles/(hr) (cu ft) as in Eq. (6-1), or in lb moles/(hr) (lb catalyst) as in Eq. (7-7); $r_{surface}$ is reaction rate, lb moles/(unit surface area) (unit time).
R_g	Gas constant 1.987 Btu/(lb mole) (°R), or 1546 ft lb/(lb mole) (°R), or 1.987 cal/(g mole) (°R), etc.
R	Radius of pores as in Eq. (7-1), or radius of pellets as in Eq. (8-1). R_0 indicates initial radius of a spherical particle, R_c is the radius of a core of unreacted solid within this particle at any time θ, while R_{cf} is the final core radius.
$\mathbf{R}$	Number of independent chemical reactions in a system, Eq. (2-16).
S	Moles of a stream per mole of feed, after reaction occurs, as in Eq. (6-15).
$\boldsymbol{S}$	Number of stoichiometric relations in a system, as in Eq. (2-6).
SV	Space velocity, defined in example 9-1.
$\Delta S°$	Standard entropy change of a chemical reaction.
t	Steady state reactor temperature as in Eq. (7-15), in absolute degrees.
T	Absolute temperature, °K or °R as the case might be. T_{FW} is the value of feed and cooling fluid to a kettle, Eq. (7-20).
U	Overall heat transfer coefficient, Btu/(hr) (sq ft) (°F).
v	Specific volume, cu ft/(lb mole).
V	Volume of system or reactor, cu ft.
$\underline{V}$	Variance of a system, as in Eq. (2-7).

W	Weight of solid, either reagent or catalyst, usually in lb.
X	Abscissa on Janecke diagram, such as Fig. 3-4.
x	Mole fraction often referring to liquid phase, as in Eq. (3-13).
y	Mole fraction, often referring to the vapor phase, as in Eq. (3-13).
Y	Moles of key component present at any time in a reacting system, often based on 1 mole of feed; also abscissa on Janecke diagram, such as Fig. 3-4.
Z	Compressibility factor, namely R_qT/pV. Dimensionless.

Greek Symbols

α	Empirical exponent of Eq. (6-5).
β	Empirical exponent of Eq. (6-5).
γ	Activity coefficient, as in Eq. (10-8) etc.; empirical exponent of Eq. (6-5).
Δ	Finite difference between final and initial quantity, such as enthalpy or free energy.
δ	Change in the number of moles of a reaction, defined by Eq. (1-7).
ξ	Defined in Eq. (1-13).
η	Effectiveness factor in Eq. (7-5); also empirical exponent of Eq. (6-10).
ρ	Density, as in lb/cu ft of catalyst.
π	Total pressure, atm.
Σ	Summation.
τ	Residence time, Eq. (6-34).
θ	Time in sec, min, or hr depending on circumstances.
ψ	Temperature interval over which reaction rate doubles, usually in °C, as in example 6-1.
μ	Empirical exponent of Eq. (6-10).

Equilibrium in Homogeneous Systems 1

Industrial chemistry is the study of the large-scale manufacture of materials whose production involves chemical reactions. The field is vast, for it includes not only the so-called industrial chemicals, but also plastics, paper, the metals, glass, synthetic fibers, and others. Industrial chemistry is therefore the study of the process industries.

In industrial chemistry, attention is focused on both the processes operated and the systems of apparatus used in manufacture. A process may be defined as an integrated series of reactions and associated operating steps, such as mixing, separation by distillation, and the like, whereby available raw materials are converted into a desired product. A process is carried out in a system of apparatus, through which materials travel in accordance with the processing plan. A schematic representation of the route so traveled is called a flow sheet, while the aggregation of equipment itself constitutes a plant. The creation, modification, and analysis of processes and their associated systems of equipment form the subject matter of this book, with illustrations and examples largely drawn from the field of industrial chemicals.

Alternatives in manufacture. Characteristically, a product can be made industrially by any of several alternative processes involving different reactions and systems of apparatus, depending on the raw materials available and the chemical route chosen. To illustrate the use of alternative raw materials with a single example, acetaldehyde is made from acetylene by hydration using an aqueous solution of mercury salts as catalyst, from ethylene by oxidation using an aqueous solution of palladium salts as catalyst, and from butane by partial oxidation. Similarly, two different chemical routes from the same raw material to the same product and therefore involving different reactions and different processing systems may be illustrated by ethanol manufacture from ethylene. The conversion to alcohol is carried out directly by passing a mixture of ethylene and steam at high pressure over a phosphoric acid catalyst. Alternatively, ethylene combines at low pressure with sulfuric acid to form ethyl hydrogen sulfate, which is then hydrolyzed to the alcohol. Finally, there are often many possible alternatives in the systems of apparatus used in carrying out conversion of a given raw material into a given product by a given set of chemical reactions. Thus, in making ammonia from hydrogen-nitrogen gas mixtures at high pressure, the reacting gases are passed through a series of adiabatic catalyst beds. Cooling between stages is necessary since this reaction is vigorously exothermal. One such method of cooling is by inserting water-cooled heat exchanger surfaces in the ducts between the reactors to generate process steam, while another involves injecting suitable quantities of cold fresh feed gas into the hot gases between reactors.

The existence of alternatives such as the ones mentioned is characteristic of industrial chemistry, and their discovery and elaboration afford extensive opportunities for innovation, ingenuity, and invention. Moreover, because of continual changes in technology and raw material costs in a dynamic economy the best processing system for manufacture at any given location today does not necessarily remain the best. Thus, the recent development of the catalytic one-step conversion of propylene, ammonia, and air into acrylonitrile has made the older syntheses from acetylene and hydrogen cyanide and from ethylene via the cyanohydrin intermediate appear much less attractive. The advent of this new acrylonitrile process illustrates the general trend in the petrochemicals industry toward the use of cheaper feedstocks and simpler but more sophisticated chemical processes. Analogous changes are, of course, occurring throughout the field of industrial chemistry.

Reactions of industrial chemistry. While the study of a processing system is conveniently started with the chemical reactions occurring, the key difficulties of the operation do not necessarily lie in the reactors. That is, a typical process might involve purification and mixing of raw materials, followed by reaction and then by recovery, separation, and purification of products. There are always problems associated with the process such as supplying or removing energy, recycling unused reagents in the product stream from the reactor, and others. It is not uncommon to find that the investment in product recovery and purification equipment in a plant is manyfold greater than for the reactors. The broad challenge is to assemble and arrange an equipment system of high efficiency which will respond sensitively and without excessive demands of attention to unavoidable variations in raw material compositions, in temperatures of coolants and heat sources, in throughput, etc.

Basically, no new chemical or physical principles are encountered in industrial chemistry. The reactions of this area, however, are often not suitable as simple illustrations for a systematic development of general inorganic or organic chemistry, and so are not discussed in detail in elementary texts. Thus, ethylene is made in enormous quantities industrially by cracking light naphthas, but because of the complexity of both the reactions and product recovery, naphtha cracking is not a good introductory example to simple dehydrogenation. Again, the reactions of industrial chemistry are often not easily performed on a laboratory scale. In the ethylene process just mentioned, serious difficulties are encountered both in controlling the residence time of the reacting stream at elevated temperature and in product recovery; consequently, simpler preparative methods are preferred for bench scale studies, such as the dehydration of ethanol with sulfuric acid. Finally, particularly in the organic chemistry field, the student is frequently not familiar with the catalytic processes being operated which sometimes involve highly unorthodox chemistry, such as in the direct oxidation of ethylene with air over a silver catalyst to ethylene oxide.

Novelty in the field of industrial chemistry lies primarily in its focus on reactions of commercial interest and the associated physical chemistry of the mixing and separation processes. The intensity with which these matters are studied in industry is often surprising to the newcomer, but such studies of general chemistry, equilibria, thermal effects, kinetics, etc., is essential for the design and operation of a processing system having lowest costs of construction and operation. These detailed technical studies are costly, but are readily justified if even small improvements in performance are attainable. A chemical plant often represents a multimillion dollar investment and it will be supplied during its lifetime with many more millions of dollars of raw material, along with associated large expenditures for labor, operating supplies, power, and fuel. The benefits of studies leading to optimum system design are thus obvious.

Initial orientation. Study of a processing system is usually begun with the chemical reactions occurring, since these help fix much of the system's stoichiometry. Three key questions are asked:

1. Over what range of temperature, pressure, and reactant feed compositions are the chemical equilibria such that a favorable concentration of the desired product is attainable in the reactor outlet streams?
2. Within this range of conditions, where do reaction rates become favorable?

3. What are the energy requirements of the operation?

The operating conditions involved must obviously be reasonably attainable. For example, as temperature is increased, undesired side reactions often become more important, corrosion becomes more severe, materials of construction rapidly lose strength, etc. Because of such difficulties, the highest temperatures encountered in industry today are around 3000°F with much lower temperatures common in most operations. Even when a reaction is technically feasible, economics are the overriding determinant of its industrial success. Thus, the equilibrium concentrations of a product become "favorable" as required in question 1 above when product recovery can be undertaken without excessive costs. At the "favorable" rates of question 2, a reasonable fraction of equilibrium conversion must be attainable without excessive residence time in the reactor. "Favorable conditions," therefore, involve operations at competitive costs whose attainability depends on the technical feasibility of the processing system. Technical feasibility, which must be demonstrated before costs and optimization studies can be undertaken, is the subject of this book and is introduced in this first chapter with a brief review of energy balances and chemical equilibrium.

Knowing the raw materials to be used, it is usually possible to anticipate the range of feed composition to the reactors in which the reactions under study are to be carried out. Regardless of the design of these reactors, the corresponding compositions of the product streams can also often be anticipated. Thus, if reaction rates are known to be high, then it is reasonable to assume that equilibrium is approximated in the product stream leaving a reactor, as in the air oxidation of SO_2 to SO_3 over a vanadium catalyst. When rates are low, then some fractional approach to equilibrium may be assumed, as in the first bed of ammonia converters being installed today, in which about 50% of the equilibrium conversion at off-gas conditions is often attained. Of course, when sequential reactions are involved such that the desired product made by one reaction is destroyed by a second reaction, then a guess as to attainable yields is more difficult to make without rate data. The compositions of streams entering and leaving a reactor can, however, often be anticipated in this manner without undertaking detailed reactor design. Preliminary process analysis is consequently simplified, and the requirements of the rest of the process are readily determined.

A key step in any process analysis, regardless of whether preliminary or not, is therefore the determination of the equilibrium compositions theoretically attainable in the reactor product streams. In this first chapter, discussion centers largely around the calculation of such equilibrium compositions for homogeneous systems, that is, for systems of one liquid or one gas phase. The reader is already familiar with such calculations when only one reaction occurs; consequently, this case is given only brief review in the next few paragraphs. In later parts of this chapter, discussion is presented of the reactor feed composition required to maximize the equilibrium concentration and yield of the desired product of a single reaction. Finally, a technique is presented for determining the equilibrium composition of a system in which several reactions occur simultaneously.

Chemical equilibrium. For the typical generalized reaction,

$$aA + bB + cC = mM + nN \tag{1-1}$$

the equilibrium constant is stated as

$$K_a = \frac{a_M^m a_N^n}{a_A^a a_B^b a_C^c} \tag{1-2}$$

where a, the activity of each reaction partner, is defined as

$$a = \frac{f}{f_{\text{ref}}} \tag{1-3}$$

Here f is the fugacity of the reaction partner in the mixture under study, while f_{ref} is this material's fugacity when in its standard state, namely when pure at 1 atm pressure and at the temperature of the mixture.* In this chapter, only those reaction mixtures are considered for which all partners are gaseous at these standard state conditions. In the region where perfect gas behavior is approximated, namely at low and moderate pressures, the fugacity and pressure of a gas are very nearly equal; consequently, at the standard state pressure of 1 atm, f_{ref} can safely be taken as unity. When the reaction partner's pressure in the mixture under study is also low enough to lie in the perfect gas region, then Eq. (1-3) becomes

$$a = \frac{p}{p_{\text{ref}}} \tag{1-4}$$

where p_{ref} is unity. As implied in this discussion, pressures used in calculations of activities are always expressed in atmospheres in this book.

Being a ratio as shown in Eq. (1-3), the activity is dimensionless; consequently, the equilibrium constant K_a stated in terms of activities in Eq. (1-2) is likewise dimensionless. For a reacting mixture in the perfect gas law region, Eq. (1-2) and (1-4) can be combined to give

$$K_a = \frac{\dfrac{p_M^m p_N^n}{p_A^a p_B^b p_C^c}}{\dfrac{(p_M^m)_{\text{ref}}(p_N^n)_{\text{ref}}}{(p_A^a)_{\text{ref}}(p_B^b)_{\text{ref}}(p_C^c)_{\text{ref}}}} \tag{1-5}$$

*While this standard state definition commonly applies to gases and solids, it is often not chosen for the solutes when dealing with real solutions. This matter is further discussed in texts on thermodynamics.

The denominator of the term on the right side of Eq. (1-5), involving the standard state pressures of the reaction partners, is numerically unity. The numerator is K_p, the equilibrium constant expressed in terms of partial pressures:

$$K_p = \frac{p_M^m p_N^n}{p_A^a p_B^b p_C^c} \tag{1-6}$$

The term K_p has the dimensions of pressure raised to the power δ, where

$$\delta = (m + n - a - b - c) \tag{1-7}$$

In accordance with common practice, pressure is expressed in atmospheres when calculating K_p in this book. Inspection shows that the numerator of the term on the right side of Eq. (1-5), namely K_p, has the same dimensions as the denominator. Since the numerical value of this denominator is unity, Eq. (1-5) could have been rewritten:

$$K_a = \frac{K_p}{1} \tag{1-8}$$

This equation holds only in the perfect gas law region.

It is surprisingly rare to encounter industrial gas reactions in which conditions are far enough outside the ideal gas range so that K_p cannot be used directly for equilibrium calculations. When in doubt, the compressibility factors of the gases at the operating temperature and pressure are calculated, with the use of fugacities rather than of the partial pressures being indicated when these compressibility factors deviate considerably from unity. For such a case, the equilibrium constant K_f is used whose form is identical with that for K_p in Eq. (1-6), except that fugacities are substituted for partial pressures.

The equilibrium composition of a homogeneous gas phase or liquid phase system, in which only one reaction occurs, is readily calculated stepwise as in the following example 1-1. Choosing any convenient number of moles of feed mixture as a basis, the moles of each constituent of the feed are listed in a mole table. The moles of an arbitrarily chosen key component present at equilibrium is next denoted as Y, and the moles of the other components in the equilibrium mixture are then expressed in terms of Y and listed in the mole table. The equilibrium constant expression Eq. (1-6) is restated in terms of these quantities, and the resultant equation is solved.

EXAMPLE 1-1

Assume that a selective catalyst becomes available, on which only the following reaction can occur:

$$CH_4 + H_2O = CO + 3H_2 \qquad (K_p)_{1100°F} = 0.41 \tag{1-9}$$

What is the mole fraction of CO in the equilibrium gas at 1100°F and 2 atm total pressure, when the feed contains 5 moles of steam/mole of methane?

Answer: Choosing CO as the key, then the mole table for this case becomes:

	Feed, moles	Product, moles
CH_4	1	$1 - Y$
H_2O	5	$5 - Y$
CO	0	Y
H_2	0	$3Y$
Total	6	$6 + 2Y$

In accordance with Eq. (1-6), the equilibrium constant for reaction (1-9) is written

$$K = \frac{p_{CO}p_{H_2}^3}{p_{CH_4}p_{H_2O}}$$

Restating this equilibrium constant in terms of the above mole table and designating the total pressure as π one gets

$$K = \frac{27Y^4\pi^2}{(1 - Y)(5 - Y)(6 + 2Y)^2}$$

Solving when π is 2 atm and K is 0.41, Y is found to be 0.6 moles, while the mole fraction of CO is $Y/(6 + 2Y)$, or 0.09.

Standard free energy changes. The standard free energy change of a chemical reaction, namely ΔF°, equals (minus sign) the reversible isothermal work of the reaction when the reactants originate in their standard states and the products end up in their standard states, all at the temperature of the reaction mixture. To be in its standard state at the mixture temperature, as stated earlier, each component must be pure and under a total pressure of 1 atm.

The relation between ΔF° and the equilibrium constant for a reaction is as follows:

$$\Delta F^\circ = -R_g T \ln K_a \tag{1-10}$$

The value of K_a for a chemical reaction is conveniently derived from Eq. (1-10), after first obtaining ΔF° by algebraically summing up the free energies of formation of the partners of the reaction under study. Values for the free energies of formation of a large number of compounds are available in the literature (29, 44, 62, 70, 130, 131, 157, 162) and when not available, methods of estimating these quantities can be used (122).

Enthalpies of reaction. The heat requirements of an operation are calculated directly from the enthalpy changes occurring by use of the First Law of Thermodynamics. The enthalpy change of a reaction is conveniently obtained from the algebraic sum of the enthalpies of formation of the reaction partners involved. It will be remembered that $\Delta H°$, like $\Delta F°$, is applicable for the change in which the reactants and products all originate and terminate in their standard states, namely as pure materials at the temperature of the system and under the standard state pressure, generally 1 atm. Often there is no significant difference between this standard enthalpy change and the actual enthalpy change involved for the isothermal change of state between the reagent mixture fed to the reactor and the product mixture. The assumption that $\Delta H°$ is equal to the actual enthalpy change of a reaction occurring in gas mixtures at lower pressures is a good first approximation, since the enthalpy changes upon mixing perfect gases at constant temperature are zero, and the change of enthalpy of a perfect gas upon isothermal expansion is zero.

EXAMPLE 1-2

A furnace is fired with methane and dry air (21% O_2, 79% N_2), the gases entering through separate ports at 25°C in a ratio of 10 moles of air/mole of methane. Mixing in the combustion space is poor, consequently there are large amounts of unburned combustibles in the off-gases, which leave the furnace at 900°K. An analysis of these stack gases after mixing is complete shows 1.0% CH_4, 4.0% O_2, 1.1% CO, 1.1% H_2, 8.5% CO_2, and 84.3% N_2. This analysis is on a "dry basis," that is, on a water-free basis. How much heat was transferred from the combustion gases to the furnace, in Btu/lb mole of CH_4 charged?

Answer: The general solution procedure involves calculation of the moles of the various product gas constituents per mole of methane fed. The total enthalpies of the air and methane fed, and of the flue gases at 900°K, are then determined, and the desired heat effect obtained from the difference of these two quantities. The moles of each component per 100 moles of dry flue gas are listed in the first column of Table 1-1. The water content can be calculated in two ways:

1. Hydrogen disappearance. Carbon is present in the flue gas as CH_4, CO, and CO_2 and per 100 moles of dry flue gas equals $(1 + 1.1 + 8.5)$ or 10.6 atoms. This carbon came from the methane feed, which supplied 4×10.6 or 42.4 atoms of hydrogen to the system. "Visible" hydrogen in the flue gas exists as CH_4 and H_2, namely $(4 \times 1 + 2 \times 1.1)$ or 6.2 atoms. Hydrogen disappearance is therefore $(42.4 - 6.2)$ or 36.2 atoms, equivalent to 18.1 moles of H_2O formed.

2. Oxygen disappearance. The oxygen in the air fed, by nitrogen balance, was $84.3 \times \frac{21}{79}$, or 22.4 moles. The "visible" oxygen in the flue gas is present as O_2, CO, and CO_2, namely $(4 + 1.1 \times \frac{1}{2} + 8.5)$ or 13.05 moles.

The total oxygen disappearance is therefore (22.4 − 13.05) or 9.35 moles equivalent to 18.7 moles of H_2O.

This agreement between water disappearance as calculated by these two independent methods is fairly good and consequently reassuring as far as reliability of the data is concerned. As a reasonable compromise, the water content associated with 100 moles of dry flue gas will be taken as (18.1 + 18.7)/2 or 18.4 moles.

Standard enthalpies of formation at 25°C for the components of the flue gas are listed in the second column of Table 1-1 while the sensible enthalpy changes at atmospheric pressure between 25–900°K are listed in the fourth column. Data on enthalpies of formation come from reference (70), in which the standard states for zero enthalpy are chosen as the pure elements in their normal states of aggregation at atmospheric pressure and 25°C.

TABLE 1-1. PER 100 MOLES OF FLUE GAS

	(1) Components, moles	(2)* Enthalpies of Formation, kcals/g mole at 25°C	(3) (1) × (2), kcal	(4)** Sensible Enthalpy, kcal/g mole	(5) (1) × (4), kcal
CH_4	1.0	−17.9	−17.9	7.5	7.5
O_2	4.0	0		4.6	18.4
N_2	84.3	0		4.4	371.0
CO	1.1	−26.4	−29	4.4	4.8
CO_2	8.5	−94.05	−800	6.7	57
H_2	1.1	9		4.2	4.6
Dry gas	100.0		−846.9		463.3
$H_2O(g)$	18.4	−57.8	−106.3	5.2	95.6
Wet gas	118.4		−953.2		558.9

*From reference (131).

**This represents the enthalpy change for each component at atmospheric pressure between 25–900°K; from reference (70).

From columns (2) and (5), the total enthalpy of 118.4 moles of wet flue gas is (−953.2 + 558.9) or −394.3 kcal. The enthalpy of wet flue gas per mole of methane fed, by nitrogen balance, is then −394.3 (7.9/84.3) or −37 kcal.

The enthalpy of the methane plus air feed equals the standard enthalpy of formation of m moles of methane at 25°C, or −17.9 kcal. By the usual energy balance, the heat effect, namely Q, is related to the enthalpy change of the gas system as follows:

$$Q = -37 - (-17.9) = -19.1 \text{ kcal/g mole } CH_4$$

The minus sign indicates that heat is removed from the reacting gases.

As an alternative to the method just described, the enthalpy change of a reaction can be determined from equilibrium data by the van't Hoff equation:

$$\frac{d \ln K_a}{dT} = \frac{\Delta H^\circ}{R_g T^2} \tag{1-11}$$

The reaction's standard enthalpy change, namely ΔH°, is zero only in rare cases, such as optical isomerizations, and is obviously a positive number for endothermal reactions and a negative one for exothermal reactions. This quantity usually changes little percentagewise with temperature, even over relatively large temperature changes; consequently, $\ln K$ will usually plot as a very nearly straight line against $(1/T)$, with a slope equal to $(-\Delta H^\circ/R)$. The van't Hoff equation, of course, is also used for determining the changes in the equilibrium constants with temperature.

The standard enthalpy change of a chemical reaction is related to the standard free energy change by the following equation:

$$\Delta F^\circ = \Delta H^\circ - T\,\Delta S^\circ \tag{1-12}$$

Here ΔS° is the standard entropy change for the chemical reaction and, like ΔH°, usually changes little with temperature. The value of ΔS° for a reaction is again conveniently determined by algebraically summing up the absolute entropies of the partners in the reaction. As for the standard free energies of formation, information on both absolute entropies and on the standard enthalpies of formation are available for many materials in the usual references (29, 44, 62, 70, 130, 131, 157, 162).

Optimum feed composition. There is always some factor which limits the rate at which reagents can be supplied to a reactor, such as pump or blower capacity, maximum flow velocity which will not blow catalyst out of the reactor, etc. At all feed rates to a reactor, including the maximum, the feed composition can usually be varied over wide limits. Assuming reaction rates to be great enough so that equilibrium is attained at all these flow rates, then several questions arise (24, 116): for a given reactor input rate, what feed composition will give the maximum moles of product per mole of feed? What feed composition will result in the maximum product concentration in the off-gas? How are these optimum feed compositions affected as the flow rate changes? The answers are simple for isothermal reactors in which a single reaction occurs: when equilibrium is attained, then regardless of flow rate, the reagents in the feed should be in the stoichiometric ratio as required by the overall chemical reaction occurring for maximized output. This will now be demonstrated.

Consider an isothermal flow reactor operating in steady state, in which the gas phase reaction (1-1) occurs. The feed stream to this reactor enters at F moles per unit time, with mole fractions of A, B, C, and inerts (such as nitrogen) of α, β, σ, and i. The off-gases, which are at equilibrium at the temperature at which they leave the reactor, contain Y moles of the desired product M per mole of feed. Designating a, b, c, m, and n as the coefficients of Eq. (1-1), then the following mole table applies per mole of feed.

TABLE 1-2. MOLE TABLE FOR REACTION (1-1)

	Feed, moles	Product, moles
A	α	$\alpha - \frac{a}{m}Y$
B	β	$\beta - \frac{b}{m}Y$
C	σ	$\sigma - \frac{c}{m}Y$
M	0	Y
N	0	$\frac{n}{m}Y$
Inerts	i	i
Total moles, ξ	1.0	$1 + \frac{\delta}{m}Y$

where $\delta = (m + n - a - b - c)$. Defining ξ as the total moles of gas product per mole of feed,

$$\xi = 1 + \frac{\delta}{m}Y \tag{1-13}$$

In the perfect gas region, the equilibrium constant becomes

$$K_p = \frac{Y^m\left(\frac{n}{m}Y\right)^n}{\left(\alpha - \frac{a}{m}Y\right)^a\left(\beta - \frac{b}{m}Y\right)^b\left(\sigma - \frac{c}{m}Y\right)^c\left(1 + \frac{\delta}{m}Y\right)^\delta} \tag{1-14}$$

and

$$1 = \alpha + \beta + \sigma + i \tag{1-15}$$

The maximum value of Y can now be determined by the Lagrangian method of undetermined multipliers. Take the logarithm of Eq. (1-14), differentiate at constant π, and set dY equal to zero:

$$\frac{a\,d\alpha}{\alpha - \frac{a}{m}Y} + \frac{b\,d\beta}{\beta - \frac{b}{m}Y} + \frac{c\,d\sigma}{\sigma - \frac{c}{m}Y} = 0 \tag{1-16}$$

Differentiate Eq. (1-15) after recognizing that the amount of inerts present, namely i, is constant at some arbitrarily chosen level:

$$0 = d\alpha + d\beta + d\sigma \tag{1-17}$$

Multiply Eq. (1-17) by the constant quantity λ, add the result to Eq. (1-16), and set the coefficients of $d\alpha$, $d\beta$, and $d\sigma$ equal to zero:

$$\frac{a}{\alpha - \dfrac{a}{m}Y} + \lambda = 0; \qquad \frac{b}{\beta - \dfrac{b}{m}Y} + \lambda = 0; \qquad \frac{c}{\sigma - \dfrac{c}{m}Y} + \lambda = 0$$

Solving,

$$\frac{\alpha}{\beta} = \frac{a}{b}; \qquad \frac{\alpha}{\sigma} = \frac{a}{c}$$

It follows, then, that the maximum moles of product per mole of feed are obtained when the reagents in the feed are present in their stoichiometric ratios. Note that this derivation is independent of whether perfect gases or perfect solutions are being handled, of the feed rate, and of the amounts of inert component or solvent present.

Feed for maximum product concentration. It is sometimes desirable to adjust the feed composition to attain maximum mole fraction of the product in the off-gas, rather than to attain maximum yield. For a single reaction at equilibrium, however, both these requirements of maximum mole fraction of product and maximum product yield are simultaneously satisfied with stoichiometric feed, as is now demonstrated. Using the nomenclature presented above, the mole fraction y of product M in the off-gases, namely Y/ξ, becomes

$$y_M = \frac{Y}{1 + \dfrac{\delta}{m}Y} \tag{1-18}$$

Differentiating with respect to α,

$$\frac{\partial y_M}{\partial \alpha} = \frac{\left(1 + \dfrac{\delta}{m}Y\right)\dfrac{\partial Y}{\delta \alpha} - \dfrac{\delta}{m}Y\dfrac{\partial Y}{\partial \alpha}}{\left(1 + \dfrac{\delta}{m}Y\right)^2}$$

When the mole fraction of M is a maximum, then $dy_M/\partial\alpha$ is zero, which by inspection of the foregoing equation occurs only when $dY/\partial\alpha$ is zero. Differentiating Eq. (1-18) in turn with respect to β and σ shows that $\partial Y/\partial\beta$ and $\partial Y/\partial\sigma$ must likewise be zero for maximum concentration of M. But these quantities $\partial Y/\partial\alpha$, $\partial Y/\partial\beta$, and $\partial Y/\partial\sigma$ are zero only when Y is a maximum, which occurs with stoichiometric feed. Of course, stoichiometric feed ratios may not be desirable in those cases in which one reagent is much costlier than the other. That is, by operating with a deficit of the costlier reagent, the improved yields obtained on this material may often more than offset the increased costs due to reduced output and more difficult recovery.

Multiple reactions. The calculation of equilibrium compositions in a system in which a single reaction occurs, as shown in example 1-1, is relatively straightforward. Example 1-1 is somewhat artificial, however, since it assumes a selective catalyst which will promote only reaction (1-9),

$$CH_4 + H_2O = CO + 3H_2 \qquad (K_p)_{1100°F} = 0.41 \tag{1-9}$$

without also catalyzing the equally important reaction (1-19),

$$CH_4 + 2H_2O = CO_2 + 4H_2 \qquad (K_p)_{1100°F} = 1.09 \tag{1-19}$$

The facts are that known catalysts promote reactions (1-9) and (1-19) simultaneously, causing both reactions to proceed. Formulation of the equations needed to determine the equilibrium compositions in systems such as this, in which more than one reaction occurs, is again straightforward. Algebraic complications in attaining a numerical solution, however, become increasingly formidable as the number of independent reactions increases.

EXAMPLE 1-3

A catalyst is available on which reactions (1-9) and (1-19) occur simultaneously. Using the same feed composition and operating conditions as for example 1-1, namely 5 moles of steam/mole of methane at 1100°F and 2 atm pressure, what is now the off-gas composition at equilibrium?

Answer: The following mole table applies when designating Y and Z as the moles of CO and CO_2 formed per 6 moles of mixed feed:

	Feed, moles	Product, moles
CH_4	5	$5 - Y - Z$
H_2O	1	$1 - Y - 2Z$
CO	0	Y
H_2	0	$3Y + 4Z$
CO_2	0	Z
Total	6	$6 + 2Y + 2Z$

The equilibrium constant expressions for reactions (1-9) and (1-19) are, respectively,

$$0.41 = \frac{Y(3Y + 4Z)^3\pi^2}{(5 - Y - Z)(1 - Y - 2Z)(6 + 2Y + 2Z)^2}$$

$$1.09 = \frac{Z(3Y + 4Z)^4\pi^2}{(5 - Y - Z)(1 - Y - 2Z)^2(6 + 2Y + 2Z)^2}$$

Solving these two equations by trial when the total pressure π is 2 atm, Y is found to equal 0.188 and Z to equal 0.594 in the equilibrium off-gas.

Solution of the two simultaneous equations resulting from the two reactions of example 1-3 is arduous. Algebraic difficulties rapidly become greater as the number of independent reactions increases further. Various methods of dealing with such equilibrium calculations have been proposed (161, 165) with a simple procedure (96), called the series reactor method, described below. The series reactor method is, of course, suited both to hand and machine calculations.

Independent reactions. The first step in determining the composition of an equilibrium system involving several reactions is to decide on the number of independent reactions to be considered. Thus, the system of example 1-3 involves reactions (1-9) and (1-19), which are clearly independent since neither can be derived from the other. If ethylene formation is also to be allowed, then a third independent reaction must be included:

$$2CH_4 = C_2H_4 + 2H_2 \qquad (K_p)_{1100°F} = 1.23 \times 10^{-6} \qquad (1\text{-}20)$$

Still other reactions might be considered, such as those giving ethane, solid carbon, and other compounds. Each new independent reaction obviously adds a new equilibrium constant equation and a new unknown to the system being studied.

In calculating a system's equilibrium composition, serious error results if a reaction is overlooked whereby a major constituent is produced. On the other hand, it is often reasonable to neglect minor constituents whose concentrations are too small to be of commercial significance. A minor constituent is recognizable as one which appears only as a product and only in reactions which have very small equilibrium constants. On this basis, ethylene production by Eq. (1-20) might very well be neglected. The propriety of such neglect can always be tested after completion of the calculations, as illustrated in later paragraphs.

"Series reactor" solution procedure. This technique for the determination of the equilibrium composition in a system of multiple reactions involves the imagined flow of the reacting system through each of a row of reactors in series. These reactors are all to be isothermal at the same temperature and to operate at the pressure of the system. Assume that the independent reactions which can occur are *A*, *B*, *C*, *D*, etc. Each reactor contains a selective catalyst which causes one of these reactions to proceed to equilibrium, but prevents all the other reactions from occurring. Numbering the reactors in sequence, then reaction *A* occurs in reactor 1, *B* in 2, *C* in 3, *D* in 4, etc. Now repeat this reaction cycle, with reaction *A* again occurring in reactor 5, *B* in 6, *C* in 7, and *D* in 8. The same cycle of reactions is repeated in the next four reactors, and then again repeated, until no further significant change in composition of the reacting stream indicates that equilibrium has been approximated.

Regardless of the series of independent reactions chosen, the composition of the mixture passing through the reactor series will always converge on the correct final equilibrium composition. This follows from the Second Law of Thermodynamics, which requires that the free energy change of an isothermal system undergoing spontaneous reaction at constant pressure is always negative, and that for a system at equilibrium at constant temperature and total pressure, no change can be found for which ΔF is negative. The change in composition in each successive reactor will therefore always be toward total equilibrium, never away from it.

The advantage of this series reactor method lies in the simplicity of the calculation for each reactor stage. It was shown in example 1-1 that a single equation in a single unknown resulted when only one reaction occurred in a reacting system, and such an equation is, of course, easily solved. By the series reactor method just described, any system of multiple reactions is reduced to a series of individual reactions occurring separately. Application of this method to the system of example 1-3, involving two independent reactions, is illustrated in the following simple example.

EXAMPLE 1-4

Determine the equilibrium composition at 1100°F and 2 atm pressure for the system whose feed composition is 5 moles of steam/mole of methane, recognizing that both reactions (1-9) and (1-19) occur.

Answer: Using an isothermal reactor series as just discussed, the indicated feed gas will pass through reactor 1, where only reaction (1-9) occurs. The product gas from reactor 1 passes through reactor 2, where reaction (1-19) occurs. The product gas from reactor 2 passes through reactor 3 for reaction (1-9), then passes through reactor 4 for reaction (1-19), reactor 5 for reaction (1-9), etc.

The composition of the gases entering and leaving each reactor is now calculated in succession, recognizing that equilibrium is attained for first one reaction, then the next, etc. Results of the calculations are presented in Table 1-3. Thus, the feed to the first reactor is 5 moles of H_2O, 1 mole of CH_4, and if *a* moles of CO form by reaction (1-9), then the moles of product gas

TABLE 1-3. EQUILIBRIUM GAS OF EXAMPLE (1-4)*

Cycle No.	1		2		7		8	
Reactor No.	1	2	3	4	13	14	15	16
Reaction occurring	(1–9)	(1–19)	(1–9)	(1–19)				
CH_4, moles	$1 - a = 0.288$	$0.228 - b = 0.072$	$0.072 - c = 0.286$	0.128	0.223	0.215	0.221	0.217
H_2O	$5 - a = 4.288$	$4.288 - 2b = 3.856$	$3.856 - c = 4.510$	3.750	3.64	3.62	3.63	3.62
CO	$a = 0.712$	$0.712 = 0.712$	$0.712 + c = 0.498$	0.498	0.194	0.194	0.188	0.188
H_2	$3a = 2.136$	$2.136 + 4b = 3.000$	$3.000 + 3c = 2.358$	2.98	2.91	2.94	2.92	2.94
CO_2	$0 = 0$	$b = 0.216$	$0.216 = 0.216$	0.373	0.582	0.590	0.590	0.594
Total moles	$6 + 2a = 7.424$	$7.424 + 2b = 7.856$	$7.856 + 2c = 7.428$	7.730	7.55	7.56	7.55	7.55

*Basis: Reactor feed is 5 moles H_2O plus 1 mole CH_4.

are as in the first column of Table 1-3. At a total pressure of 2 atm, the equilibrium constant for reaction (1-9) becomes

$$0.41 = \frac{(3a)^3(a)(2)^2}{(1 - a)(5 - a)(6 + 2a)^2}$$

Solving, a equals 0.712 moles. The feed from reactor 1 enters reactor 2, where b moles of CO_2 are formed by reaction (1-19), whose equilibrium constant is as follows:

$$1.09 = \frac{(2.136 + 4b)^4(b)(2)^2}{(0.228 - b)(4.288 - 2b)^2(7.424 + 2b)^2}$$

Solving, b equals 0.216. This gas now enters reactor 3, where c moles of CO form by reaction (1-9), with an equilibrium constant as follows:

$$0.41 = \frac{(0.712 + c)(3.00 + 3c)^3(2)^2}{(0.072 - c)(3.856 - c)(7.856 + 2c)^2}$$

Solving, c equals -0.214, indicating reaction (1-9) reverses a trifle in reactor 3. The product gas now enters reactor 4, where the d moles of CO_2 in the product gas formed by reaction (1-19) is found to be 0.373. Proceeding similarly, off-gas compositions are calculated for the successive reactors. To save space, results of the third through the sixth cycle are omitted from Table 1-3. It is evident from inspection that the gas composition has stabilized in the eighth cycle. As expected, the test shows that the off-gases from this cycle simultaneously satisfy equilibrium constants of both reactions (1-9) and (1-19).

The justification for ignoring the presence of ethylene can be tested. From reaction (1-20), $P_{C_2H_4}$ equals $1.23 \times 10^{-6}(P_{CH_4}/P_{H_2})^2$. Substituting partial presssures from the table, it follows that $P_{C_2H_4}$ in this sytem is only 1.38×10^{-7}, which is negligible.*

It might be argued that there is no great advantage for this series reactor method, as illustrated in example 1-4 and as contrasted with the more straightforward approach of example 1-3. The advantages of the series reactor method become greater when three or more independent reactions are involved.

**Note:* This system is readily tested for the presence of solid carbon. That is, carbon might form as follows:

$$C + 2H_2 = CH_4 \qquad K_{1100°F} = 0.129$$

Obviously, carbon cannot be present in any system at 1100°F in which $(P_{CH_4}/P^2_{H_2})$ is less than 0.129. From values in Table 1-4, this ratio has the value $(0.014/0.428^2)$, which is less than 0.129; consequently, no solid carbon forms.

Dependent versus independent reactions. In systems of multiple reactions, it is always possible to write dependent as well as independent reactions. Thus, for the hydrogen-producing system of example 1-3, the following reaction can be written along with reactions (1-9) and (1-19):

$$CH_4 + CO_2 = 2CO + 2H_2 \qquad (K_p)_{1100°F} = 0.154 \qquad (1\text{-}21)$$

Reaction (1-21) is dependent in the sense that it is obtained by adding reactions (1-9) and (1-19). In an equilibrium system, the equilibrium constants of all such dependent as well as independent reactions are, of course, simultaneously satisfied. The choice of which reactions are independent and which dependent is entirely arbitrary. Thus, in a system in which reactions (1-9), (1-19), and (1-21) occur, any two of these three are independent and the third is dependent.

When ethylene is also recognized as being produced along with hydrogen, then three independent reactions are involved, which are restated below for convenience.

TABLE 1-4. HYDROGEN-PRODUCING REACTIONS

$CH_4 + H_2O = CO + 3H_2$	$(K_p)_{1100°F} = 0.41$	(1-9)
$CH_4 + 2H_2O = CO_2 + 4H_2$	$(K_p)_{1100°F} = 1.09$	(1-19)
$2CH_4 = C_2H_4 + 2H_2$	$(K_p)_{1100°F} = 1.23 \times 10^{-6}$	(1-20)

An alternative group of independent reactions, derivable by simple addition or subtraction from the reactions of Table 1-4, are as shown in Table 1-5.

TABLE 1-5. ALTERNATIVE HYDROGEN-PRODUCING REACTIONS

$CH_4 + H_2O = CO + 3H_2$	$(K_p)_{1100°F} = 0.41$	(1-9)
$CO + H_2O = CO_2 + H_2$	$(K_p)_{1100°F} = 2.65$	(1-22)
$2CH_4 = C_2H_4 + 2H_2$	$(K_p)_{1100°F} = 1.23 \times 10^{-6}$	(1-20)

Still another combination of independent reactions is presented in Table 1-6.

TABLE 1-6. ALTERNATIVE HYDROGEN-PRODUCING REACTIONS

$CH_4 + 2H_2O = CO_2 + 4H_2$	$(K_p)_{1100°F} = 1.09$	(1-19)
$CH_4 + CO_2 = 2CO + 2H_2$	$(K_p)_{1100°F} = 0.154$	(1-21)
$2CO + 4H_2 = C_2H_4 + 2H_2O$	$(K_p)_{1100°F} = 7.3 \times 10^{-6}$	(1-23)

Other groups of three independent reactions involving these same reaction partners exist for this system. It follows that there is wide latitude in the choice of the independent equations to be used in calculating any system's equilibrium composition.

Parallel, sequential, and mixed reactions. Before considering the selection of a particular reaction group for use in "series reactors" to determine an equilibrium composition, it is helpful to distinguish between what are designated here as parallel and sequential reactions. Parallel reactions are those which require the same reactants, as with reactions (1-9) and (1-19), both of which start with CH_4 and H_2O. In sequential reactions, the products of one reaction are consumed as reactants in a following reaction. Thus, reactions (1-21) and (1-23) of Table 1-6 are sequential, in that the CO and H_2 obtained as products in reaction (1-21) are consumed as reactants in reaction (1-23). Finally, in a mixed reaction, the reactant of an earlier reaction combines with a product of the same or another earlier reaction. Reaction (1-21) of Table 1-6 is mixed since a reactant and a product of the earlier reaction (1-19), namely CH_4 and CO_2, appear on its left side. In the same sense, reaction (1-22) of Table 1-5 is mixed with respect to reaction (1-9).

Convergence. To use this technique effectively, a reaction sequence must be chosen whereby convergence is rapid, as in example 1-4. Two situations can arise in which convergence is slow, one with sequential reactions and the other with parallel reactions. As illustrated below, slow convergence can be avoided by selection of a proper reaction sequence.

1. Sequential reactions. An improperly chosen series of sequential reactions for the production of hydrogen from methane and steam is illustrated in Table 1-7.

TABLE 1-7. ALTERNATIVE HYDROGEN-PRODUCING REACTIONS

Reaction	Equilibrium constant	
$2CH_4 = C_2H_4 + 2H_2$	$(K_p)_{1100°F} = 1.23 \times 10^{-6}$	(1-20)
$C_2H_4 + 2H_2O = 2CO + 4H_2$	$K_{1100°F} = 1.4 \times 10^5$	(1-24)*
$CO + H_2O = CO_2 + H_2$	$K_{1100°F} = 2.65$	(1-22)

Inspection shows that the concentration of C_2H_4 in the gases leaving reactor 1 is minute because of the very small value of K for reaction (1-20). As a consequence, the sequential reaction (1-24) would be "starved" of ethylene, and so, despite the very large value of K for reaction (1-24), very little hydrogen could form per cycle in reactor 2. Moreover, since very little CO is made in reactor by reaction (1-24), reaction (1-22) would be similarly starved of CO in reactor 3. As a result, the reacting gas would have to pass through many cycles before equilibrium could be approximated. The reaction series of

**Note:* This is the reverse of reaction (1-23).

Table 1-7 is therefore poorly chosen for operation in the reactor series. To avoid such problems of slow convergence, all the sequential reactions chosen (except the last) should have equilibrium constants greater than about 0.1, and preferably greater than unity. The reactions of both Tables 1-3 and 1-4 fulfill these requirements, hence either group would be more suitable than the reactions of Table 1-5.

2. Parallel reactions. Improperly chosen parallel reactions can also bring about reactant starvation with consequently slow convergence. Consider the combustion of CS_2 with excess oxygen (say, 5 moles of O_2/mole of CS_2) to form a product stream at 800°F containing the following constituents: CO_2, CS_2, O_2, SO_2, and SO_3. The reaction sequence chosen is presented in Table 1-8 with reactions (1-25) and (1-26) being parallel in the sense that the same reagents, namely CS_2 and O_2, appear on the left side of each equation. Only two of the reactions of Table 1-8 are independent, since each reaction can be derived from the other two by addition or subtraction.

TABLE 1-8. CARBON DISULFIDE COMBUSTION

$CS_2 + 3O_2 = CO_2 + 2SO_2$	$K_{800°F} = 3.8 \times 10^{74}$	(1-25)
$CS_2 + 4O_2 = CO_2 + 2SO_3$	$K_{800°F} = 2.55 \times 10^{79}$	(1-26)
$SO_2 + \frac{1}{2}O_2 = SO_3$	$K_{800°F} = 260$	(1-27)

If reaction (1-25) is carried out in the first converter, little CS_2 will survive in the product gas because of the large value of K for this reaction. Because of this lack of CS_2, very little SO_3 can form in reactor 2, despite the large size of the equilibrium constant for reaction (1-26). In other words, reactor 2 is starved of reagent CS_2 because most of the CS_2 originally present was consumed in reactor 1. When dealing with parallel reactions, therefore, the earlier reactions should have small equilibrium constants, preferably less than unity. To avoid difficulty with the parallel reactions of Table 1-8, reaction (1-25) should be followed not by reaction (1-26), but by the sequential reaction (1-27). Note that no starvation now occurs, since reaction (1-27) is well provided with SO_2 by reaction (1-25), and with oxygen from the excess in the feed.

In most situations, more than one reaction series can be found which allow rapid convergence, and any one of these series would be suitable for the determination of the system's equilibrium composition. Moreover, the presence of dependent or "redundant reactions" in a series-reactor system is permissible, providing they do not cause a starvation step to develop, but are disadvantageous in involving one more step per cycle of reactors, and therefore more calculation.

Often, there is uncertainty as to whether a reaction for making a constituent should have been neglected. When in doubt, it is simple to test for the propriety of such neglect by the method already illustrated in the footnote to example 1-4.

Rules of procedure. It is now possible to formulate a set of rules for avoiding reagent starvation with consequently slow convergence, as follows:

1. Decide on the number of independent reactions to be recognized as occurring in the equilibrium system under study.

2. Formulate various possible groups of independent reactions, and select that reaction group for the reactor series which converges rapidly. The following should be observed:

   (a) To the extent possible, make the compounds present by sequential reactions having an equilibrium constant no smaller than 0.1, and preferably greater than unity.

   (b) When one of the reagents in the feed is in stoichiometric excess with reference to the first reaction occurring, write a later reaction of the mixed type in which this excess reagent reacts with a product of an earlier reaction. This earlier reaction will desirably have a K value greater than unity.

   (c) It may not be possible to produce all or any of the compounds present by sequential reactions having K values greater than unity. Make these compounds by parallel reactions having K values less than unity. Insert these reactions in the series at those points where the reactants for these parallel reactions are not yet reduced to starvation levels.

Rates of chemical reaction. While this first chapter is concerned primarily with chemical equilibrium, a few paragraphs are included here on reaction rates to provide background for the problems with this chapter. It is, of course, a familiar fact that chemical equilibrium is often not attained because mixing, mass transfer to a catalyst surface, or the chemical reaction itself is slow relative to residence time in the reactor. Mixing and mass transfer can sometimes be improved by changed reactor design, while reaction rates almost invariably respond to increased temperature.

The beneficial effects of increased temperature on reaction rate is illustrated by an old rule of thumb, which states that the rate of a chemical reaction doubles for every 10°C increase in temperature. This rule indicates correctly that rates tend to vary exponentially with temperature, but is often incorrect in that doubling of rates may occur at temperature increments much larger or smaller than 10°C, depending on the case. Because of this powerful influence of temperature on rates, it is not surprising that for many reactions, chemical reaction rate becomes negligible below some temperature called the "kindling" temperature, at which the reaction is said to "freeze."

Thus, mixtures of oxygen and fuel gases like hydrogen or methane can be kept for many years at room temperature without detectable reaction; hence the slow step here is obviously reaction rate, not mixing. As the temperature of such a mixture is raised, however, a kindling temperature is reached at which reaction rate becomes rapid enough to result in detectable (in this case violent) combination. At higher temperatures, mixtures of oxygen and fuel cannot be prepared since these gases react immediately upon contact. The rate-determining step which controls the amount of reaction occurring is now the rate at which the gases are allowed to come together and mix rather than the rate of chemical combination.

Reaction rate is influenced not only by temperature, but also by reactant concentration, proximity to equilibrium, and the presence of impurities and catalysts. Impurities may be introduced in the raw materials, in the materials of construction used for the apparatus, or from some other adventitious source. In sensitive catalytic or biochemical processes, raw materials from a new supplier must always be tested before use in the plant. The problem may become more severe with reactions involving solids, whose reactivities tend to vary greatly with the amount and type of surface exposed.

Unfortunately, the science of kinetics has not yet progressed to the point where it is possible to predict the rate of an untried chemical reaction, even when pure reagents are used. It is, therefore, necessary for the industrial chemist to resort to laboratory and pilot plant measurement of rates, making this study under conditions which reflect plant conditions as closely as possible.

Catalyzed systems. Catalysis in industry is usually heterogeneous, in the sense that the catalyst exists in a phase different from that of the main reagent stream. Gas reactions are usually catalyzed by solids, although the catalyst may also be in liquid form, as with the aqueous palladium chloride–copper chloride solution used in the Wacker process for acetaldehyde from ethylene and air.

Some catalysts used today are very effective, others less so, and the degree to which equilibrium is approached varies, depending on contact times. The discovery of an effective catalyst for a given reaction is difficult, since catalysts are often highly specific, and only a few materials (which must exist in proper physical as well as chemical form) will catalyze any given reaction. There is real question, of course, as to whether some reactions can be catalyzed at all.

Both the discovery and preparation of catalysts are of key importance in industrial chemistry, and an increasing research effort is today being devoted to this field as it slowly changes from an art to a science. The industrial benefits of a good catalyst are clear, since it makes possible the carrying out of a reaction which does not occur at all otherwise, or only so slowly that very large residence times and enormous reactors would be required. A catalyst may make possible operations at lower pressures and temperatures than would otherwise be possible. Again, a catalyst is often highly selective, and may accelerate the desired reaction to the exclusion of other undesired reactions, even though all these have favorable equilibrium constants.

The equilibria of catalyzed reactions are theoretically unaffected by the presence of catalysts, since these materials do not participate in the reaction, and so should remain unchanged indefinitely. In actual operations, catalysts rarely show unlimited life, being subject to deterioration and loss by physical or chemical attack. As an illustration of this point for solid catalysts, the copper chloride used to catalyze the oxidation of HCl to chlorine is slowly lost by volatilization, vanadium oxide catalysts used in the air oxidation of SO_2 to SO_3 are "poisoned" by arsenic and vanadium, while the silver catalyst for the oxidation of ethylene to ethylene oxide loses its activity by sintering. Similarly, liquid phase catalysts often lose their efficacy due to accumulation of tar and by-products, as in the solutions of mercuric salts used in the classical oxidation of acetylene to acetaldehyde. Such catalyst changes usually take place slowly, however, and in general do not involve sufficient participation in the primary reaction to affect the equilibrium calculations.

Heterogeneous Reaction Equilibria 2

A reaction is said to be homogeneous when it occurs in a single phase, either gaseous, liquid, or solid, within which all reactants and products remain in solution. In heterogeneous reactions, on the other hand, two or more phases are present, the reaction partners are somehow divided between these phases, and reaction occurs primarily at the interfaces. Examples of heterogeneous reactions are as follows:

1. Solid-gas reactions, such as the reduction of metal oxides with hydrogen or carbon monoxide, the combustion of solid fuel, or the oxidation of gaseous $SiCl_4$ to form gaseous Cl_2 and solid silica particles.
2. Liquid-gas reactions, such as the removal of ammonia vapor from coke oven gas by scrubbing with sulfuric acid.
3. Liquid-solid reactions, such as the leaching of many ores, or the precipitation of pigments like titanium dioxide from solution.

In theory, all solids and liquids exert vapor pressures and, although often extremely small at lower temperature, these vapor pressures are always finite. At those lower temperatures corresponding to low vapor pressures, the rate of sublimation of a solid diminishes with decreasing temperature even more rapidly than its vapor pressure (72). Usually the rate of sublimation of a solid falls substantially to zero when its equilibrium sublimation pressure lies under about 0.01 mm of mercury; therefore, in heterogeneous systems at low temperature, little material is available for reaction in the gas phase. For example, in the hydrogen reduction of iron oxide to metal at 500°F, the gas phase may be considered to contain only hydrogen and water, since the vapor pressures of iron and iron oxide at this temperature, while finite, are so small as to be completely negligible. As a consequence, the reduction reaction occurs heterogeneously at the iron oxide surface in contact with hydrogen, but not to any significant extent within the gas phase. When making equilibrium calculations, however, it is fair to assume that the gas phase is saturated with the vapors of all solids (and liquids) present.

Heterogeneous versus homogeneous operation. In theory, heterogeneous systems can usually be made homogeneous, and the reverse is also feasible. Thus, the direct chlorination of chloroform to carbon tetrachloride at atmospheric pressure and 200°C is a homogeneous gas phase reaction. An isothermal increase in pressure results in the formation of liquid droplets within the vapor phase, whereupon the system becomes heterogeneous. When condensation is complete, then the system has been converted from a homogeneous gas phase to a homogeneous liquid phase system. As illustrated here, unless the system's pressure is above the critical pressure for all the components and possible mixtures, an isothermal increase in total pressure will always ultimately result in the formation of liquid or solid phases, and thus convert a homogeneous into a heterogeneous system.

Especially in organic reactions, there is often the option of operating homogeneously at low pressure, or heterogeneously at higher pressure. The pros and cons of heterogeneous versus homogeneous operation will be discussed in a later chapter. Frequently, of course, there is no choice, as in systems involving solids like carbon, which have such low sublimation pressures at all reasonably attainable temperatures that heterogeneous operation cannot be avoided. Similarly, heterogeneous operations are not possible when, as in the ammonia synthesis, reaction kindling temperatures lie above the critical temperatures of all components within the system.

Activity of liquids and solids. When a material has a vapor pressure of less than 1 atm at a given temperature, then it necessarily exists as a liquid or solid in its corresponding standard state. Thus, solid iodine has a vapor pressure of almost 3 mm of Hg at 25°C, hence it exists as a crystalline solid at 25°C under its standard state pressure of 1 atm. A vapor pressure is, of course, still exerted by the solid in this standard state, as is shown by the finding that iodine crystals sublime when placed in an open dish at 25°C. An increased total pressure, as of air on iodine crystals, always results in a minute increase in vapor pressure, as required by the Poynting equation.

Up to total pressures of 20 atm and more, however, the increase in vapor pressure with total pressure is negligible in most industrial situations.

To be in its standard state with unit activity, a solid or liquid must be pure and under a total presure of 1 atm at the temperature in question. The fugacity of a solid in its standard state equals p^0, its vapor pressure under these conditions. As with vapor pressure, the increase in the activity of a solid with total pressure is very small and is often neglected. In other words, the activity of a condensed phase is usually taken as unity in equilibrium calculations, provided the total pressure is not very high.

EXAMPLE 2-1

What is the activity of carbon at 100 atm and 80°F, when carbon's activity at 1 atm and 80°F is taken as unity? Carbon is assumed to be incompressible and to have a density of 150 lb/cu ft.

Answer: The isothermal relation between the activity and pressure of an incompressible solid is

$$(\pi - p)V = RT \ln a$$

where π and p are the final and initial pressures, and V is the volume per gram atom of material. Substituting

$$\ln a = \frac{(\pi - p)V}{RT} = \frac{(100 - 1)(14.7)(144)(\frac{12}{150})}{(1546)(540)} = 0.02$$

and solving, the activity of carbon is about 1.02. Thus, even a high total pressure makes little change in the activity of a liquid or a solid (solid carbon in this case).

A material which is solid or liquid in its standard state will exist in the equilibrium system under study as a vapor when p, the partial pressure of this material, is less than the material's vapor pressure p^0. The activity of this material is then

$$a = \frac{p}{p^0} \tag{2-1}$$

Heterogeneous equilibrium. Depending on the temperature being considered, any or all of the partners of reaction (1-1) may exist as solids in their standard states, namely, pure and under a pressure of 1 atm. When A and M of reaction (1-1) are both solids, then this reaction is rewritten as

$$aA(c) + bB(g) + cC(g) = mM(c) + nN(g) \tag{2-2}$$

where the designations $A(c)$ and $M(c)$ indicate the crystalline state. The equilibrium constant for reaction (2-1) is

$$K_{\text{het}} = \frac{a_M^m a_N^n}{a_A^a a_B^b a_C^c} \tag{2-3}$$

The activities of the gaseous reaction partners are again replaced by their partial pressures, from Eq. (1-4). As long as A and M are present at equilibrium in the reacting system as pure solids, then the activity of each is unity with the standard state choice just indicated (this again neglects the small Poynting effect). Equation (2-3) is therefore rewritten:

$$K_{\text{het}} = \frac{p_N^n}{p_B^b p_C^c} \tag{2-4}$$

When conditions of total pressure and composition are such that A and M are present in the equilibrium system not as solids but as vapors, with partial pressures below their sublimation pressures at this temperature, then their activities are smaller than unity and are represented by Eq. (2-1). Eq. (2-4), of course, no longer applies, and instead, by combination of Eqs. (2-3), (2-1), and (1-4), the following relation must be used:

$$K_{\text{het}} = \frac{\left(\frac{p_M}{p_M^0}\right)^m p_N^n}{\left(\frac{p_A}{p_A^0}\right) p_B^b p_C^c} \tag{2-5}$$

The subscript on K_{het} is included in the foregoing equations to indicate that this constant is for a heterogeneous reaction. From here on, use of this subscript will be discontinued, since the type of reaction to which an equilibrium constant applies will be clear from the context. For all of these heterogeneous reactions, enthalpy changes are again related to their equilibrium constants and free energy changes by Eqs. (1-11) and (1-10).

Equilibrium composition. As with homogeneous systems, the equilibrium composition of a heterogeneous system can always be calculated when information is available on the initial feed composition, total pressure, and equilibrium constants for the reactions occurring at the system temperature. These calculations are somewhat more complex than for homogeneous systems, especially when it is important to know not only the composition of the solution (gas or liquid) phase, but also the relative quantities of the solution and solid phases present. The approach to such a problem is illustrated in the following example.

EXAMPLE 2-2

An equimolal mixture of hydrogen and methane is placed in a cylinder equipped with a movable piston, and allowed to come to equilibrium at 600°K and a total pressure of 1.1 atm. (a) What is the composition of the gas phase? (b) How much solid carbon is present? (c) Keeping the temperature at 600°K, the piston is moved inward to compress the system until all but a trace of the solid has disappeared. What is the system pressure now? (d) Compress to a total pressure of 3.0 atm. What is the composition of the system? What is the activity of carbon in this system?

Note:

$$C(c) + 2H_2 = CH_4 \qquad K_{600°K} = 1.0 \text{ atm}^{-1}$$

The vapor pressure of carbon at 600°K is vanishingly small.

Answer: (a) The equilibrium constant for the (heterogeneous) reaction whereby methane is formed from hydrogen and solid carbon is $(p_{CH_4})/(a_C p_{H_2}^2)$. In the presence of solid carbon, a_C is unity. For 2 moles of total mixed gas feed to the reactor, the equilibrium system contains $(1 - X)$ moles of CH_4, $(1 + 2X)$ moles of H_2, and X moles of solid carbon. Since the total moles of gas equal $(2 + X)$ and the total pressure is π, then

$$K = \frac{(1 - X)(2 + X)}{\pi(1 + 2X)^2}$$

Solving when π equals 1.1 atm and K equals unity, X equals 0.15 atoms. The mole fraction of H_2 in the gas is $(1 + 2X)/(2 + X)$, or 0.605.

(b) The solid carbon present is 0.15 atoms/2 moles of initial mixed feed.

(c) As the pressure is increased, hydrogen and carbon react to form more methane. When the amount of carbon present reaches zero, then from the above equation, π equals 2 atm.

(d) Since the last trace of carbon has disappeared to form methane when the pressure has reached 2 atm, as just shown, no further methane formation can occur upon further compression. That is, the vapor pressure of carbon at 600°K is vanishingly small, consequently, no carbon is available in the vapor for further reaction. It follows that, but for a minute trace of carbon vapor (of absolutely no industrial significance), the system contains only a gas phase consisting of 1 mole of CH_4 and 1 mole of H_2, each with a partial presssure of 1.5 atm. The activity of the carbon vapor is calculated directly from the equilibrium constant expression

$$a_C = \frac{p_{CH_4}}{K_{p_{H_2}^2}} = \frac{1.5}{1.5^2} = 0.67$$

Thus, the partial pressure of carbon in the system is only 0.67 times that of the already minute saturation vapor pressure of carbon at 600°K.

Examination of the foregoing example shows that a phase may be caused to appear or disappear in heterogeneous systems by an isothermal change in pressure. That is, by compressing the system of example 2-1 isothermally at 600°K to above 2 atm, solid carbon disappears, but reappears upon expansion to pressures below 2 atm. In the sense that a solid phase, namely carbon, disappears upon compressing this system, this heterogeneous reaction can be made to go to completion. It must be emphasized that the homogeneous reaction in the gas phase, however, has not gone to completion, since there is still carbon vapor present, regardless of the total pressure.

THE PHASE RULE

The generalizations of the Phase Rule (38) as these relate to the commonly encountered heterogeneous reactions of industry are presented in the following paragraphs. After this introductory discussion, it is shown that most industrial gas-solid reactions fall into one of five classes, and examples of each are considered. The gas-liquid reactions of industry usually involve only one liquid phase, and generalizations relating to such systems are likewise presented. Consideration of liquid-solid systems is postponed until Chapter 3.

A closer examination of the conditions under which gases, liquids, and solids can coexist at equilibrium benefits from a review of the Phase Rule for two reasons:

1. The Phase Rule makes possible the classification of diverse and seemingly unrelated heterogeneous reactions into a few categories. Thus, most of the gas-solid reactions of industrial importance can be grouped into five classes, as described below. Reactions in any one of these groups show important similarities.
2. The derivation and application of the Phase Rule calls for the same type of thinking as is involved in the analysis of systems encountered in industrial chemistry, as becomes evident from the examples presented.

The Phase Rule applies only to systems at equilibrium, therefore to systems which in an unchanging environment can undergo no further change with time. The equilibrium systems considered here are at uniform temperature and pressure throughout. In developing the Phase Rule, use is made of familiar terms, such as constituent, phase, component, variable, etc., whose definitions must be carefully respected. These definitions are now briefly reviewed.

Constituent. A constituent is any compound, ion, or element found by analysis to be present in any of the phases in the system under study.

The number of constituents in a system is designated by N. As already pointed out, all constituents except ions are assumed present at least tracewise in the gas phase of a gas-solids system, and similarly all constituents are present in the liquid phase of a liquid-solids system. Each material present is recognized as a constituent, even though its concentration in the gas or liquid phase is extremely small.

Phase. A phase is that part of any equilibrium system in which the concentrations of all constituents, expressed as mass per unit volume, are constant. Each phase is separated from neighboring phases by boundary surfaces across which the concentration of at least one constituent undergoes a stepwise change. Thus, in a system comprising water in a closed container at 1 atm and 100°C, the vapor is separated from the liquid phase by an interface across which the concentration (g/cc) undergoes an abrupt change. A single phase in a system may exist in several parts, as in the form of separate droplets of water. The number of phases in a system is denoted by $\boldsymbol{P}$.

A phase is considered to exist in a system at equilibrium, if it is present to any degree, regardless of its relative mass. Consider a two-phase system involving liquid and gaseous water in equilibrium at exactly 100°C and 1 atm pressure. Now pass a thin membrane between these two phases, so that the gas phase contains only gaseous water at 100°C and 1 atm. From the point of view of the Phase Rule, this gas phase system still contains two phases since, while not actually present, liquid water could exist here at equilibrium. The same reasoning is applicable to the liquid phase of this imaginary experiment. In other words, a phase is assumed to exist if it could be present without altering the system's temperature, pressure, and composition.

Intensive variables. A system is at equilibrium in a given unchanging environment when it undergoes no further reaction or change with time. Each arbitrarily chosen part of such a system is at equilibrium within itself, and also with respect to every other part of this system. Thus, a system can be at equilibrium, regardless of its mass or size. The temperature and the total pressure of all parts of a system at equilibrium are the same throughout, and the concentration of each constituent, expressed as moles per unit volume, is constant within any given phase. In the gas phase, the partial pressure can obviously be calculated for each component from its concentration.* It is generally more convenient to use partial pressures than concentrations when describing a gas phase composition, and this practice is followed here.

**Note:* This assumes that the perfect gas laws apply, which is equivalent to saying that the compressibility factor is close to unity. In the real gas region, fugacities must be used in place of partial pressures. It should be noted that mole fractions are not suitable Phase Rule variables, since fixing only temperature and mole fractions does not define the system's fugacities or partial pressures.

These variables, namely temperature, total pressure, and concentrations (or partial pressures in the gas phase) are called "intensive," since they are independent of the system's mass. Having fixed these variables, then all the other intensive variables, such as optical index, specific conductivities for heat and electricity, dielectric constant, etc., are also fixed. Thus, a phase is characterized by temperature, total pressure, and concentrations (or partial pressures).

Obviously a complete description of a system, such as is commonly required in industrial chemistry, requires information on not only intensive variables, but also on its extensive variables, namely those depending on the size of the system. Examples of extensive variables are total mass and total volume, and also the volume and mass of each phase. The relation of extensive variables to the Phase Rule variables will be discussed in later sections of this chapter.

Relations between phases. When a given constituent is present as a solid as well as in the gas phase, then its concentration at a given temperature in the gas phase is fixed by its sublimation pressure. Thus, in a gas-solid system at $-10°C$ containing ice as a solid phase, the partial pressure of water in the gas phase equals the sublimation pressure of ice. When two different pure solid phases are present, then the sublimation pressure of each is fixed by temperature. Thus, the number of such vapor pressure relations is determined by the number of solid phases present. If there are $\boldsymbol{P}$ phases in a system, of which one is a gas phase and the rest are pure solids, then ($\boldsymbol{P}$-1) vapor pressures are fixed in the gas phase.

Intrinsic stoichiometric relations. It is evident from the discussion in the preceding paragraph that the ratio between vapor pressures of any two pure solid phases present is constant as the system volume is varied isothermally. The existence of such ratios is a direct consequence of the fact that vapor pressures of pure solids are temperature functions alone (again neglecting the Poynting relation).

A second type of constant ratio between partial pressures or concentrations, independent of system volume changes, is sometimes found in heterogeneous systems, and such ratios are called intrinsic stoichiometric ratios. These ratios do not involve vapor pressures, but are the consequence of a chemical reaction occurring, and are always expressed in terms of partial pressures or concentrations in the solution (gas or liquid) phase. Thus, when zinc oxide is mixed with carbon and heated in an initially evacuated container, the two following reactions proceed simultaneously to equilibrium:

$$\mathrm{ZnO(c) + C(c) = Zn(g) + CO(g)}$$
$$\mathrm{2ZnO(c) + C(c) = 2Zn(g) + CO_2(g)}$$

In this system, an intrinsic stoichiometric relation exists between the partial pressure of the reaction products, since the atoms of "gasified" zinc equal the atoms of "gasified" oxygen, or

$$p_{Zn} = p_{CO} + 2p_{CO_2}$$

This equation holds regardless of the ratio of solid ZnO to solid carbon in the feed. It obviously also holds, regardless of changes in the system volume, or the extent to which the reaction has proceeded. If the number of intrinsic stoichiometric relations is designated as $\boldsymbol{S}$, then in this system $\boldsymbol{S}$ is unity.

Systems made by charging a single material to a reaction vessel sometimes have a value for $\boldsymbol{S}$ of unity. Thus, when solid pure ammonium chloride is fed into an initially evacuated cylinder fitted with a frictionless piston, then gaseous HCl and NH_3 form by dissociation as the system volume is increased and so three constituents are recognizable:

$$NH_4Cl(c) = NH_3(g) + HCl(g)$$

The partial pressure of HCl equals the partial pressure of NH_3, regardless of the temperature and volume of the system, consequently for this system, $\boldsymbol{S}$ is unity.

An important change occurs now when the feed to the cylinder–piston arrangement just described is equivalent to HCl plus NH_3 in a mole ratio other than unity. This could involve feeding solid NH_4Cl plus gaseous NH_3 or solid NH_4Cl plus gaseous HCl. Obviously, as the relative mass of solid NH_4Cl and total gas phase is changed by altering the system volume or temperature, the gas composition will also change. It follows that $\boldsymbol{S}$ is zero for any system in which the mole ratio of HCl/NH_3 in the feed is not unity.

Constant intrinsic stoichiometric ratios are encountered in liquid as well as in gaseous solutions. A stoichiometric relation is always found in electrolyte solutions because the total equivalents of positive ions and negative ions must be equal to preserve electrical neutrality. Thus, in a solution of KCl and Na_2SO_4, denoting the moles of an ion present by brackets [K^+], then

$$[K^+] + [Na^+] = [Cl^-] + 2[SO_4^{--}]$$

Equilibrium relations. An equilibrium constant relating at least some of the partial pressures or concentrations of a system can be written for each independent chemical reaction occurring. The number of such independent reaction constants is designated by the symbol $\boldsymbol{R}$.

Components. The number of components in a system, namely $\boldsymbol{C}$ is defined as follows:

$$\boldsymbol{C} = \boldsymbol{N} - \boldsymbol{R} - \boldsymbol{S} \tag{2-6}$$

Consideration of this equation indicates that the number of components is the minimum number of materials required to duplicate the system under consideration in any phase ratio. While the definition in the preceding sentence is useful, the number of components present in a system is most easily determined from Eq. (2-6).

Variance. The Phase Rule variables are temperature, pressure, and concentrations. Partial pressures may be used in place of concentrations for the gas phase, if the ideal gas laws apply. The "variance" of a system, also called its "degrees of freedom," is then the minimum number of these Phase Rule variables which, upon being arbitrarily fixed, cause all the other intensive variables of the system to be fixed. Thus, a pure substance in the gaseous state has two degrees of freedom, or is said to be bivariant, in that only after fixing any two of the three intensive variables are all the intensive variables of the system fixed. A system's variance is designated by the symbol $\boldsymbol{V}$.

Phase rule for gas-solid systems. The Phase Rule is usually written as follows:

$$\boldsymbol{V} = \boldsymbol{C} + 2 - \boldsymbol{P} \tag{2-7}$$

A restricted derivation is now presented, applying to systems containing a gas phase and $(\boldsymbol{P} - 1)$ different pure solid phases, making a total of $\boldsymbol{P}$ phases present. Each solid phase is a single pure constituent rather than a solid solution of two or more constituents; consequently, the only intensive variables relating to composition which need be specified are in the gas phase. Thus, the system is characterized by the temperature, and by the partial pressure of each constituent, including, of course, the vapor pressure of each solid present. Since there are $\boldsymbol{N}$ constituents, there are $(\boldsymbol{N} + 1)$ independent intensive variables. The total pressure is also an intensive variable, but is not counted since it is not independent, being equal to the sum of the partial pressures.

Relations between these $(\boldsymbol{N} + 1)$ independent intensive variables are of three kinds:

1. The saturation vapor pressure of each of the pure solid phases is a function of temperature alone.* Since there are $(\boldsymbol{P} - 1)$ solid phases in this system, there are $(\boldsymbol{P} - 1)$ such relations.
2. The equilibrium constant of a chemical reaction is a function of temperature alone. Since there are $\boldsymbol{R}$ independent reactions in the system, there are $\boldsymbol{R}$ such equilibrium constant relations.
3. Stoichiometric relations between partial pressures or concentrations

*This neglects the Poynting effect, which is generally negligible for industrial systems.

are found to exist in some systems, as discussed earlier. The number of such relations is $\boldsymbol{S}$.

The variance of this system is now merely its total number of intensive variables, namely $(\boldsymbol{N} + 1)$, minus the above listed relations between these variables, namely $\{(\boldsymbol{P} - 1) + \boldsymbol{R} + \boldsymbol{S}\}$, or

$$\boldsymbol{V} = \boldsymbol{N} + 1 - \{(\boldsymbol{P} - 1) + \boldsymbol{R} + \boldsymbol{S}\}$$

Rearranging:

$$\boldsymbol{V} = \{\boldsymbol{N} - \boldsymbol{R} - \boldsymbol{S}\} + 2 - \boldsymbol{P} \tag{2-8}$$

The term $\{\boldsymbol{N} - \boldsymbol{R} - \boldsymbol{S}\}$ equals the number of components as defined by Eq. (2-6); consequently Eq. (2-8) is identical with the Phase Rule Eq. (2-7). Obviously, when the experimenter fixes $\boldsymbol{V}$ independent variables, then all other independent variables of the system of $\boldsymbol{P}$ phases are also fixed. It is not possible arbitrarily to fix more than $\boldsymbol{V}$ independent variables in any system.

CLASSES OF GAS-SOLID REACTIONS

The gas-solid reactions of industry are usually inorganic, and most of these fall into one of the five types discussed below.

1. Solid = Solid + Gas. Reactions involving the conversion of a solid to a solid plus gas are fairly common. Thus, decomposition of solid salts of oxy-acids to a gaseous acid anhydride plus a solid oxide, such as the decomposition of $Al_2(SO_4)_3$ to Al_2O_3 and SO_3, or of $CaCO_3$ to CaO plus CO_2, fall into this category. Other examples are (a) the decomposition of salt hydrates like $Na_2SO_4 \cdot 10H_2O$ to the anhydrous salt plus water vapor; (b) the conversion of acid salts to neutral salts and the corresponding acid, as in the decomposition of sodium bicarbonate to sodium carbonate, CO_2, and water vapor; (c) the decomposition of a solid copper chloride–ethylene complex to ethylene and copper choloride.

Decomposition of limestone serves to illustrate the behavior of systems of this type:

$$\mathrm{CaCO_3 = CaO + CO_2} \tag{2-9}$$

Solid calcium carbonate and solid calcium oxide show no mutual solubility, and since the activity of each pure solid is unity, the equilibrium constant for the foregoing heterogeneous reaction becomes

$$K = p_{\mathrm{CO_2}} \tag{2-10}$$

In a system containing solid $CaCO_3$, solid CaO, and gaseous CO_2, the number of phases present is 3. The constituents are obviously $CaCO_3$, CaO, and CO_2, making N equal to 3. Only one chemical reaction occurs, namely reaction (2-9), so $\boldsymbol{R}$ is unity. Since the gas phase contains nothing but CO_2, there are no partial pressures between which a constant ratio can be maintained, consequently $\boldsymbol{S}$ is zero. Substituting into Eq. (2-6), the number of components is found to be 2, and so by Eq. (2-7), the system's variance is unity.

Experimental study of limestone decomposition shows the expected monovariant behavior, as is evident from line *A-A* in Fig. 2-1 presenting the equilibrium decomposition pressure of CO_2 at various temperatures when both CaO and $CaCO_3$ exist as solid phases. The monovariance of this three-phase system is evidenced by the fact that, when a single intensive variable such as temperature is fixed, then all other intensive variables are fixed. As expected, line *A-A* applies as long as all three phases are simultaneously present at equilibrium, regardless of the relative quantities of these phases.

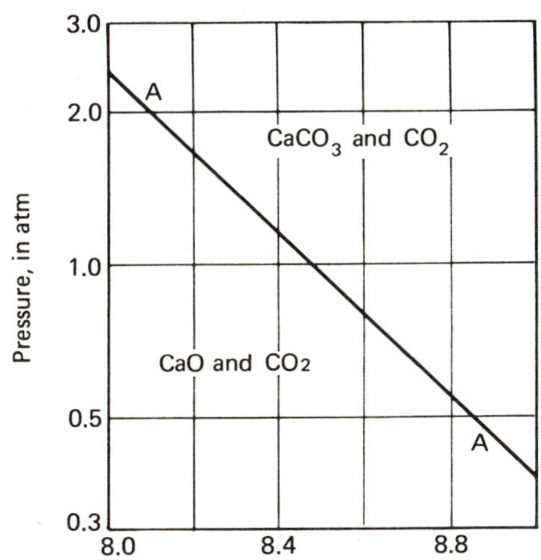

Fig. 2-1. Dissociation pressure of $CaCO_3$ (reference 61, vol. VII, p. 297): $CaCO_3 = CaO + CO_2$. All three phases are present on line *A-A*. Temperature is in °K.

Further reflection will show that, under the proper conditions, systems containing only two of these three phases can also exist at equilibrium. Thus, upon heating to a temperature such that the equilibrium decomposition pressure of line *A-A*, Fig. 2-1, exceeds the CO_2 pressure maintained in the system, then all $CaCO_3$ will be completely decomposed to CaO and CO_2. Therefore, all points lying in the region below line *A-A* are equilibrium points at which only CaO and CO_2 can coexist. By inspection, we see that all these points are divariant, since both temperature and pressure must now be fixed in order to fix all the intensive variables of this system. As expected, the Phase Rule analysis is consistent with these experimental findings in showing such a two-phase system to be divariant. That is, $\boldsymbol{N}$ is 2 as before, $\boldsymbol{R}$ is unity and $\boldsymbol{S}$ is again zero, but $\boldsymbol{P}$ is now only 2, hence, the variance is 2 from Eq. (2-7).

By similar reasoning, it follows that the region above line *A-A* contains only two phases at equilibrium, namely solid $CaCO_3$ and gaseous CO_2. This region is again divariant, in that both temperature and pressure must again be stipulated to fix all the intensive variables of this system. Systems containing only the two solid phases without a gas phase are similarly divariant. These systems, which cannot be so conveniently represented in Fig. 2-1, are not of commerical importance.

It is now clear that Fig. 2-1 may be viewed not only as a graphical representation of the variation of the dissociation pressure of calcium carbonate with temperature, but also as a phase diagram showing regions of stability of the different phases at equilibrium. Obviously, to cause reaction (2-9) to proceed quantitatively to the right, it is necessary to operate at conditions lying below line *A-A*, namely at such a temperature that the corresponding decomposition pressure of $CaCO_3$ as defined by line *A-A* is greater than the actual pressure existing within the system. Contrariwise, to cause reaction (2-9) to proceed quantitatively to the left, one operates somewhere in the region above line *A-A*.

2. Solid + Solid = Solid + Gas. The reduction of metal oxides with solid carbon to form metal plus the carbon oxides are typical examples of reaction between solids to form solid plus a gas. Another example is the reaction occurring in the so-called Pidgeon process for magnesium production:

$$Si(c) + 4MgO(c) = 2Mg(g) + Mg_2SiO_4(c) \tag{2-11}$$

The equilibrium constant for reaction (2-11) obviously equals the pressure of magnesium vapor. By inspection, $\boldsymbol{N}$ is 4, $\boldsymbol{R}$ is unity, and $\boldsymbol{S}$ is zero, since the gas phase contains only magnesium vapor. It follows from Eq. (2-6) that $\boldsymbol{C}$ is 3, and from Eq. (2-7) that the variance is unity as long as three solid phases plus a gas phase are present. Figure (2-2) shows the dissociation pressure of magnesium at various temperatures for reaction (2-11), when all three solid phases are present. As expected, these experimental findings again show this system to be monovariant. Further analysis of this diagram is similar to that of Fig. 2-1 discussed in earlier paragraphs.

3. Solid = Gas. All simple sublimations of solids, such as that of solid iodine, are examples of the direct conversion of a solid into a gas. Of interest here, however, are those cases in which chemical reaction also occurs, as in

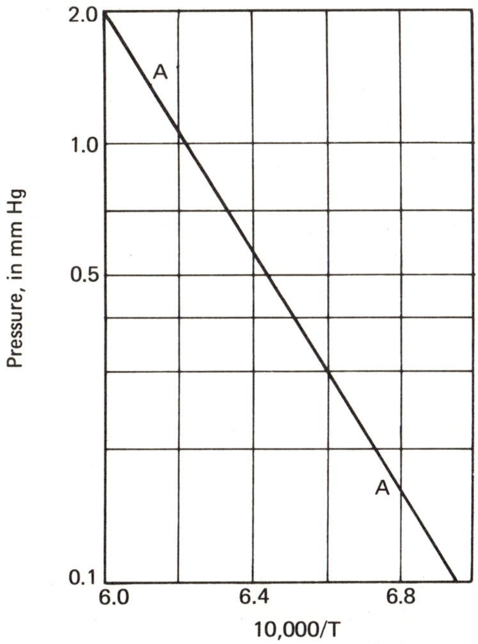

Fig. 2-2. Dissociation pressures of magnesium by reaction (2-11) (80): $Si(c) + 4MgO(c) = 2Mg(g) + Mg_2SiO_4(c)$. All phases are present on line *A-A*. Below line *A-A* only magnesium vapor, solid Mg_2SiO_4, and either solid Si or solid MgO are present.

the previously mentioned decomposition of solid ammonium chloride to ammonia and hydrogen chloride:

$$NH_4Cl(c) = NH_3(g) + HCl(g) \tag{2-12}$$

The equilibrium constant for this reaction is

$$K = \frac{P_{NH_3} \cdot P_{HCl}}{a_{NH_4Cl}} \tag{2-13}$$

Assuming solid NH_4Cl to be present, then since the activity of solids is unity, Eq. (2-13) may be rewritten

$$K = P_{NH_3} \cdot P_{HCl} \tag{2-14}$$

Now consider the system obtained by placing solid NH_4Cl in an initially evacuated vessel. Dissociation will proceed as in reaction (2-12), and if the vessel volume is small enough so that the solid is not gasified completely, then both a solid and a gas phase will be present, making $\boldsymbol{P}$ equal to 2. In this system, $\boldsymbol{N}$ is 3 for the three constituents, and $\boldsymbol{R}$ is unity for the dissociation reaction. When using solid NH_4Cl as feed, $\boldsymbol{S}$ is unity because the partial pressure of NH_3 equals that of HCl regardless of the temperature and volume of the system. Thus, the number of components is unity by Eq. (2-6), and the variance is unity by Eq. (2-7) as long as $\boldsymbol{P}$ is 2.

As expected, experimental findings likewise indicate that the two-phase system made by placing solid NH_4Cl in an initially evacuated box is monovariant. Thus, the total dissociation pressure π of solid NH_4Cl can be approximated by the experimentally derived relation [(61), Vol. III, Erster Teil, p.144]:

$$\log \frac{\pi}{2} = 6.62 - \frac{4240}{T} \tag{2-15}$$

where π is in atm and T in °K. Vapor density measurements show the gas to contain no detectable, undissociated NH_4Cl molecules, consequently,

$$\pi = p_{\mathrm{HCl}} + p_{\mathrm{NH_3}} \tag{2-16}$$

Since the reaction vessel was initially evacuated and charged with only solid NH_4Cl, p_{HCl} and $p_{\mathrm{NH_3}}$ remain equal. It follows that the equilibrium constant for reaction (2-12) can be directly obtained from Eqs. (2-14), (2-15), and (2-16):

$$\tfrac{1}{2} \log K_{\mathrm{het}} = \tfrac{1}{2} \log (p_{\mathrm{NH_3}})(p_{\mathrm{HCl}}) = 6.62 - \frac{4240}{T} \tag{2-17}$$

and that fixing one of the unknown variables in Eq. (2-17) will fix the other two and so fix the system.

The system just described, formed by adding solid NH_4Cl to an evacuated vessel, could also have been made by adding appropriate amounts of gaseous HCl and gaseous ammonia in an equimole ratio. When enough of these gases have been added to form solid ammonium chloride, then this system is again monovariant. However, when the mole ratio of NH_3/HCl charged to the vessel is not unity, then $\boldsymbol{S}$ becomes zero. By the Phase Rule, such a system is divariant when both a solid and gas phase are present. Experiment is again consistent with this finding. That is, the equilibrium constant Eq. (2-17) still applies, yet since the ratio of $p_{\mathrm{NH_3}}$ to p_{HCl} is not fixed, but varies with changes in system volume or temperature, two of the unknowns in Eq. (2-17) must be specified to fix the third.

4. Solid + Gas = Gas. A familiar example of this reaction type is the so-called Boudouard reaction:

$$CO_2(g) + C(c) = 2CO(g) \tag{2-18}$$

For this system, N is obviously 3, while S is zero and R is unity, hence C is 2 by Eq. (2-6). When the solid phase is present, then by Eq. (2-7) the variance is 2. When the solid phase disappears, the variance becomes 3.

The heterogeneous equilibrium constant for reaction (2-18) is (p_{CO}^2/p_{CO_2}), and the variation with temperature of this constant is shown by line *A-A* of Fig. 2-3. The region above line *A-A* is not attainable under equilibrium conditions, since any attempt to increase the partial pressure of CO will result in the decomposition of this gas by reaction (2-18), satisfying the heterogeneous equilibrium constant for this reaction. In the region below line *A-A* in Fig. 2-3 only a gas phase is present, with no solid phase. Solid carbon can exist in contact with a gas phase only when the conditions on line *A-A* are satisfied.

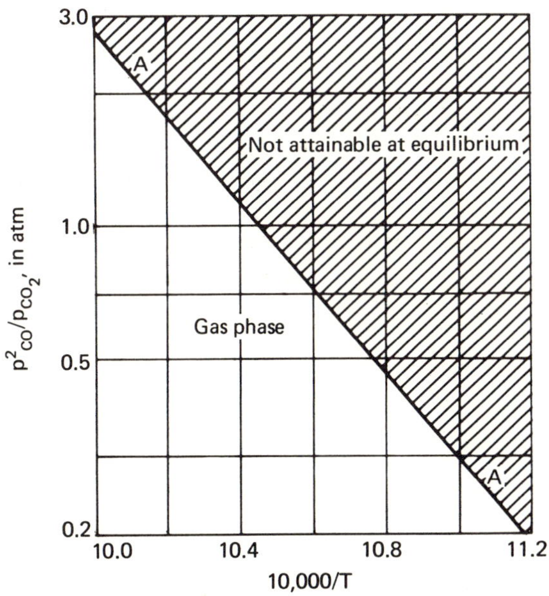

Fig. 2-3. Decomposition of carbon monoxide (22): $C + CO_2 = 2CO$. Gas and solid phase are both present on line *A-A*, with only the gas phase present below this line.

It should now be obvious that, as predicted by the Phase Rule, the two-phase system (gas phase plus solid carbon) represented by line *A-A* is divariant, in that two of the three intensive variables T, p_{CO}, and p_{CO_2} must be fixed to fix the third one. The gas phase region below line *A-A* is trivariant, in that all three intensive variables must be fixed to define the system.

Inspection of this diagram makes it clear that pure carbon monoxide gas is unstable at normal temperatures and pressures. Carbon monoxide is, nevertheless, found to exist under the conditions of the surroundings, and indeed is available for purchase in pressure cylinders, because the rate of reaction (2-18) is substantially zero at normal temperatures.

5. Solid + Gas = Solid + Gas. There are many examples of this type of reaction, especially in the field of process metallurgy. Thus, the reduction of nickel oxide with hydrogen proceeds as shown:

$$NiO(c) + H_2(g) = Ni(c) + H_2O(g) \tag{2-19}$$

In this system, N is 4, R is 1, and S is zero, hence, C is 3 by Eq. (2-6). Nickel and nickel oxide are insoluble in each other, and so P is 3 when both are present in addition to the gas phase, hence the system's variance is 2. When three phases are present, namely a gas phase plus two solid phases, then it follows from the Phase Rule Eq. (2-7) that the system is fixed when any two intensive variables are fixed, such as temperature and the partial pressure of water. Recognizing that each solid constituent has an activity of unity, then the equilibrium constant for reaction (2-19) has the form (p_{H_2O}/p_{H_2}), and the variation of this constant with temperature is shown graphically as curve *A-A* in Fig. 2-4. When the value of the ratio (p_{H_2O}/p_{H_2}) at the system's temperature lies on line *A-A*, then both solid Ni and solid NiO are present, and in any phase ratio whatever.

It is, of course, also possible to have only two phases present at equilibrium in this system, namely, a solid phase and a gas phase. Thus, when the

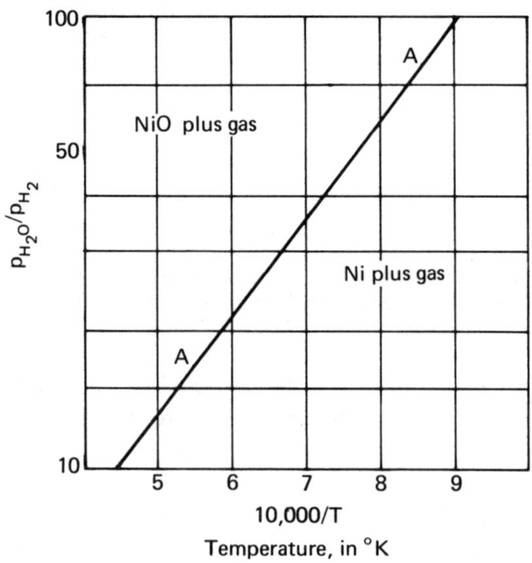

Fig. 2-4. Equilibrium constant vs. temperature for NiO reduction with hydrogen (22): $NiO + H_2 = Ni + H_2O$. All three phases are present only on line *A-A*.

ratio (p_{H_2O}/p_{H_2}) is deliberately adjusted to a value lying above line *A-A* in Fig. 2-4, reaction (2-19) is forced to go to completion to the left. This means that all the solid nickel metal present reacts to form NiO; consequently, only solid NiO can exist above line *A-A* in Fig. 2-4. By analogous reasoning, it follows that only nickel metal can exist below line *A-A*. Phase Rule analysis shows the NiO region above line *A-A* and the Ni metal region below line *A-A* to be trivariant. That is, ***C*** is now 3 but ***P*** is 2, hence, ***V*** is 3 by Eq. (2-7). Thus, below line *A-A*, three intensive variables, such as temperature, the partial pressure of water, and the partial pressure of hydrogen, can be varied arbitrarily and independently without necessarily causing a phase to appear or disappear.

6. Reaction sequence. Reduction of the oxides of elements having several valences, such as iron, titanium, copper, etc., typically proceeds through a series of sequential reactions. Thus, hydrogen reduction of magnetite (Fe_3O_4) to metal proceeds through FeO as in reactions (2-20)

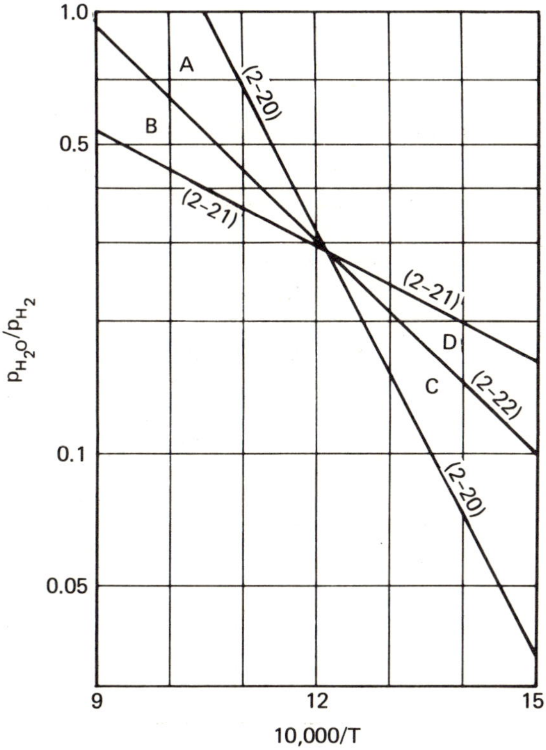

Fig. 2-5. Reduction of iron oxides with hydrogen (22):

$$Fe_3O_4 + H_2 = 3FeO + H_2O \qquad (2\text{-}20)$$
$$FeO + H_2 = Fe + H_2O \qquad (2\text{-}21)$$
$$\tfrac{1}{4}Fe_3O_4 + H_2 = \tfrac{3}{4}Fe + H_2O \qquad (2\text{-}22)$$

and (2-21) below, or else direct by reaction (2-22):

$$Fe_3O_4 + H_2 = 3FeO + H_2O \tag{2-20}$$
$$FeO + H_2 = Fe + H_2O \tag{2-21}$$
$$\tfrac{1}{4}Fe_3O_4 + H_2 = \tfrac{3}{4}Fe + H_2O \tag{2-22}$$

Each of these heterogeneous reactions has an equilibrium constant of the form p_{H_2O}/p_{H_2}, whose variation with temperature is shown in Fig. 2-5. These reactions all fall into class 5, as just discussed, hence each heterogeneous reaction goes completely to the left when the ratio p_{H_2}/p_{H_2O} lies in the region above its equilibrium line, while when this ratio lies below its equilibrium line, each heterogeneous reaction goes completely to the right. The regions of stability of the various solid phases are readily determined. Consider the area of Fig. 2-5 marked *A*. Here reactions (2-21) and (2-22) go to the left, and (2-20) to the right, hence the surviving solid phase is FeO, regardless of whether the starting solid phase was Fe_3O_4, FeO, or metallic iron. Similarly, the region marked *B* is one in which reactions (2-20) and (2-22) go to the right, while reaction (2-21) goes to the left, hence the surviving solid

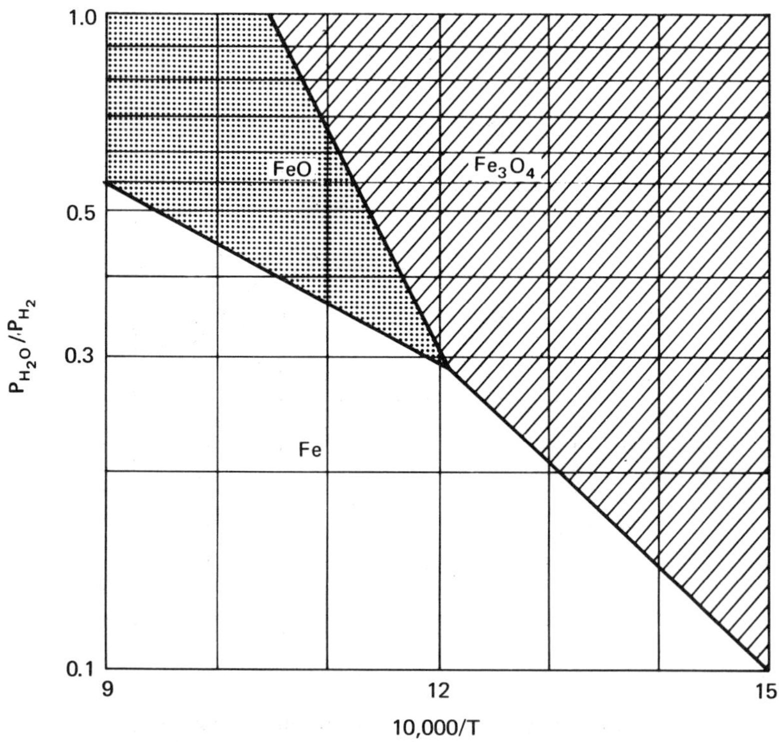

Fig. 2-6. Approximate phase diagram for the system Fe — O — H, derived from Fig. 2-5.

is again FeO. Below these lines, all reactions go to the right, and only metallic iron is stable, while above the lines all reactions go to the left and only Fe_3O_4 is stable. By similar reasoning, field C is a metallic iron region, and field D an Fe_3O_4 region. The resultant diagram is as shown in Fig. 2-6, with the equilibrium lines dotted in those regions where they no longer represent phase boundaries. This diagram is still not quite correct, in that it does not allow for solid solution formation. That is, magnetite is slightly soluble in FeO, so that the lower oxide phase formed by reduction of Fe_3O_4 has the composition $Fe_{0.95}O$ and is called Wustite. The correct final equilibrium diagram for the hydrogen reduction of the iron oxides is shown in reference (92).

A Phase Rule analysis indicates that the iron oxide system represented in Fig. 2-5 contains three components. That is, recognizing the possible presence of Fe, FeO, Fe_3O_4, H_2, and H_2O as constituents, then N is 5. There are two independent chemical reactions but no stoichiometric relations, hence, C is again 3 by Eq. (2-6). Reasoning as with the Ni–NiO–H_2 system of the preceding paragraphs, all these three-phase systems are divariant, and are represented by the lines of Fig. 2-5. The two-phase systems are trivariant, and are represented by the areas between the lines of Fig. 2-5.

While a reaction sequence is illustrated here by oxide reduction, reactions of many other compounds of elements capable of existing in several valence states, such as the halides, the nitrates, etc., behave similarly.

GAS-LIQUID-SOLID SYSTEMS

Derivation of the Phase Rule for a system containing a gas phase, a liquid solution phase, and $(P - 2)$ pure solid phases follows along the lines of the derivation just presented for gas-solid systems. To characterize the system, assuming all constituents to be present at least tracewise in both the gas and solution phases, it is necessary to know temperature, total pressure, N partial pressures in the gas phase and N concentrations in the liquid phase, or a total of $(2N + 2)$ intensive variables. However, the density of the liquid phase is fixed by temperature, pressure, and composition, hence, there are only $(N - 1)$ independent liquid phase concentrations. Similarly, the total pressure is the sum of the partial pressures, and consequently is not an independent variable. The independent intensive variables, including temperature and pressure, therefore, total to $2N$. Relations between these variables are as follows:

1. Each of the $(P - 2)$ pure solids present has a definite solubility in the liquid phase and a definite vapor pressure in the gas phase, hence, there are $2(P - 2)$ such independent solubility and vapor pressure relations. In addition, there are obviously $(N - P + 2)$ constituents

present which do not exist as solids. For each of these, the partial pressure in the gas phase is related to its mole fraction in the liquid by the usual type of vapor-pressure equation, consequently, there must be $(N - P + 2)$ such relations.

2. There are R independent chemical reactions, each involving a relation between concentrations (or partial pressures) in the form of an equili brium constant.
3. There are S independent constant stoichiometric ratios between partial pressures or concentrations in a phase.

The system's variance again equals the total number of intensive variables, minus the number of relations between them as listed above, or

$$V = 2N - 2(P - 2) - (N - P + 2) - R - S$$

Simplifying, this equation again resolves itself directly into the familiar Phase Rule Eqs. (2-7) and (2-8).

A more general derivation (38) would show that these latter two Phase Rule equations apply regardless of the number of solution phases, of whether all constituents are present in each solution phase, of whether a gas phase exists, and of whether the solid phases are pure. The Phase Rule as stated in Eq. (2-7) always applies to equilibrium systems.

EXTENSIVE VARIABLES

In industrial operations as in preparative chemistry, it is common practice to fix the initial quantities of reagents used, along with temperature and pressure. In other words, the extensive variables involving the system's total mass, rather than the intensive variables having to do with its final equilibrium composition, are fixed at the outset. For example, when operating in a simple batch reactor at constant pressure, the weight of each component in the charge is known. Students of chemistry assume intuitively that, when equilibrium is attained in such a system, then the total pressure plus the masses and compositions of all phases present are fixed, if temperature is fixed. This assumption is correct, as will be now demonstrated:

Given a system in which the following are known: temperature, pressure, and the mass of each of C components, then the number of independent unknowns in this system are as follows:

1. The number of intensive variables which are not fixed equal the system's variance as shown by the Phase Rule:	$C + 2 - P$
2. The number of unknown phase masses:	P
Total Unknowns:	$C + 2$

The number of relations between these variables are:

1. The mass of each component is known, hence, the number of such known masses is C
2. Temperature and pressure are fixed, hence, the number of known variables is 2

Total Relations: $C + 2$

As just demonstrated, the number of unknowns in such a system equals the number of relations between these unknowns, hence there is only one value for each unknown possible, and the system is fixed.

This derivation obviously applies to a steady-state flow system as well as to a batch system at constant pressure, since flow may be viewed as comprising a series of slugs of constant mass. The extension of this thinking to a batch system at constant volume instead of constant pressure is again obvious.

Liquid-Solid Equilibria Involving Ions 3

The precipitation of more or less insoluble materials formed from ionized reagents in aqueous media represents one rather important type of liquid-solid reaction. A typical example from industry is the Solvay process, in which sodium bicarbonate is precipitated from a sodium chloride solution by introduction of ammonium bicarbonate. Rates are generally very high for reactions involving ions so that equilibrium is rapidly attained unless delay results from slow mixing of reagents, unfavorable nucleation rates, etc. Apart from such difficulties, yield is limited by equilibrium conditions, and an analysis of phase diagrams is consequentially essential. Most industrial precipitations involve two inorganic salts, or are readily converted into such systems, and are discussed here under the following two headings:

1. Two salts with a common ion. Thus Cl^- is the common anion in a mixture of KCl and NaCl, while K^+ is the common cation in a mixture of KCl and KNO_3.
2. Two salts having no common ion. Such an example would be a mixture of NaCl plus KNO_3. As discussed later, any such salt pair having no common ion is called a reciprocal salt pair.

A second type of liquid-solid reaction, which is relatively new and of growing importance, involves the use of ion exchange resins to replace one ion in solution with another. Ion exchange reactions are discussed later in this chapter.

TWO SALTS WITH COMMON ION

Equilibrium phase compositions in systems of two salts with a common ion are conveniently presented on triangular diagrams. An important feature of these diagrams is that the Inverse Lever law applies.

> When two mixtures of mass W_1 and W_2 and compositions represented by points R_1 and R_2 on a triangular diagram (see Fig. 3-1) are added together, then the composition of the resultant mixture is represented by point M lying on the straight line drawn between R_1 and R_2, so that

$$\frac{\text{Distance } R_1M}{\text{Distance } MR_2} = \frac{W_2}{W_1} \tag{3-1}$$

For those interested, a derivation of this generalization is presented in the appendix to this chapter, where the general properties of triangular diagrams are further discussed.

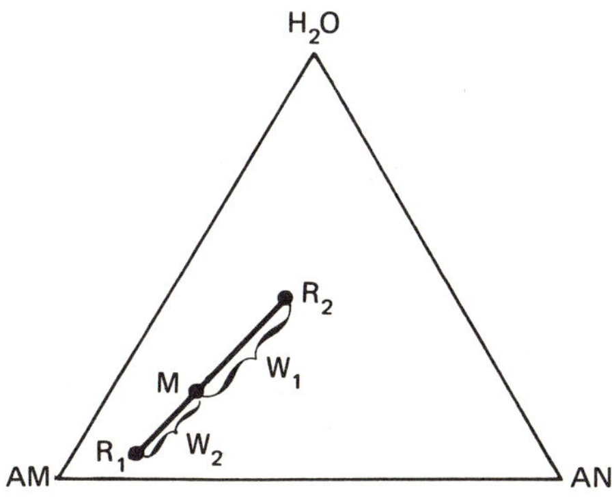

Fig. 3-1. Triangular diagram illustrating the Inverse Lever law.

Typical salt system. The phase behavior at 25°C and atmospheric pressure of the system $NaNO_3$–KNO_3–H_2O, containing NO_3^- as a common ion, is shown in Fig. 3-2. This diagram is relatively simple, since no hydrates, double salts, or solid solutions form. Point *m* in Fig. 3-2 represents a saturated solution containing only $NaNO_3$, point *n* is a saturated solution containing only KNO_3, and point *p* represents a solution saturated with both of these salts. Line *pm* represents compositions of all solutions saturated with $NaNO_3$ but not with KNO_3, while line *pn* represents compositions of all solutions

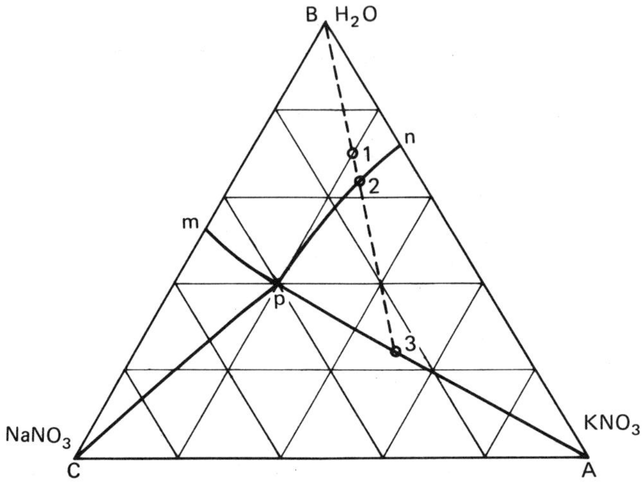

Fig. 3-2. System: water – KNO_3 – $NaNO_3$ (138). See Example 3-1. Numbers in weight per cent.

saturated with KNO_3 but not with $NaNO_3$. Area *Bnpm* is a one-phase region of unsaturated solutions, area *mpC* is a two-phase region of solid $NaNO_3$ and solutions saturated with this salt, area *pnA* is a two-phase region of solid KNO_3 and solutions saturated with this salt, while area *ApC* is a three-phase region of solid KNO_3, solid $NaNO_3$, and saturated solution corresponding to point *p*.

All salt systems contain at least one drying-up point, representing the composition of the last droplet of solution which exists in equilibrium with the indicated solid phases as evaporation to dryness is carried out. Inspection of Fig. 3-2 shows that solutions of only $NaNO_3$ in water dry up at point *m*, while solutions of only KNO_3 in water dry up at point *n*. All solutions containing both $NaNO_3$ and KNO_3, regardless of proportions, dry up at point *p*, since both these nitrates must be present as solids as water evaporates out of the last droplet of liquid. A moment's reflection shows that the composition of this last droplet must be the same as that of the first drop of saturated solution formed when a small amount of water is brought into contact with relatively large amounts of these same solids.

EXAMPLE 3-1

Given the solubility diagram of Fig. 3-2 at 25°C for the system $NaNO_3$–KNO_3–H_2O, (a) locate a solution containing 10% $NaNO_3$, 20% KNO_3, and the rest water. Is this solution saturated? (b) Water is evaporated at 25°C from this solution until solid first starts to precipitate. What is now the composition of this solution and of the solid, and what percent of the water originally present was evaporated? (c) How many pounds of pure KNO_3 can be recovered from 100 lb of initial solution by evaporation only, when operating isothermally at 25°C?

Answer: (a) This composition, plotted as point 1 in Fig. 3-2 falls into the region of unsaturated solutions, namely field *Bnpm*. (b) As water is evaporated, the solution composition moves down along the extension of the line connecting point 1 and apex *B*, since the ratio of $NaNO_3$ and KNO_3 in solution remains unchanged on this line. Precipitation of KNO_3 first starts at point 2, which represents the composition of the saturated solution containing 24% KNO_3 and 64% H_2O. For an initial system at point 1 weighing 100 lb, 70 lb are water. At point 2, by the Inverse Lever law, the water evaporated equals

$$100 \cdot \frac{\text{Distance } 1 - 2}{\text{Distance } B - 2} = 100 \cdot \frac{6}{36} = 16.7 \text{ lb}$$

The same answer is, of course, obtained by use of an algebraic material balance. That is, if W is the water evaporated, then

$$70 = W + 30 \cdot \frac{0.64}{1 - 0.64}$$

hence W equals 16.7 as before. (c) Assume the solution to remain supersaturated as evaporation proceeds so that no precipitation occurs, then the solution composition progresses along the extension of the dotted line *B*-2 into the two-phase region *pnA*. The composition of this supersaturated solution can progress to point 3, lying on line *Ap*, before any precipitation would result in contamination with $NaNO_3$. Consider this two-step process: Evaporate water from 100 lb of initial solution, finally yielding a supersaturated solution at point 3 weighing

$$100 \cdot \frac{\text{Distance } B - 1}{\text{Distance } B - 3} = 100 \cdot \frac{30}{76} = 39.5 \text{ lb}$$

Allow this supersaturated solution to reach equilibrium at 25°C, thus separating into precipitated KNO_3 and saturated solution at point *p*. The weight of KNO_3 crystals produced is

$$39.5 \cdot \frac{\text{Distance } p - 3}{\text{Distance } p - A} = 39.5 \cdot \frac{32}{81} = 15.5 \text{ lb}$$

This result can, of course, again be obtained by use of material balance.

In inorganic systems, precipitation rates are usually high and so no supersaturation occurs. Crystals of KNO_3 will therefore form during evaporation as soon as point 2 in Fig. 3-2 is crossed, while the liquid phase composition will travel to the left along line *np* from point 2 toward point *p*. Since equilibrium compositions are state properties, however, any convenient path may be chosen for calculation, such as the two-step process above.

Phase rule analysis. Figure 3-2 applies to the system KNO_3–$NaNO_3$–H_2O under 1 atm pressure when no gas phase is present. This diagram also applies when the system is open to the atmosphere, since the compositions of saturated solutions are not significantly changed by the presence of the small amounts of air dissolved in these solutions. For simplicity's sake, since dissolved air does not significantly affect salt solubilities, air is usually assumed not to be present. Since the vapor pressure of water over these solutions at 25°C is small, in the absence of air under a total pressure of 1 atm no vapor phase would exist. Total pressure on a solution also affects solubilities of the components present. Relatively great changes in pressure are however required to make even small changes in these solubilities. For all practical purposes, solubilities of inorganic solids in water are not significantly altered by increasing the total pressure from the lowest pressures attainable up to many atmospheres.

The number of components in this system can be calculated in several ways, depending on the degree of sophistication shown in the selection of constituents. In each case, however, the component count is found to be 3, as shown by the three alternative approaches below:

1. Each point within the triangle of Fig. 3-2 represents a system made up of the three materials H_2O, $NaNO_3$, and KNO_3, hence, there are 3 components.
2. Assuming complete ionization in the solution and in the solid phases, the following four constituents are present: H_2O, K^+, Na^+, and NO_3^-, hence $\boldsymbol{N}$ is 4. There is no chemical reaction; therefore, $\boldsymbol{R}$ is zero. Since all ionic solutions are electrically neutral, the number of positive ions equals the number of negative ions, hence $\boldsymbol{S}$ is unity. Thus, from Eq. (2-6), $\boldsymbol{C}$ is 3.
3. Assuming incomplete ionization, the following constituents are present: H_2O, K^+, Na^+, and NO_3^-, plus unionized KNO_3 and $NaNO_3$, hence $\boldsymbol{N}$ is 6. Independent equilibrium constants can be written for the following two reactions:

$$NaNO_3 = Na^+ + NO_3^-$$
$$KNO_3 = K^+ + NO_3^-$$

   hence $\boldsymbol{R}$ is 2. Charge equality between positive and negative ions must still exist, hence $\boldsymbol{S}$ is unity. It again follows from Eq. (2-6) that $\boldsymbol{N}$ equals 3.

As just illustrated, when using improved analytical techniques, more constituents become identifiable. In a given system, however, the component count does not change since new stoichiometric relations and chemical reactions are also involved. As a consequence of Eq. (2-6), therefore, C remains unaltered.

When the temperature and total pressure of a system are fixed, then the Phase Rule Eq. (2-7) becomes

$$V = C - P \tag{3-2}$$

This equation applies to the systems of Fig. 3-2, since these are all at 25°C and a total pressure of 1 atm. It is evident that each point within region *Bmpn* represents an unsaturated solution, and such a one-phase system is divariant by Eq. (3-2) since two concentrations can be changed arbitrarily and independently without necessarily causing a solid phase to precipitate. On the other hand, a system in region *mpC* is monovariant by Eq. (3-2) since two phases are now present, namely, solid $NaNO_3$ plus a saturated solution lying on line *mp*. The concentration of only one constituent in the saturated solution can be arbitrarily changed, since all the other concentrations will undergo associated changes as determined by line *mp*. Finally, the variance of area *ApC* is zero by Eq. (3-2), since there are three phases present here, namely, solid KNO_3, solid $NaNO_3$, and saturated solution at p which can exist in contact with these solids.

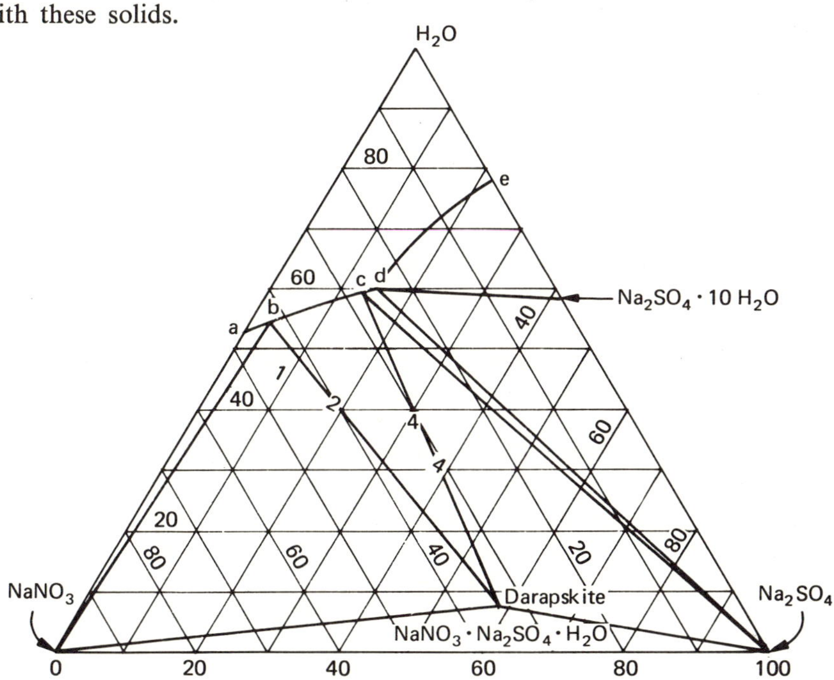

Fig. 3-3. The system $NaNO_3 - Na_2SO_4 - H_2O$ at 25° C, in weight per cent (138).

Diagram construction. Relatively complex equilibrium diagrams are not uncommon, such as for the system $NaNO_3$–Na_2SO_4–H_2O as shown in Fig. 3-3. To construct any of these diagrams, a reasonable number of sets of analytical data are required, each set being the composition of a saturated solution and the composition of the solid in equilibrium therewith. Upon initial consideration, it appears that the composition of a solid phase can be determined only after freeing it from the mother liquor clinging to its surfaces. This difficult problem of mother liquor removal is cleverly avoided by the "method of wet residues" (135). The wet residue is the wet solid which remains after draining off most of the mother liquor. This mother liquor is analyzed and the composition plotted on the diagram. The average analysis of the wet residue is next determined, and this is conveniently done by mixing the wet residue with enough water to form an unsaturated solution and analyzing. Knowing the weights of wet residue and of added water in this unsaturated solution, the average composition of the wet residue is readily computed. By the Inverse Lever law, a straight line through the two points representing the composition of the mother liquor and the composition of the wet residue must somewhere, upon extension, include a point representing the composition of the original dry solid. The dry solid composition point itself lies at the intersection of several of these lines. Locating the compositions of the solid phase in equilibrium with saturated solutions by the method of wet residues is illustrated in example 3-4.

In theory, it is possible to construct a triangular diagram, knowing only the solubilities of the pure salts in water. That is, for the saturated solution of any salt AM, the following relation holds:

$$P_{AM} = (a_A^+)(a_M^-)$$

where P_{AM} is the solubility product, while a_A^+ and a_M^- are the activities of the two ions. In extremely dilute, saturated solutions formed by insoluble salts such as AgCl, the ion molality can safely be substituted for the ion activity. If m_s is the solubility of such a salt, in units of molality, then for a salt of monovalent ions P_{AM} becomes equal to m^2. With salts showing reasonable solubility, this substitution cannot be made without seriously sacrificing accuracy, since the activity coefficients are now no longer unity. When it is nevertheless assumed that the activity of an ion equals its molality, namely the number of moles of this ion present per 1000 g of solvent, then the phase diagram can be very roughly calculated, knowing the solubility of the pure salts, as illustrated in example 3-2 below:

EXAMPLE 3-2

At 25°C, the molality of a saturated aqueous solution of $NaNO_3$ is 10.8 g moles/1000 g of water, that of KNO_3 is 3.82, while a solution saturated with both salts is 14.0 molal in $NaNO_3$ and 5.02 in KNO_3.

(a) Making the assumption (known to be poor) that molalities equal ionic activities, calculate the composition of the solution saturated with both salts from the known solubilities of these salts individually.

(b) Calculate the amount of KNO_3 which can be dissolved at 25°C into a solution containing 5.05 g moles $NaNO_3$/1000 g water (this solution is not saturated with $NaNO_3$), and compare with the experimental value of 3.94 g moles of KNO_3.

Answer: (a) Using the symbols $[Na^+]$, $[K^+]$, and $[NO_3^-]$ to designate molalities in this solution, then solubility products of the two salts become:

$$[Na^+][NO_3^-] = 10.8^2$$
$$[K^+][NO_3^-] = 3.84^2$$

By charge balance

$$[K^+] + [Na^+] = [NO_3^-]$$

Solving, $[K^+]$ is 1.28 and $[Na^+]$ is 10.3 g moles/l. Agreement with experimental data is not good. (b) Using the symbols of the preceding paragraphs, then the solubility product for KNO_3 is again

$$[K^+][NO_3^-] = 3.84^2$$

By charge balance

$$5.05 + [K^+] = [NO_3^-]$$

Solving, $[K^+]$ equals 2.1 moles/1000 g water, hence 2.1 moles of KNO_3 will dissolve in this system. Agreement with experiment is again poor.

It is clear from example 3-2 that the assumption of unity for the activity coefficients of ions in strong solutions is not good, consequently, use of the foregoing calculational procedure can give only a qualitative indication of the form of the triangular diagram.

TWO SALTS, NO COMMON ION

Two salts having no common ion, such as KCl and $NaNO_3$, undergo a double decomposition as follows:

$$KCl + NaNO_3 = NaCl + KNO_3 \tag{3-3}$$

Because one pair of salts is converted into another pair by such double decomposition, they are called reciprocal salt pairs. The equilibrium behavior of these systems is conveniently presented on a Janecke diagram (63, 121),

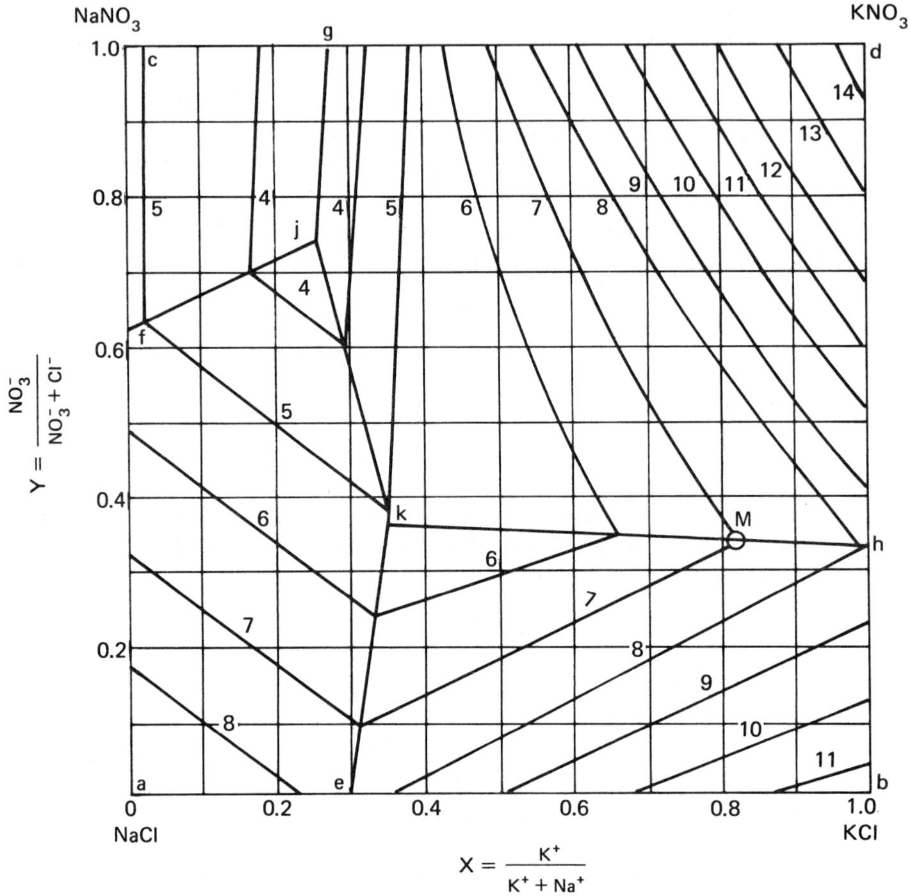

Fig. 3-4. Janecke diagram for the system NaCl — KNO_3 — H_2O at 250° C. The contour lines represent moles of water per mole of total salt in solution when that solution is saturated (123). *X* and *Y* are mole fractions.

whose construction is illustrated in Fig. 3-4. The ordinate represents the fraction of the total moles of anions existing as NO_3^-, while the abscissa shows the fraction of the cations present which exist as K^+, these quantities being on a water-free or "dry" basis. Obviously, any dry mixture of the four salts of reaction (3-3) can be represented on the Janecke square, with systems containing only one salt falling at the corners.

All lines in Fig. 3-4 refer specifically not to dry salt mixtures, but to the dissolved salts in saturated solutions, with the moles of water per mole of total dissolved salt being represented by the contour lines. The solid phases with which these solutions are saturated are indicated by the heavy boundary lines on the diagram. Thus, all solutions within area *afjke* are in equilibrium with solid NaCl, in area *bekh* with KCl, in area *cfjg* with $NaNO_3$, and in area *dgjkh* with KNO_3. All solutions on the dividing lines between these areas are in equilibrium with two solid salts, consequently solutions along *fj* are in equilibrium with both solid $NaNO_3$ and solid NaCl, along *jk* they are in equilibrium with solid KNO_3 and solid NaCl, etc. Finally, three solid phases can coexist at only two points, namely, at *j* and *k*. Inspection shows that the solution at *j* is saturated with $NaNO_3$, NaCl, and KNO_3, while that at *k* is saturated with KNO_3, KCl, and NaCl. There is clearly no saturated solution which can exist in contact with both solid KCl and solid $NaNO_3$, called the unstable salt pair.

EXAMPLE 3-3

(a) What is the composition of the saturated solution at point *M* in Fig. 3-4? (b) What is the composition of the saturated solution at point *j*? (c) Water is added to a saturated solution with composition at point *M* until the moles of water per total moles of dissolved salt are 20. Where does this solution point lie in Fig. 3-4?

Answer: (a) At point *M* in Fig. 3-4, $Y = 0.35$ and $X = 0.82$, while the moles of water per mole of dissolved salt are 7. Per mole of dissolved salt, there are therefore 0.35 moles of NO_3^-, 0.65 moles of Cl^-, 0.82 moles of K^+, 0.18 moles of Na^+, and 7 moles of water in the saturated solution. (b) By similar reasoning, per mole of dissolved salt, the saturated solution at point *j* contains 0.76 moles of NO_3^-, 0.24 moles of Cl^-, 0.27 moles of K^+, 0.73 moles of Na^+, and about 3.5 moles of water. (c) When water is added to a saturated solution at *M*, its values of *Y* and *X* remain unchanged. The solution, however, is no longer saturated, since it now contains 20 moles of water per mole of dissolved salt. Figure 3-4 might be considered an irregular surface with coordinates in three dimensions, with the *X*- and *Y*-axes as shown, while the *Z*-axis representing moles of water per mole of dry salt being vertical to the paper. Thus, adding water to the saturated solution at point *M* leaves *X* and *Y* unchanged, but increases *Z* into the unsaturated region lying above the surface formed by the saturated solutions.

Inverse lever law. Just as for triangular diagrams, the Inverse Lever law applies to the "dry" Janecke diagram, in that if N_1 moles of dry mixture at any point in diagram R_1 are combined with N_1 moles of dry mixture at any other point R_2, then the resultant mixture point *M* lies on a straight line connecting R_1 and R_2 such that

$$\frac{\text{Distance } R_1M}{\text{Distance } MR_2} = \frac{N_2}{N_1}$$

The proof of this law is analogous to that presented in the Appendix for triangular diagrams.

EXAMPLE 3-4

(a) A saturated solution containing exactly 1 mole of NaCl in water is added to a saturated solution of exactly 1 mole of KNO_3 in water, all at 25°C. Describe the resultant mixture. (b) Dry KNO_3 is added to a saturated solution of NaCl. Prepare a graph showing as abscissa the moles of KNO_3 added per mole of NaCl initially in solution and as ordinate the moles of water per mole of total salt. Operations are again at 25°C.

Answer: (a) From Fig. 3-4, a solution saturated with pure NaCl contains 9 moles of H_2O per mole of salt, while a saturated solution of pure KNO_3 contains 14.5 moles of water per mole of salt. The proposed mixture of these solutions would contain 1 mole each of K^+, Na^+, Cl^-, and NO_3^-, and 23.5 moles of H_2O.

Note that this same mixture could have been made in many different ways, as by mixing 1 mole of $NaNO_3$ in its saturated solution with 1 mole of KCl in its saturated solution, and adding water, etc. By the Inverse Lever law applied to the dry salt content, the mixture lies midway on the diagonal between the NaCl and KNO_3 corners, where X and Y are each 0.5. Since the water actually present (11.75 moles per mole of salt mixture) is greater than that of the saturated solution at this point (5.7 moles H_2O/mole salt), this final solution is not saturated. (b) Let n be the moles of KNO_3 added per initial mole of NaCl and w the moles of water per total mole of salt in solution, then since initially there are 9 moles of water in the saturated solution per mole of NaCl,

$$w = \frac{9}{1 + n} \tag{a}$$

The values of X and Y on the Janecke diagram corresponding to addition of n moles of KNO_3 per mole of NaCl are then

$$X = Y = \frac{1}{1 + n} \tag{b}$$

The dotted line of Fig. 3-5 shows the change of w with n as calculated from Eq. (a). If S represents the moles of water in a saturated solution, then the solid line in Fig. 3-5 shows the variation of S with n as determined from Fig. 3-4. By inspection, the solution immediately becomes unsaturated when solid KNO_3 is first added, and remains so as more KNO_3 is added until n equals 0.7. Saturation is reached at this point where $X = Y = 0.445$, and $S = w = 5.6$. Further additions of solid KNO_3 will cause no further change in the saturated solution composition, but merely increase the quantity of solid KNO_3 in contact with this solution.

It is evident that the entire Janecke diagram could be predicted, knowing solubilities of the individual pure salts, by the method of example 3-2. Unfortunately, the success of these predictions is again no good because of the unwarranted assumption that ionic activity coefficients are unity, or else equal for all salts present.

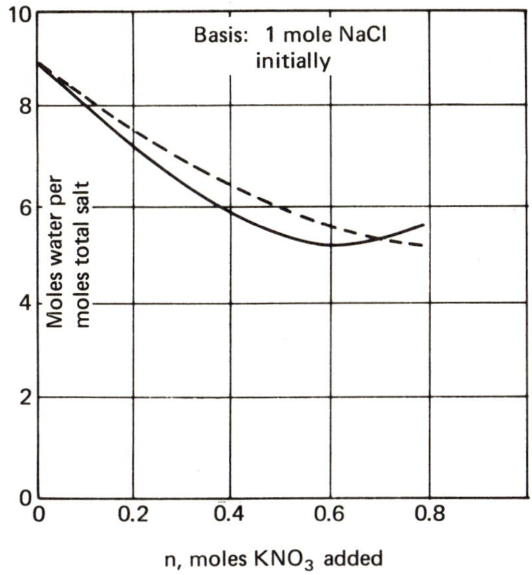

Fig. 3-5. For Example 3-4. Solid line shows saturated solution compositions on diagonal *ad* of Fig. 3-4. Dotted line shows solution compositions as solid KNO_3 is added.

Drying-up points. Consider the generalized reciprocal salt system of the following reaction between salts of monovalent ions:

$$AM + BN = AN + BM \tag{3-4}$$

The diagrams of Fig. 3-6 are typical of such systems when no hydrates, double salts, or solid solutions form. In each diagram, the stable salt pair is *AM* plus *BN*, since these solids can coexist at conditions represented by line *jk*. The pair *AN* plus *BM* is unstable since there is no region on these diagrams where these two salts can coexist at equilibrium.

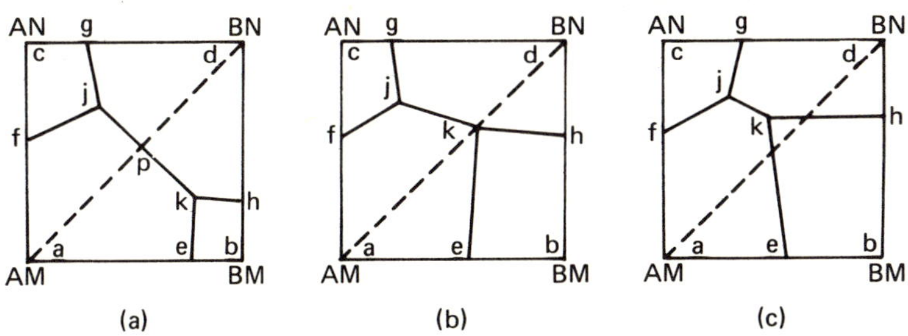

Fig. 3-6. Typical Janecke diagrams. The ordinates and abscissas for each are as follows:

$$Y = (N^-)/(M^- + N^-) \qquad X = (B^+)/(A^+ + B^+)$$

where moles positive ions equal moles negative ions.

It is noteworthy that, as a consequence of the Inverse Lever law, any point lying above diagonal *ad* in all three diagrams can be made by mixing *AN*, *AM*, and *BN*. No points below the diagonal, however, can be made by mixing only these three salts. Similarly, all points below the diagonal can be made by mixing *AM*, *BM*, and *BN*, but no points above the diagonal can be made in this way. As a further consequence of the Inverse Lever law, any point lying below diagonal *ad* in all three diagrams can be formed by mixing *AM* and *BN*, and then removing *AN*. This removal of *AN* is equivalent to adding *BM*, since from Eq. (3-4)

$$BM = AM + BN - AN \tag{3-5}$$

By similar reasoning, any point above diagonals *ad* on Fig. 3-6(a), (b), and (c) can be made by mixing *AM* and *BN*, and then either removing *BM* or adding *AN*.

The behavior upon evaporation of the saturated solutions of Figs. 3-6(a), (b), and (c) can now be considered. These solutions are assumed at the outset to contain no solid phases. It is evident that all such solutions lying above the dotted diagonals dry up at point *j*, where the last droplet will be in contact with solids *AM*, *AN*, and *BN*. Similarly, all saturated solutions below the diagonals dry up at point *k*, in contact with solids *AM*, *BM*, and *BN*. The fact that point *k* in Fig. 3-6(c) lies above the diagonal seems at first surprising. This behavior is consistent with the fact that, when a droplet of water is brought into contact with a large excess of solids *AM*, *BM*, and *BN*, reaction (3-4) goes to the right to a limited degree, whereupon a little *AM* and *BN* go into solution, forming additional solid *BM* and leaving *AN* in solution.*

In addition to points *j* and *k*, point *p* is obviously also a drying-up point in Fig. 3-6(a), since this point, representing a solution saturated with *AM* and *BM*, cannot move during evaporation to dryness. In Fig. 3-6(b), area *ebhk* has grown so that point *k* now lies on the diagonal, consequently, points *k* and *p* coincide. Because of this circumstance, Fig. 3-6(b) has only two drying-up points, namely, *j* and *k*, as compared to the three drying-up points of Fig. 3-6(a). Inspection shows that point *p* has entirely disappeared in Fig. 3-6(c), since there is now no point on the diagonal which remains fixed as evaporation to dryness proceeds.

EXAMPLE 3-5

Point *m* in Fig. 3-7 represents a saturated solution at 50°C, for which X is 0.6, Y is 0.8, and w is 4.2. This is isothermally evaporated to dryness. Trace the path, and calculate the amount of water evaporated and the amounts of

**Note:* If this droplet is now separated from the three solid phases and evaporated to dryness, its drying-up point is found to be point *j*, for which the moles of water per mole of dissolved salt are smaller than point *k*.

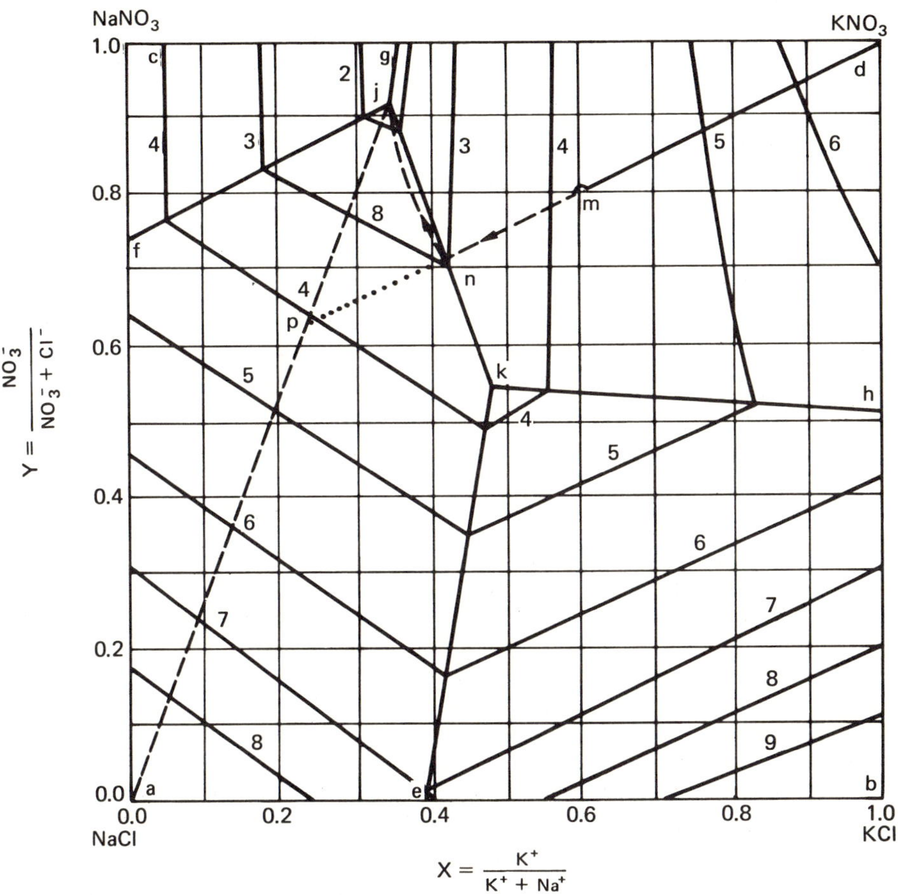

Fig. 3-7. Janecke diagram for the system $NaNO_3 - KNO_3 - H_2O$ at 50° C (123). The contour lines show the moles of water per mole of dissolved salt in saturated solutions. X and Y are mole fractions, dry basis.

each solid phase present when (a) solid NaCl first starts to form and (b) solid $NaNO_3$ first starts to form.

Answer: (a) The actual path of the saturated solution composition during evaporation is shown by the heavy dashed lines *mn* and *nj* of Fig. 3-7. Immediately after the start of evaporation, KNO_3 starts to precipitate, while, as required by the Inverse Lever law, the solution composition moves along the dashed extension of line *dm* toward *n*. When *n* is reached, the solution is saturated with both NaCl and KNO_3 and contains 3 moles of water per mole of dissolved salt. The KNO_3 precipitated per mole of total salt initially in solution when *n* is reached is

$$1 \times \frac{nm}{nd} = 1 \times \frac{0.6 - 0.42}{1 - 0.42} = 0.31 \text{ moles}$$

The water evaporated is $\{4.2 - 3(1 - 0.31) = 2.13\}$ moles per mole of salt initially in solution. (b) One of many possible ways of solving this problem is as follows: extend dotted line *mn* until it intersects the line drawn between points *a* and *j* at *p*, where by inspection X equals 0.23. Now evaporate the original solution, starting at *m* and ending at *p*, assuming supersaturation so that no NaCl but only KNO_3 has precipitated at point *p*. The moles of KNO_3 precipitated per mole total of salt initially in solutions is

$$1 \times \frac{pm}{pd} = \frac{0.6 - 0.23}{1 - 0.23} = 0.48$$

The moles of salt in supersaturated solution at *p* is then 1 − 0.48 or 0.52 moles. Now allow the supersaturated solution at *p* to come to equilibrium, and then evaporate as needed until the saturated solution is at *j*, where X is 0.34. The NaCl precipitated is

$$0.52 \times \frac{jp}{ja} = 0.52 \times \frac{0.34 - 0.23}{0.34 - 0} = 0.166 \text{ moles}$$

Hence, when point *j* is reached, per moles of initial salt in solution, there are present 0.166 moles of solid NaCl, 0.48 moles of solid KNO_3, and 0.354 moles of salt mixture (1.9 moles of H_2O mole of dissolved salt) in saturated solution with composition as of point *j*. The total evaporated water is (4.2 − 0.354 × 1.9), or 3.53 moles.

Any evaporation of the solution at point *j* results in no further composition change for the solution.

Phase rule analysis. An air-free system containing water plus two reciprocal salts, namely, AM plus BN, is a four-component system, as is evident from the following alternative methods of reasoning:

1. Assume the salts do not ionize. The constituents present are water plus nonionized salts AM, AN, BM, and BN, hence N is 5. The occurrence of reaction (3-4) makes $\boldsymbol{R}$ equal to unity and, since $\boldsymbol{S}$ is zero, $\boldsymbol{C}$ is 4.
2. Assume all four salts are fully ionized both in the solid state and in solution, hence, the constituents present are A^+, B^+, M^-, N^-, and H_2O, so N is 5. No chemical reaction occurs, so $\boldsymbol{R}$ is zero, but $\boldsymbol{S}$ is unity because of the equality of positive and negative charges, hence, $\boldsymbol{C}$ is again 4.
3. Each of the four salts is assumed partly ionized, as follows:

$$AM = A^+ + M^- \tag{3-6a}$$

$$BM = B^+ + M^- \tag{3-6b}$$

$$AN = A^+ + N^- \tag{3-6c}$$

$$BN = B^+ + N^- \tag{3-6d}$$

There are now nine constituents, namely, H_2O, A^+, B^+, M^-, N^-, AM, BM, AN, and BN. Because of reactions (3-6a, b, c, d), ***R*** equals 4 [note reaction (3-4) is not independent]. As before, ***S*** is unity because the number of positive ions equals the number of negative ions. It follows again that ***C*** is 4.

Thus, by each of these three different initial assumptions, it is found that an aqueous solution of a reciprocal salt pair contains four components.

Upon fixing the temperature and pressure, application of the Phase Rule Eq. (3-2) to the four-component system just described shows that these become invariant when four phases are present (a solution plus three solid phases at points *j* and *k* of Figs. 3-4, 3-6, and 3-7). Similarly, these systems are monovariant when three phases are present (a solution phase plus two solid phases along *hj*, *kj*, and other lines of the diagram). Finally, these systems are divariant when two phases are present (a solution phase plus a solid phase, as within areas *afjke*, *cgjf*, etc.).

ION EXCHANGE

An interesting technique for replacing one cation in solution with another, or one anion with another, involves the use of ion exchange resins. These resins, also called polyelectrolytes, are insoluble high molecular weight solids or liquids carrying exchangeable anions or cations. One important type of ion exchanger is polystyrene containing sulfonic acid groups, made by polymerizing styrene in the usual way and then sulfonating. A small chain segment of this polymer may be represented as follows:

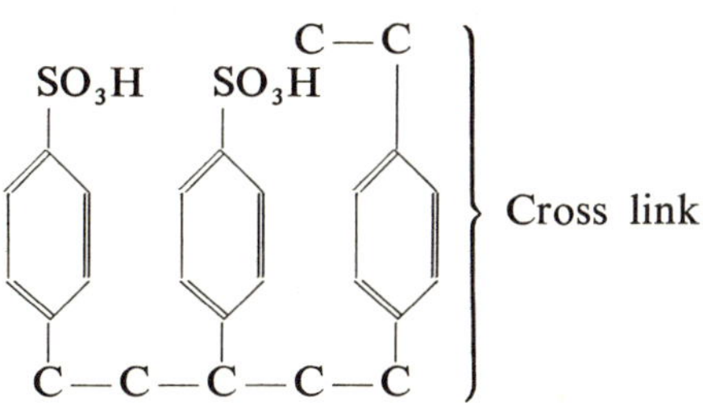

When desired, cross-links of the type shown above are introduced by mixing divinyl-benzene with the styrene before polymerization. As chain lengths and the degree of cross-linkage are increased, the exchanger is converted from a liquid to a solid.

Since polystyrene-sulfonic acid, like all polyelectrolytes, must remain electrically neutral, its H^+ ions cannot be removed by washing with an ion-free liquid such as water. However, when this exchanger is brought into contact with an aqueous solution of a sodium salt such as sodium chloride, then some of the sodium ions are found to migrate rapidly from the solution into the resin, displacing an equivalent amount of H^+ ions from the resin into the solution. The reaction occurring is as follows:

$$HR + Na^+ = NaR + H^+ \tag{3-7}$$

where R represents the appropriate segment of the polyelectrolyte associated with each ionic group, while HR and NaR, etc., represent the electrically neutral resin exchanger in the hydrogen and sodium forms, respectively. The foregoing is an equilibrium reaction, since the ratio of Na^+/H^+ in the resin is always the same for a given ratio and concentration level of the Na^+ and H^+ ions in the final aqueous phase, regardless of from which side the final condition is approached.

Further study shows that anions (in the example just presented, these are the chloride ions) are unable to penetrate to any significant degree into this exchanger, since they are excluded by the negative charges of the sulfonic acid groups rigidly attached to the polystyrene network. Because of the presence of these ionic groups, polyelectrolytes are hydrophylic, readily imbibing water and swelling when in contact with water or an aqueous solution. The resins, when cross-linked, show no solubility in water.

Anion exchangers as well as cation exchangers are being synthesized and used industrially today. Anion exchangers are made by introducing quaternary ammonium, phosphonium, or tertiary sulfonium groups into a polystyrene network. These anion exchangers show behavior which is complementary but analogous to that of the cation exchangers.

Equilibrium. Exchange reactions, such as reaction (3-7) involving the monovalent cations $A^+ + B^+$ may be generalized as follows:

$$RA + B^+ = RB + A^+ \tag{3-8}$$

The equilibrium constant for this reaction is written as follows:

$$K_{A^+}^{B^+} = \frac{(\bar{B}^+)(A^+)}{(\bar{A}^+)(B^+)} \tag{3-9}$$

where A^+ and B^+ are ionic activities in solution, while $\bar{A}^+$ and $\bar{B}^+$ are ionic activities in the polyelectrolytes. When for convenience, the approximation is made that ionic activities equal ionic concentrations, then K of Eq. (3-9) is found to vary somewhat with the ionic ratios and concentrations at a given temperature. With activities expressed as concentrations, then K is called the selectivity coefficient, since it represents the ratio of B^+/A^+ in the resin to that in the aqueous phase, as rearrangement of Eq. (3-9) makes clear

$$K_{A^+}^{B^+} = \frac{\bar{B}^+/\bar{A}^+}{B^+/A^+} \tag{3-10}$$

As a first approximation, the selectivity is often assumed independent of temperature and concentrations.

For a resin-solution system involving reaction (3-8) at equilibrium, it is often convenient to define y as the fraction of the total cations in the resin which is $\bar{B}^+$, while x is the fraction of the total cations in solution which is B^+. Substituting into Eq. (3-10),

$$K_A^B = \frac{(y)/(1 - y)}{(x)/(1 - x)} \tag{3-11}$$

The typical variation of y and x at equilibrium for three constant values of the selectivity is shown in Fig. 3-8.

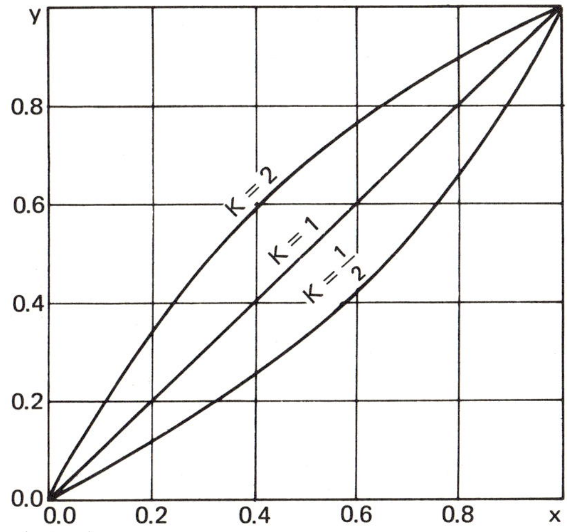

Fig. 3-8. Equilibrium relations between ions in solution and in ion exchange resins, based upon Eq. (3-10).

$y = \bar{B}^+/\bar{A}^+$ in resin

$x = B^+/A^+$ in solution

Inspection shows that the selectivity as defined for reaction (3-8) by Eqs. (3-9) and (3-11) is not only dimensionless but numerically independent of whether concentrations are in molal or molar units. However, for a reaction involving ions of different valence,

$$2AR + C^{++} = CR_2 + 2A^+ \tag{3-12}$$

the selectivity becomes

$$K_A^C = \frac{(\bar{C}^{++})/(\bar{A}^+)^2}{(C^{++})/(A^+)^2} \tag{3-13}$$

A moment's reflection will show that the numerical value of K_A^C in Eq. (3-13) now depends upon whether concentrations are in molal or molar units, since the density of the polyelectrolyte phase differs from that of the aqueous phase. Thus, for systems involving ions of different valence, it is important to use consistent concentration units when making calculations involving the selectivity.

Modern resins contain from 1 to 10 equivalents of active ionic groups per 1000 g of material, hence may be said to have a normality ranging approximately from 1 to 10. Typical values of the selectivity coefficient for several resins are shown in Table 3-1. It is usually found that divalent ions are held more strongly than monovalent ions, and that trivalent ions in turn are held more strongly than divalent ones. Not evident from this table is the fact that selectivities change little with temperature, because ΔH values for reactions (3-8) and (3-12) are small, being generally less than 2 kcal/g equivalent of ions exchanged.

TABLE 3-1. SOME AVERAGED SELECTIVITY COEFFICIENTS*

Resin	Ion A	Ion B	K_A^B
Sulphonated polystyrene (8–10%DVB)	H^+	Li^+	0.8
	H^+	Na^+	2.0
	H^+	Ca^{++}	42
Benzyldimethylammonium polystyrene (8%DVB)	Cl^-	F^-	0.1
	Cl^-	Br^-	2.5
	Cl^-	OH^-	0.5
	Cl^-	I^-	18

*From J. A. Kitchener, *Ion Exchange Rates*, London: Methuen & Co. Ltd., 1957, page 23.

Solid, ion-exchange resins are usually produced by bead polymerization in the form of small spheres, available commercially in several sizes in the diameter range of from at 0.05–1.0 mm. These diameters are for dry material, and become somewhat larger when wet because of swelling due to water imbibition. Solid ion-exchange resins in bead form can be put through many thousand cycles without deterioration. Liquid resins are subject to greater possible losses in use than solid resins. These losses result from trace solubility in water, from entrainment and emulsion formation during use, and, of course, from leaks and spills.

Ion-exchange operations. When a solid exchanger is used to replace B^+ ions in solution by A^+ ions, as in reaction (3-8), conventional operation involves passing the solution slowly downwards through a fixed bed of the appropriate resin initially in AR form. Exchange rates are usually relatively rapid as compared to throughput rates, and so composition history through the bed is determined by equilibria rather than kinetics. When equilibria are favorable, exchange takes place within the bed in a narrow zone of constant depth, and this zone moves slowly down through the column as an increased amount of solution is processed, until the resin is spent. The resin above this reaction zone is all in the BR form, and below is all in the AR form. For such an operation, the effluent contains only B^+ ions as cations until the moment of "breakthrough," which occurs when the reaction zone reaches the bottom of the bed. Operations are now discontinued and, after washing, the resin is regenerated by treatment with a solution containing only B^+ as cations.

Reaction zones in fixed exchanger beds are not always narrow and of constant depth. In many cases, this zone widens rapidly as the solution passes down through the exchanger column, consequently, breakthrough occurs long before all of the A^+ ions in the resin have been replaced by B^+ ions. As shown in the review of Baddour (6), a narrow reaction band of constant width having stable and sharp boundaries always develops when the exchanger prefers B^+ to A^+, that is, when the selectivity K_A^B is greater than unity. On the other hand, a rapidly widening band and early breakthrough develops when the selectivity is less than unity. Therefore, for a given pair A^+ and B^+, since K_B^A is the reciprocal of K_A^B, when displacement of A^+ by B^+ proceeds efficiently with a narrow reaction zone because K_A^B is greater than unity then the displacement of B^+ by A^+ proceeds inefficiently with a widening reaction zone because K_B^A is less than unity.

Occasionally, selectivities vary with the total concentration of ions in solution in such a way that both efficient ion replacement from solution and efficient resin regeneration become possible. Thus, when replacing calcium with sodium ions in dilute solution, the selectivity coefficient $K_{\mathrm{Na^+}}^{\mathrm{Ca^{++}}}$ is found to be greater than unity, consequently, calcium can be efficiently replaced by sodium when passing hard water through a sodium resin bed. At high ionic concentrations in the aqueous phase, however, the selectivity $K_{\mathrm{Ca^{++}}}^{\mathrm{Na^+}}$ is greater than unity for some exchangers, consequently, the resin containing Ca^{++} ions can again be efficiently regenerated with a strong NaCl solution. Magnesium ions show an analogous behavior.

Continuous operation. Continuous ion-exchange operation, in which the resin moves countercurrently to the solutions being processed, is relatively easy to arrange with a liquid resin, but is mechanically more difficult with solid resins. The mathematics of continuous operations, however, are relatively simple. Imagine a six-plate column as in Fig. 3-9 for carrying out reaction (3-7), with solution containing only B^+ as cation fed to the top plate, and resin containing only A^+ as cation fed to the bottom plate. The idealized tower is so arranged that the solution passes down through the column while the exchanger material passes up through the column from plate

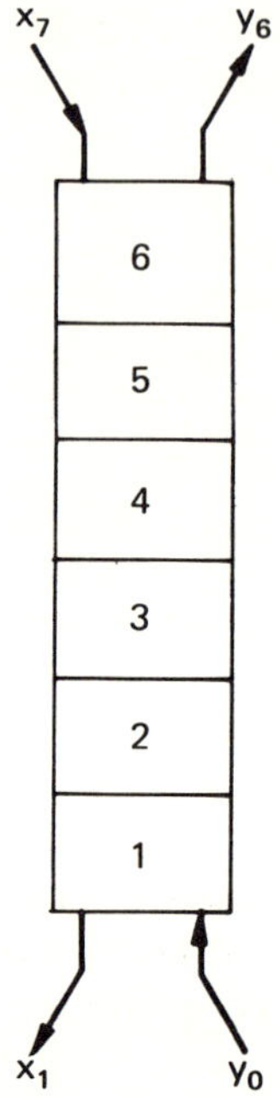

Fig. 3-9. A six-plate column operating continuously. Plate numbers and flow directions are shown.

to plate. The solution and resin on each plate are assumed thoroughly mixed, so that they leave each plate at equilibrium. The similarity of this operation to a fractionating tower or a countercurrent extraction system is apparent.

The plates in the column of Fig. 3-9 are numbered, starting with the bottom plate as number 1, as shown. Thus, y_1 and x_1, respectively, refer to the compositions of the resin and the solution leaving the bottom plates; y_2 and x_2 refer to the second plate, up to y_T and x_T for the Tth plate. The compositions of the resin and liquid feeds to the column are then y_0 and x_{T+1}. By material balance on B^+ over the whole tower is

$$R(y_T - y_0) = L(x_{T+1} - x_1) \tag{3-14}$$

where R is the total number of equivalents of cations in the resin passing up through the tower per unit time, while L is the total number of equivalents of cations in the liquid solution passing down through the tower per unit time. Making R and L equal, then for the whole tower,

$$(y_T - y_0) = (x_{(T+1)} - x_1) \tag{3-15}$$

For plate 1, a material balance on B^+ becomes

$$(y_1 - y_0) = (x_2 - x_1) \tag{3-16}$$

where y_1 and x_1 represent the equilibrium compositions of the resin and liquid leaving this plate, while x_2 is the composition of the liquid entering from plate 2 and y_0 is the composition of the fresh resin feed. Similar equations can be written for each of the other plates in the tower. This operation can obviously be represented on the familiar McCabe-Thiele type diagram, as shown in Fig. 3-10 for the case in which K_A^B of Eq. (3-10) is 2, and in which there are six plates in the tower. The "operating line" of this column presented in Fig. 3-10 is, of course, a graphical presentation of Eq. (3-15).

It is clear from inspection of Fig. 3-10 that, even if many more plates were added to the column of Fig. 3-9, most of the ion exchange would still occur in the middle six plates. Composition changes on the remaining plates, while small, would be such that the solution leaving at the bottom of the column would contain at most only a trace of B^+.

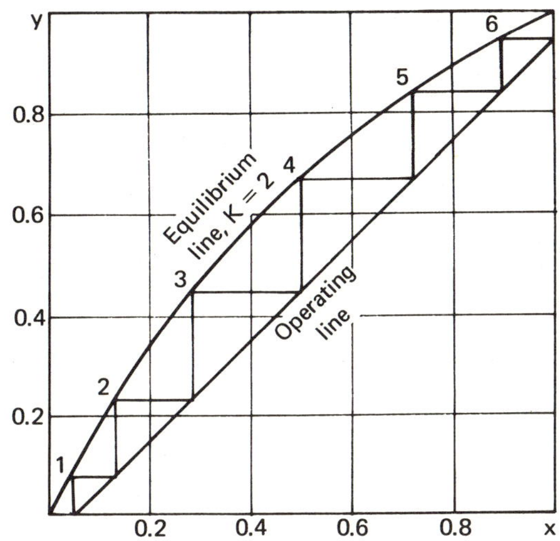

Fig. 3-10. A six-plate countercurrent tower for ion exchange, with $K_A^B = 2$; $L/R = 1.0$.

$y = \bar{B}^+/\bar{A}^+$ in resin

$x = B^+/A^+$ in solution

Operation of this six-plate column with a system for which the selectivity is less than unity is shown in Fig. 3-11, for which K_A^B is 0.5 and L/R is 1.0. Inspection shows that complete exchange cannot even be approximated in this case. Even an infinite number of plates would obviously show A^+ and B^+ present in both the liquid and the resin leaving the tower. By increasing the ratio L/R to 2.0, the operation of Fig. 3-12 results. Inspection shows that the product resin leaving the top of the tower is now largely changed from the A-form to the B-form, but the liquid is only partly stripped of B^+.

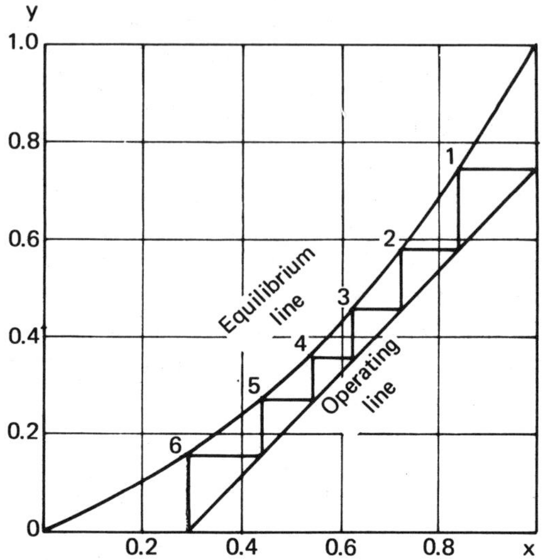

Fig. 3-11. A six-plate countercurrent column for ion exchange with $K_A^B = \frac{1}{2}$; $L/R = 1.0$.

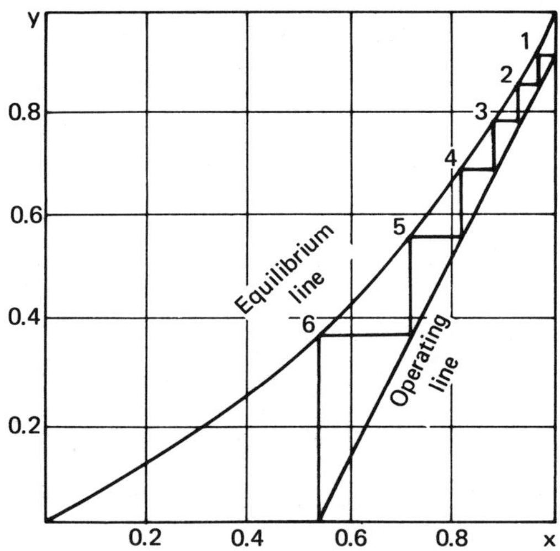

Fig. 3-12. A six-plate column for ion exchange with $K_A^B = \frac{1}{2}$; $L/R = 2$.

It follows from the foregoing that, in steady-state countercurrent operations, a complete exchange of A^+ ions with B^+ ions in the resin and a simultaneous complete exchange of B^+ ions with A^+ ions in the liquid is possible only, if K_A^B is greater than unity.

Moving bed system. Consider a moving bed system operating in a steady-state, in which a bed of solid resin is made to travel up through a tower counter to a liquid flowing down which completely fills the small, void space of the tower. When K_A^B is 2 and L/R is fixed at unity, then the equilibrium and the operating lines for this system are obviously similar to those of the plate tower of Fig. 3-10. When K_A^B is 0.5 and L/R is fixed at 2.0, then the equilibrium and operating lines are the same as those of the plate tower of Fig. 3-12.

APPENDIX

Properties of triangular diagrams. The weight or moles per cent of *A*, *B*, and *C* in any three-component mixtures are here denoted by *a*, *b*, and *c*, respectively. To represent such a composition on an equilateral triangular diagram, the corners of the triangle are marked *A*, *B*, and *C* as in Fig. 3-13. Now divide side *AC* into 100 parts, and measure off distances *a*, *b*, and *c* as shown, with distance *c* measured from point *A*, distance *a* measured from point *C*, and distance *b* lying between *a* and *c*. The point between *a* and *b* is *x*, and that between *b* and *c* is *y*. Draw dotted lines through *x* and *y* paral-

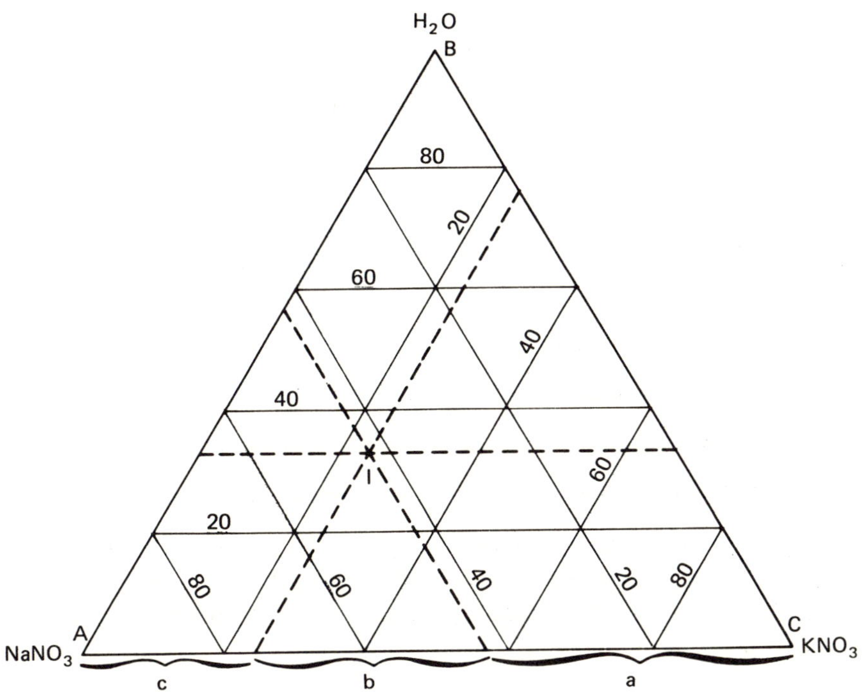

Fig. 3-13. A triangular diagram illustrating the method for locating point *I* for a mixture containing 43% *A*, 33% *B*, and 24% *C* by weight.

lel to the sides of the triangle. Their point of intersection, namely I, represent the desired mixture of A, B, and C.

Distances a, b, and c are to be found on any one of the three sides of the triangle in Fig. 3-13 as is evident from inspection. The location of a, b, and c on each of the triangle sides follows directly from the properties of equilateral triangles. Consideration of this method of plotting shows that every possible mixture of A, B, and C can be represented by a point on an equilateral triangle, and every point on the triangle is a possible mixture of A, B, and C.

Inverse lever law. Given two mixtures R_1 and R_2 of A, B, and C, prove that M, which is a mixture of R_1, and R_2, has a composition which lies on a straight line connecting R_1 and R_2 on a triangular diagram, and that distances R_1M and MR_2 are related by Eq. (3-1). To prove this, locate compositions R_1 and R_2 in Fig. 3-14, draw a straight line between these points, and locate a point marked M' in accordance with Eq. (3-1). It must now be shown that the composition at M' on the diagram is identical with the one which is obtained in the actual mixing.

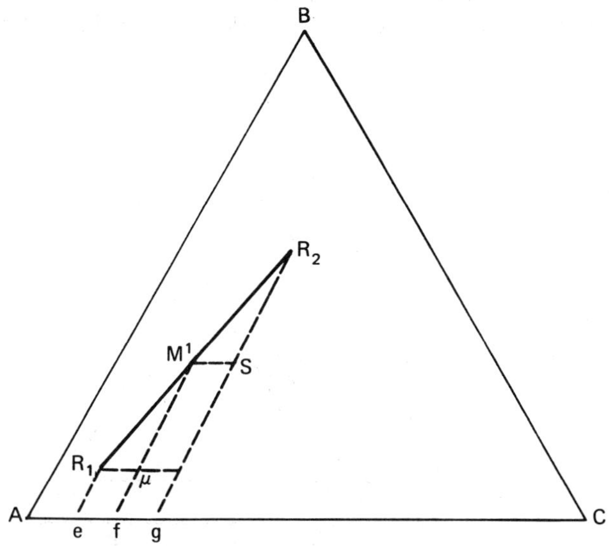

Fig. 3-14. Illustrating the derivation of the Inverse Lever law.

Draw dotted line through points R_1, M', and R_2 parallel to AB and AC, as shown. By simple geometry, it is evident that eR_1, fM', and gR_2, respectively, equal b_1, b_M, and b_2, namely, the per cent of component B in R_1, M,′ and R_2, hence:

$$\frac{\text{Distance } M'\mu}{\text{Distance } R_2S} = \frac{b_{M'} - b_1}{b_2 - b_M} \tag{3-17}$$

But triangles $R_1\mu M'$ and $M'sR_2$ are clearly similar, hence:

$$\frac{\text{Distance } M'\mu}{\text{Distance } R_2S} = \frac{\text{Distance } R_1M'}{\text{Distance } M'R_2} \tag{3-18}$$

But this diagram was so constructed that

$$\frac{\text{Distance } M'\mu}{\text{Distance } R_2S} = \frac{W_2}{W_1} \tag{3-19}$$

Combining Eqs. (3-17) and (3-19) and simplifying:

$$W_1b_1 + W_2b_2 = b_M(W_1 + W_2) \tag{3-20}$$

But Eq. (3-20) is the material balance on component B when R_1 and R_2 are mixed. It follows that the composition at M' equals the one obtained in the actual mixing, and that the Inverse Lever law is correct. Obviously, similar constructions of dotted lines to sides AB and BC would lead to material balances on components A and C.

Management of Materials 4

A chemical plant is an assembly of processing equipment through which material and energy travel batchwise or continuously. Aside from heat exchangers and materials handling devices like pumps, blowers, and conveyors, this equipment is used either for materials mixing, reaction, or separation, with an enormous diversity of design encountered in each of these three areas. Material is desirably made to move through these various operating stages so that there is an optimum yield of "on-specification product." Raw material costs and other operating costs, as well as investment in plant facilities, must be kept at levels resulting in minimum total costs of production. The attainment of such minimum costs is the primary objective of materials management.

Process flow sheet. A flow sheet is a graphical representation of the equipment in a plant and the path followed through this equipment by the material being processed, with the movement of energy often also indicated. A flow sheet is usually in the form of a line and block diagram, as illustrated by the figures in this chapter, with quantities of material and energy in the various streams sometimes shown. There has been little success in developing a standard set of symbols for general use on flow sheets representing different types of equipment, and usual practice still involves indicating each equipment stage by a square, suitably labeled as in Fig. 4-1.

As already pointed out in Chap. 1, system planning often starts with the reactor, in which it is frequently convenient to assume that equilibrium is approximated. Alternatively, as with sequential reactions, a reasonable product composition is assumed. Apparatus and sequences for mixing and separation are then chosen, for which typical efficiencies are assumed, so that intial perspective on overall material and energy needs can be attained. If large amounts of energy are rejected, then various methods for utilizing this energy within the process are tested. Given such initial perspective on performance, the system of apparatus is modified, recognizing that in most cases equilibrium will not be in fact attained in the reactor, and that separations will perhaps not be as efficient as assumed. System performance is then recalculated, using better estimates of the reaction rates, mass, and heat transfer coefficients, etc. The processing system evolves in this fashion by a series of increasingly accurate and careful estimates. Depending on the ingenuity and inventiveness of the planner and the complexity of the case considered, not one but more often two and three alternative schemes are usually found. Final choice between such alternatives is always a matter of economics.

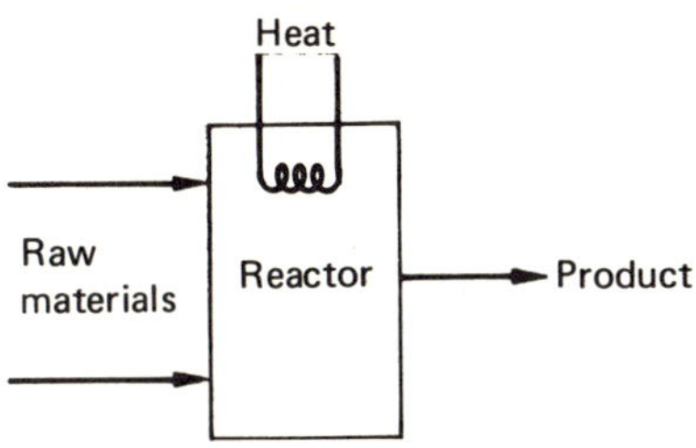

Fig. 4-1. A simple reactor for homogeneous reaction.

While evolution of a system usually starts with the reactor, the reactor does not necessarily continue to dominate planning. For example, separation of pure butadiene from the product stream of a butane dehydrogenation unit is difficult, requiring a complex and extensive recovery system. In cases like this, reactor investment and operating costs may be only small fractions of the total plant costs. Equipment for product separation may not only require a much larger investment than the reactor, but may represent far greater design and operating problems. The fact remains, however, that the reactor is a convenient starting point for initial system orientation, except in those rare processes, such as sea-water desalination by distillation, in which no reactor exists because only physical separations are involved.

Flow sheet types. Regardless of whether operations are batch or continuous, flow sheets of processes involving a single reactor can be classified into one of the five categories described below. Flow sheets for more complex processes involving two or more reactors may be viewed as combinations of these five basic types.

Type 1. Reactor for homogeneous reactions, no accessory equipment. Some processing systems, for which a typical flow sheet is shown in Fig. 4-1, involve no equipment other than a reactor, to which reactants are fed directly without purification, and from which the product stream emerges as a single phase, ready for use.

Energy is added or removed from the reactor as needed. Thus methyl methacrylate is polymerized in batches in the presence of a catalyst in a kettle equipped with an agitator and a jacket for cooling water or steam for heating. With some formulations, the cooled product is ready for use. Another example is an ordinary gas-burning steam boiler for generating process steam. The fuel is injected into the fire box, burned with air, and the products of combustion are cooled by vaporizing water in the boiler tubes prior to leaving the furnace via the stack. Even though the same flow sheet applies to these two examples, reactor construction is obviously very different.

Type 2. Reactor without accessory equipment, heterogeneous reaction. The foregoing processing system must be modified when the reactants and products remain in different phases as in Fig. 4-2. Examples are as follows:

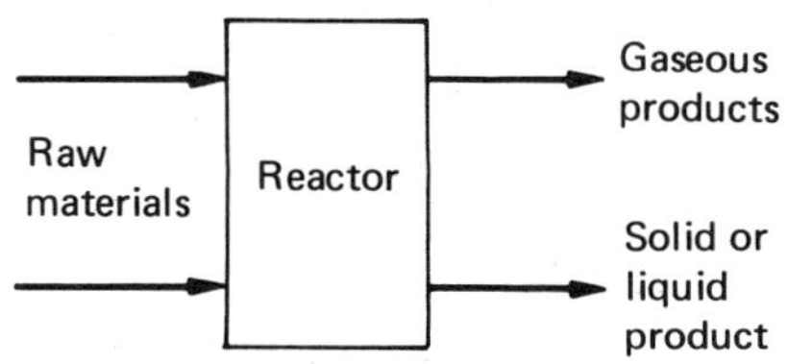

Fig. 4-2. A flow sheet for a simple heterogeneous reactor.

1. Lime burning involves decomposition of chunks of natural calcium carbonate by heating with hot combustion gases. The chunks of solid lime which form have roughly the same size as the limestone fed, and are separated from the gaseous product without difficulty. A typical flow sheet for this operation may not indicate whether a rotary kiln, a shaft kiln, or a fluid bed kiln is used as reactor. Similarly, the fact that operations are countercurrent within the kiln is again often not evident from the flow sheet as usually presented.
2. Acetylene gas and solid calcium hydrate form upon mixing calcium carbide (a solid) with liquid water. Separation of the acetylene gas produced from the pasty reacting mass is automatic, and operations are again represented by the flow sheet of Fig. 4-2.

Type 3. Reactor with single product stream, product separation required. In many processes, the desired product must be somehow separated and recovered from the mixture or suspension leaving the reactor. Any suitable separation scheme may be used, such as condensation, absorption, adsorption, filtration, etc. In the flow sheet of Fig. 4-3, recovery is by absorption in a suitable solvent followed by stripping. Examples of processes operating this type of flow sheet are as follows:

1. Ethylene oxide is made by passing ethylene and air continuously through a silver catalyst at about 250°C. The ethylene oxide vapor in the off-gases is recovered as in Fig. 4-3 by water scrubbing and purified by stripping and distillation.
2. Carbon black is made by injecting an oil high in aromatics into a flame of natural gas and air (in quantities less than those needed for complete combustion) in a furnace designed so that mixing is not rapid. Part of the natural gas "cracks" as a consequence, and the finely divided carbon produced is recovered from the product gases by cyclones and bag filters.

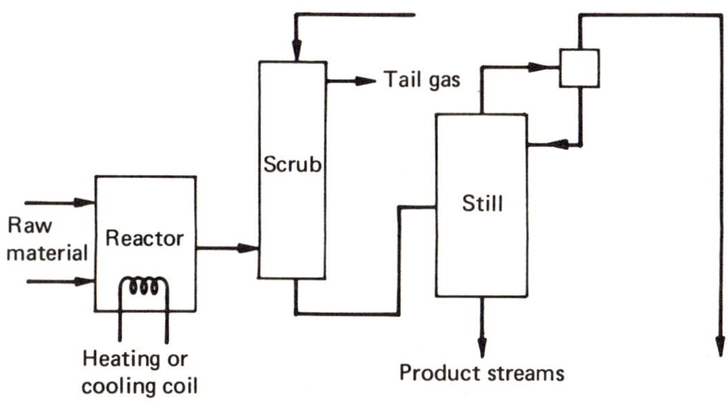

Fig. 4-3. A flow sheet with one reactor and a single product stream with product separation.

Type 4. Single reactor, multiple separations from feed and product stream. This flow scheme represents an elaboration of Type 3 above, in that the raw materials are somehow purified before entering the reactor, after which the product stream(s) is again separated into various fractions. In the illustrative flow sheet for this case shown in Fig. 4-4 raw material purification is by distillation, but of course, any other means of separation involves the same type of block diagram. There are many examples of operations described by this flow sheet, as when titanium tetrachloride is first purified by distillation, then burned with air in a suitable reactor to form finely divided TiO_2 particles and chlorine gas, after which the TiO_2 is recovered by filtration and the chlorine is reused.

Type 5. Reactor with recycle. As further discussed in later paragraphs, recycle is practiced in a large variety of situations, as in the typical flow sheet

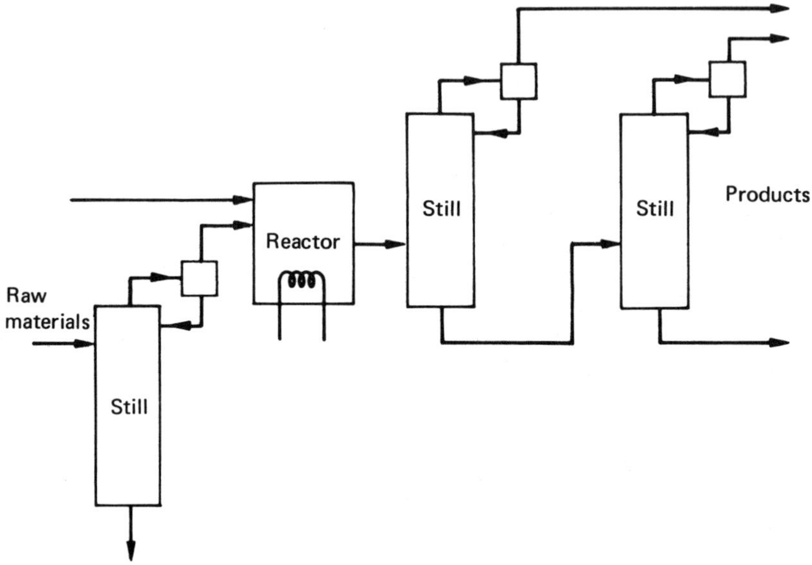

Fig. 4-4. A flow sheet with a single reactor and multiple separations.

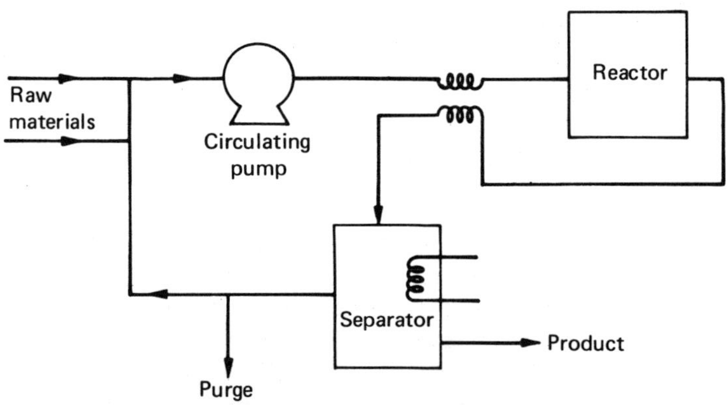

Fig. 4-5. A flow sheet with recycle.

shown in Fig. 4-5. Thus, when a reaction does not go to completion because of unfavorable kinetics or equilibrium, the unconsumed material is separated from the product stream and returned to the reactor feed stream. When material is recycled, in this fashion, a purge must be provided as shown to prevent accumulation of impurities. Examples of the use of recycle are given below.

1. Ammonia is made by passing a hydrogen-nitrogen mixture at elevated pressures over a suitable catalyst and recovering the product from the reactor off-gases by cooling and condensation or by water scrubbing. Because of equilibrium limitations encountered at the operating conditions usually used, only a fraction of the hydrogen and nitrogen reacts to form ammonia, consequently the unreacted gases are recycled. A purge is required to eliminate argon, methane, and any excess hydrogen or nitrogen fed to the system.
2. Acrylonitrile is made by passing a mixture of HCN and excess acetylene through an aqueous solution of cuprous chloride. The acrylonitrile and any unreacted HCN are removed from the product gases by water scrubbing, and the excess acetylene is recycled to the reactor. A purge gas stream is provided in the recycle loop to prevent accumulation of impurities.

Complex flow sheets. It has already been stated that most complex flow sheets may be considered as combinations of the five schemes just discussed. A few of the reasons for the existence of such combinations are as follows:

(*a*) *Reaction series.* Frequently a process requires a number of reactions to be carried out in series. Such reactions often cannot be made to occur in a single reaction vessel, since the conditions favoring one reaction may not favor the others. A familiar illustration is the contact process for sulfuric acid in which several different reactors are operated in series. The flow sheet used, as shown in Fig. 4-6, is obviously a combination of the flow sheets of Types 1 and 4. The first reactor of Fig. 4-6 is a burner in which dry air is caused to react with sulfur at over 1000°C to form SO_2. These burner gases are then cooled in a waste heat boiler and passed through a series of three or four adiabatic reactors, called converters, equipped with interstage cooling. These converters, in which SO_2 is catalytically combined with the molecular oxygen present in the gas to form SO_3, typically operate between 400–650°C. The gases from the last converter are cooled and passed to a scrubber in which the SO_3 is absorbed in 98% sulfuric acid. This scrubber is again really a reactor, in which the absorbed SO_3 combines with the free water in the recycled acid to form sulfuric acid.

The three different types of reactors in this operation, namely, the sulfur burner, the SO_3 converters, and the scrubbing towers, are operated separately because of the different requirements of the reactions occurring. Theoretically, these reactions could be carried out simultaneously at room temperature, where the equilibrium constant of each is favorable. Sulfur, however, will not burn at satisfactory rates except at about 1000°C, which is close to its adiabatic combustion temperature. Conversion of SO_2 to SO_3 must be kept

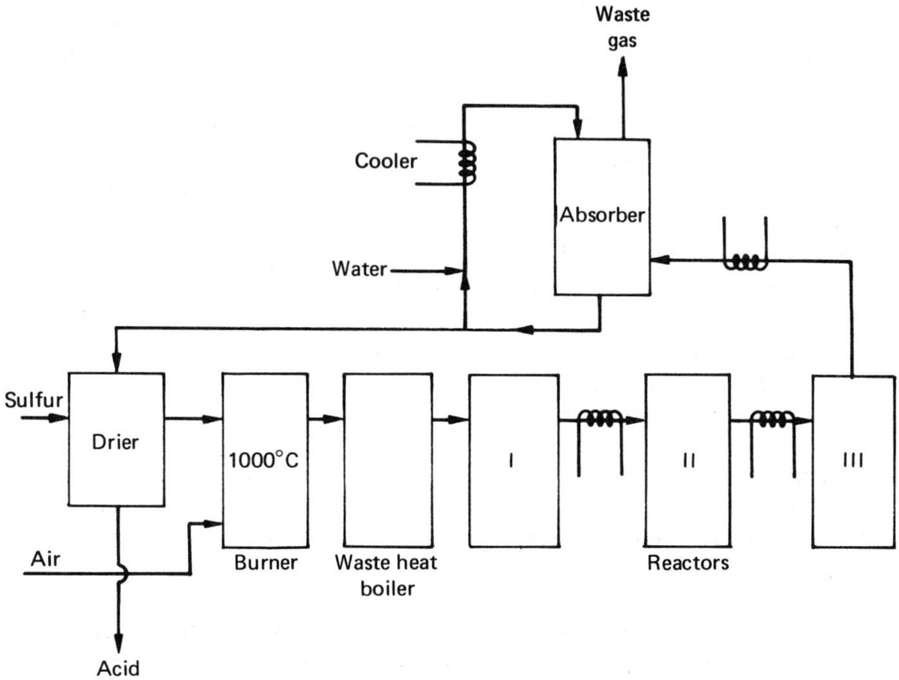

Fig. 4-6. A complex flow sheet showing the production of sulfuric acid by the contact process.

at as low a temperature as possible, since equilibria of SO_2 oxidation rapidly become unfavorable above about 650°C. On the other hand, temperatures below about 400°C are not practicable, since the catalysts available today will not operate satisfactorily at such lower temperatures. Again, the SO_3 absorption tower is operated below 150°C in order to keep down the back pressure of both SO_3 and H_2O over the sulfuric acid scrubbing liquor, and to avoid corrosion.

(*b*) *Joint products.* Joint products are unavoidably made in many chemical reactions. Often, the ratio in which these products appear is fixed, and cannot be altered. Thus, in the production of caustic soda and chlorine by the electrolysis of an aqueous solution of sodium chloride, 1 g mole of chlorine gas is produced for every 2 g formula weights of NaOH obtained. It is rare for market demand to be in balance with this ratio, hence the manufacturer seeks to find new outlets for the less salable of the two products. Several decades ago, chlorine was the more difficult product to dispose of, so that the caustic soda-chlorine producer entered into various chlorination operations. Today, caustic soda is in relative abundance, causing the producer to embark upon a whole series of operations to consume it. The net result has been a very extensive expansion of manufacturing facilities and an increasingly complex and extensive plant flow sheet. The new overall plant flow sheet, however, is again a combination of the five basic elements described earlier.

(*c*) *Multiple product separations.* A product stream containing a number of components is produced in some processes. Recovery of the individual components in relatively pure form may require many steps, sometimes involving a selective chemical reaction. Thus, many of the complications of an oil refinery flow sheet develop because the reactor product streams contain many components, which must (to a degree) be separated. A simpler example is the traditional by-product coke plant shown in Fig. 4-7, whereby coal is destructively distilled to yield solid coke, liquid products, and gas. Thus, the vapors leaving the coke ovens at 1800°F pass to a primary condenser where tar and water together with some ammonia condense. The cooled gases are now bubbled through a saturated solution of ammonium sulfate in sulfuric acid where the remaining gaseous ammonia reacts to form ammo-

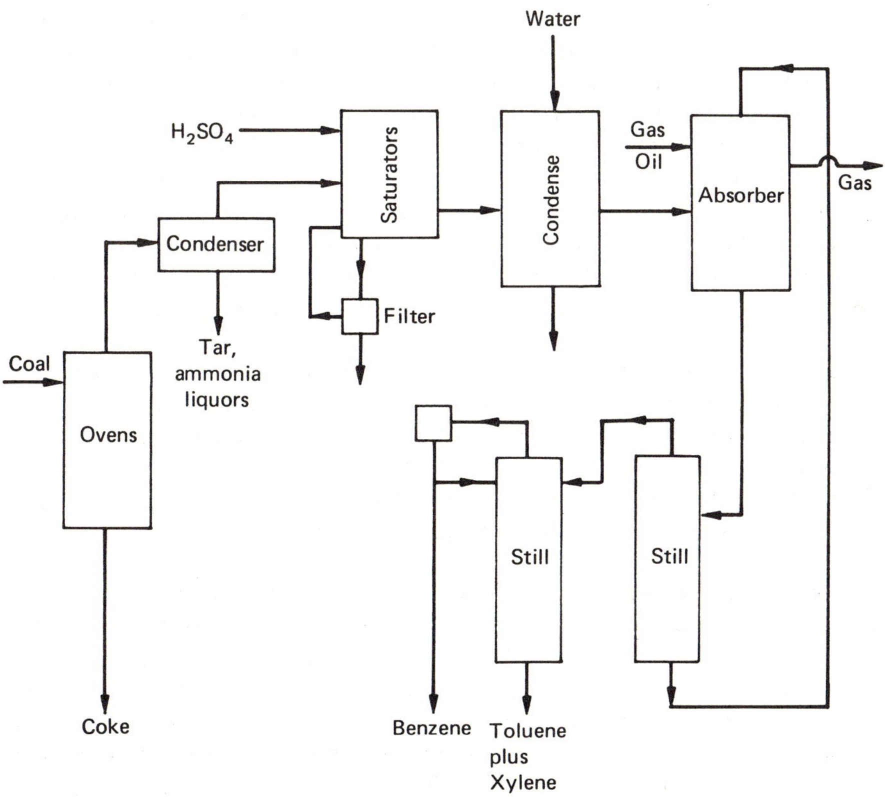

Fig. 4-7. A flow sheet showing a by-product coke plant.

nium sulfate, which precipitates in the saturator liquor. The gases are further cooled by a cold water scrub, whereupon naphthalene separates out. Benzene, toluene, and xylenes in the gases are finally removed by absorption in a "gas oil." The flow sheet involved is obviously again a combination of several of the five basic types discussed earlier.

RECYCLE

As already indicated, recycle involves the return of part or all of a process stream from a later to an earlier stage in the process. The general object of recycle in a chemical process is to minimize materials consumption per unit of product.

Heat energy as well as material is often returned from a later to an earlier state of a process. As further discussed in Chap. 5, this is accomplished by use of a heat exchanger, wherein heat moves between two process streams. Within the scope of the definition of recycle just presented, heat exchangers could be viewed as accomplishing the recycle of heat. The term recycle, however, is normally restricted to matter, and this usage is continued here.

Recycle of matter is common not only in processes for making chemicals, but also in apparatus whose primary purpose is the conversion of heat to work, or the transfer of heat from a lower to a higher temperature level. Thus, the water serving as working fluid in a steam power plant is totally and continuously recycled between the condenser and the boiler. Again, in a simple refrigerating system, the refrigerant such as ammonia or one of the Freons is totally recycled between the low temperature evaporator and the high temperature condenser. Processes of this sort involve a closed loop within which the recycled material remains sealed, never coming into physical contact with any material being processed. Such operations are not considered here, since discussion in this book is limited to the use of recycled materials in the processing of chemicals, which means that the recycled material comes into direct contact with the stream being processed, and interacts with it.

Recycle in a chemical process is undertaken for one of three reasons: for secondary reagent recovery and reuse, for yield improvement, or for control of reactor conditions. These are now discussed.

Secondary reagent recovery. A secondary reagent is here defined as one which reacts with the raw materials of a process to form an intermediate, which is later decomposed to regenerate the secondary reagent and a desired product. The secondary reagent is recycled, and so, except for losses, is not consumed in the operation. A typical example of such a secondary reagent is the caustic soda used in the Bayer process for converting impure aluminum ore, namely bauxite, to a highly purified aluminum hydroxide, later to be converted to alumina. Bauxite is composed of $Al(OH)_3$ so intimately associated with silica, iron oxides, titanium dioxide, and the like that separation by physical means is not possible. In the Bayer process, extraction is accomplished by treatment of the bauxite with strong, aqueous, hot caustic soda to form a strong solution of sodium aluminate,

$$Al(OH)_3 + NaOH = NaAlO_2 + 2H_2O \tag{4-1}$$

The impurities in the bauxite are largely insoluble in caustic soda and remain behind. After clarification, the liquors are cooled and diluted with water, reversing the foregoing reaction and causing a large part of the alumina in solution to precipitate as pure crystalline $Al(OH)_3$, which is recovered by filtration. The dilute mother liquors, which contain the caustic soda used in the extraction, together with unprecipitated alumina, are reheated and reconcentrated and then returned to the bauxite extraction vessels for reuse. Note that the overall reaction of a Bayer process plant may be represented as

$$Al(OH)_3\text{, crude} = Al(OH)_3\text{, pure} \tag{4-2}$$

Except for losses in the washed filter cake wastes and for spills, no caustic soda is consumed in this operation since none appears in any product stream.

Gaseous secondary reagents. Reactions wherein a solid material is gasified by reaction with a secondary gaseous reagent under one set of conditions, followed by the redeposition of the solid with regeneration of the secondary reagent under another set of conditions, are also operated in recycle loops. Thus, the production of highly purified metallic nickel by the Mond process is based upon the reversibility of the following reaction:

$$Ni(c) + 4CO(g) = Ni(CO)_4(g) \tag{4-3}$$

When passing CO at 70°C and atmospheric pressure over crude nickel containing copper, cobalt, etc., as impurities, the above reaction proceeds to the right and volatile nickel carbonyl forms, leaving the impurities behind. This reaction reverses to deposit pure nickel when the carbonyl-containing gases are passed through a bed of nickel shot held at about 180°C. The secondary reagent, which is CO in this example, is recycled for reuse.

Transport reactions of this general type, sometimes called chemical vapor deposition (CVD) reactions, are fairly common (137) Thus, extensive efforts have been made to develop an economic nonelectrolytic aluminum metal process based upon the reversibility of the following reaction:

$$2Al + AlCl_3(g) = 3AlCl(g) \tag{4-4}$$

This process, in which chlorine remains within the recycle system in the form of the volatile intermediates AlCl and $AlCl_3$, is further described in Prob. 4-1. Again, the chemicals of the electronics industry are often produced by gas transport, as in the formation of epitaxial gallium arsenide, described in Prob. 5-5. In all these operations, the secondary reagents used must, in the interest of economy, be recycled.

This technique, so far illustrated with inorganic reactions, is also used in the organic field. Thus, terephthalic acid is purified by conversion to the methyl ester which, unlike the acid itself, can be volatilized without decomposition. The distilled ester is then saponified and the secondary reagent, namely methanol, is recycled.

Purge. Raw materials always contain impurities. In any flow sheet, if no avenue for escape is provided, then these impurities accumulate within the system and will soon stop production. This difficulty is perhaps especially apt to occur in a recycle loop because of the care used to contain the secondary reagents within the system. Provision must, therefore, always be made to purge any impurities or inerts introduced with the raw materials.

In steady-state operations with purge, calculations often are best started by recognizing that all the impurities in the feed must leave in the purge or in suspension or in solution in product stream. This is illustrated in the following very simple recrystallization problem with recycle, in which the solvent, namely water, may be regarded as the secondary reagent.

EXAMPLE 4-1

Relatively pure KNO_3 is to be made from a crude salt mixture analyzing 99% KNO_3 by weight, the rest being NaCl, by the continuous operation shown in the flow sheet of Fig. 4-8. The crude salt mixture is dissolved in recycled mother liquor plus fresh water as shown, yielding a solution saturated with KNO_3 at 70°C, and also containing some dissolved NaCl. This

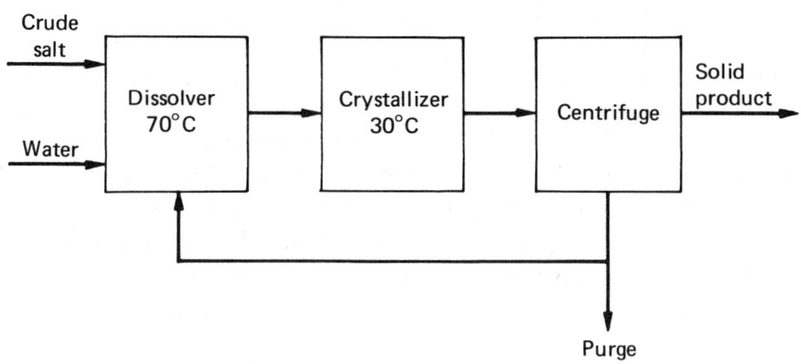

Fig. 4-8. A flow sheet for the production of pure KNO_3 from a crude salt mixture, as in Ex. 4-1.

liquor is sent to a continuous crystallizer where relatively pure KNO_3 precipitates, leaving a mother liquor saturated with KNO_3 at 30°C, containing also 2.5% by weight of NaCl in solution. The solid KNO_3 is removed by centrifuging, while the mother liquor is in part recycled and in part purged as shown. For 100 lb of crude salt mixture:

(a) How much fresh water is added to the system?
(b) How many pounds of pure KNO_3 are produced?
(c) How many pounds of saturated mother liquor leave the centrifuge?

Note: The solubility of KNO_3 in 100 lb of water is 45.8 lb at 30°C and 138 lb at 70°C independent of the NaCl in solution over the limits considered here.

Answer: (a) The obvious basis to choose for calculation is 100 lb of crude salt feed. The feed contains 1 lb of NaCl as impurity, which must leave the system in the purge, since the NaCl content of the KNO_3 product is negligible. The purge liquor, which obviously contains all this NaCl as a 2.5% solution, therefore weighs 1/0.025 or 40 lb. In addition to 1 lb of NaCl, this purge liquor contains (40 − 1)(45.8)/(145.8) or 12.2 lb of KNO_3 and (40 − 1 − 12.2) or 26.8 lb of water. The fresh water added per cycle is, therefore, 26.8 lb. (b) The pure KNO_3 produced equals the KNO_3 in the feed minus that in the purge, namely (99 − 12.2) or 86.8 lb. (c) The W lb of water in the mother liquor is related to the KNO_3 precipitated as follows:

$$86.8 = (1.38 - 0.458)W$$

Solving, W is 94.2 lb. This mother liquor, therefore, contains 94.5 lb of water, (94.2)(45.8)/(100) or 43.1 lb of KNO_3, and S lb of NaCl, where

$$(94.2 + 43.1 + S)(0.025) = S$$

Solving, S equals 3.5 lb. The total weight of mother liquor leaving the centrifuge is, therefore, (94.2 + 43.1 + 3.3), or 140.6 lb.

Composite reactions. The examples of simple recycle reactions cited so far in this chapter all have one characteristic in common, namely that in each, the secondary reagent participates in a reversible reaction which can readily be made to go quite far to completion in either direction. Reaction reversal is attained by alterations in pressure, temperature, and concentrations, these changes being reasonably within the scope of commerical plant operations. Recovery of the secondary reagent is not as simple, however, when the reaction making the intermediate cannot be reversed without going to extreme conditions. As an example, consider the production of white titanium dioxide pigment from the mineral rutile (crude TiO_2) by the new "chloride" process. Rutile is caused to react with chlorine and coke to form titanium tetrachloride, which is then carefully purified by distillation, vaporized, and "burned" at about 1000°C with oxygen to form finely divided pigment particles:

$$TiCl_4(g) + O_2 = TiO_2(c) + 2Cl_2 \tag{4-5}$$

The pigment particles are filtered out of the chlorine-rich product gas which is then recycled to react with more crude rutile, again producing $TiCl_4$. By this operation, massive discolored natural TiO_2 is converted to tiny particles of controlled size, high purity, and excellent white color. Unfortunately, reaction (4-5) cannot be reversed under any reasonably attainable conditions, since its equilibrium constant remains large up to very high temperatures. By adding coke to the crude rutile, however, this difficulty is overcome, since the equilibrium constant for the combination of TiO_2 with chlorine and carbon is favorable at elevated temperatures:

$$TiO_2(c) + 2C + 2Cl_2 = TiCl_4(g) + 2CO \tag{4-6}$$

In this process, therefore, reaction (4-5) is reversed despite an unfavorable equilibrium constant by combination with a second reaction [the interaction of oxygen with carbon to form CO by reaction (4-6)] having a very favorable equilibrium constant.

Such combinations of two complementary reactions, respectively, having unfavorable and favorable free energies, to yield an overall reaction having a favorable free energy are further discussed in Chap. 10. Obviously, this procedure requires the use both of a secondary reagent (chlorine in this case), which is regenerated, and also of another reagent (carbon) which is consumed. Thus the sum of reactions (4-5) and (4-6) results in the following overall reaction:

$$TiO_2(\text{crude}) + 2C + O_2 = TiO_2(\text{pure}) + 2CO \tag{4-7}$$

The chloride process for titanium pigment is complicated by occurrence of side reactions, since the chlorine in this operation reacts not only with the titanium in the ore, but also with impurities such as iron, manganese, etc. The resultant undesired chlorides are separated from $TiCl_4$ by distillation, but represent a source of chlorine loss to the process.

Catalyst recycle. A special case of recycle of a secondary reagent is represented by catalyst recycle. While by definition a catalyst does not participate in the chemical reaction which it accelerates, a catalyst nevertheless often suffers chemical or physical deterioration from prolonged use. Thus in petroleum cracking, the particles of treated clay used as catalyst become coated with a layer of carbon whose thickness increases as reaction proceeds. The catalyst surfaces thus become increasingly inaccessible to the reactants, until reaction finally ceases. Again in the hydration of acetylene to acetaldehyde, when bubbling the acetylene gas through an aqueous mercuric sulfate solution, the Hg^{+2} ions which act as catalysts are slowly lost by reduction to Hg^{+} and then to metal. Simultaneously, tarry polymers form which interfere with proper gas-liquid contacting.

Both in the catalytic cracking operation and in the acetylene hydration process just cited, a stream of catalyst is continuously withdrawn from the reactor, regenerated, and returned to the unit. Many similar examples of continuous catalyst regeneration exist. Such continuous regeneration, of course, is not practiced in fixed bed units when dealing with catalysts of relatively long life. Under these conditions, the entire batch of catalyst is replaced at appropriate intervals.

Recycle for yield improvement. In many operations, either the equilibrium constant, the reaction rate, or both are sufficiently unfavorable so that the product stream leaving the reactor contains much unconverted raw material. By recovery and recycle of these unused reactants, overall conversion can always, in theory, be improved to 100%, regardless of the magnitude of the equilibrium constant. Thus, in the synthesis of ammonia from the elements, a stoichiometric feed of 3 moles of H_2 to one of N_2 (and no inert material) shows an equilibrium conversion on a "once through" basis of only 38% at 200 atm and 400°C. By recycle of the unconverted nitrogen and hydrogen, an overall plant yield of well over 90% is attained. Complete utilization of the raw materials is not posssible in practice because of loss of nitrogen and hydrogen from the recycle loop, both by solution in the liquid ammonia product and in the gas purge stream operated to remove argon and other inert substances.

Recycle flow sheet. A typical flow sheet involving recycle for yield improvement is shown in Fig. 4-5. The reactor is fed with a mixture of recycle and fresh feed. The product stream leaving the reactor passes to the separator in which the product is removed from the unconverted reagents and inert substances, the latter being in part recycled and in part purged, as shown. The separator itself may be a scrubber, an absorber, a still, a decanter, or any other suitable device.

EXAMPLE 4-2

The following reaction involving perfect gases is to be carried out isothermally at 500°C in the reactor of the flow sheet of Fig. 4-5:

$$A + B = 2M \qquad K = 1.0 \text{ at } 500°\text{C} \tag{4-8}$$

The fresh feed gas to the plant analyzes 20% N_2, 40% A, and 40% B. The product gases leaving the reactor are at equilibrium and all of the M in these gases is condensed in the separator. The overall yield of M for this operation is 90% of theory, meaning that 72 moles of M uncontaminated with any other material are withdrawn from the separator per 100 moles of fresh feed to the system. (a) How many moles of purge gas are produced per 100 moles of fresh feed? (b) What is the mole per cent of nitrogen in the purge? (c) What is R, namely, the moles of gas recycled per 100 moles of fresh feed to the system?

Answer: (a) In this steady-state operation, the purge must contain all of the nitrogen which entered with the feed, along with unconverted A and B. By material balance on the total system, therefore, the purge contains 20 moles of N_2, 4 moles of A, and 4 moles of B, or a total of 28 moles of gas. (b) The purge contains (100)(20/28), or 71.5% inert material. (c) Inspection of the flow sheet shows that the recycle leaving the separator, involving R moles per 100 moles of feed, must have the same composition as the purge, calculated in part (a) above. Designating the moles of M recovered in the separator per 100 moles of feed as "m", then on the basis of 100 moles of feed:

Moles

	Recycle	Fresh Feed	Total Gas into Reactor	Total Gas Leaving Reactor
A	$\frac{4R}{28}$	40	$\frac{4R}{28} + 40$	$\frac{4R}{28} + 40 - \frac{m}{2}$
B	$\frac{4R}{28}$	40	$\frac{4R}{28} + 40$	$\frac{4R}{28} + 40 - \frac{m}{2}$
M	0	0	0	m
Inert	$\frac{20R}{28}$	20	$\frac{20R}{28} + 20$	$\frac{20R}{28} + 20$
Total moles	R	100	$R + 100$	$R + 100$

The equilibrium constant for reaction (4-8) is then

$$K = \frac{m^2}{\left(\frac{4R}{28} + 40 - \frac{m}{2}\right)^2}$$

The value of R is found from this expression to be 476 moles per 100 moles of feed, after substituting unity for K and 72 moles [from part (a)] for m.

Recycle ratio control. Some factor, such as the pressure drop through the reactor or the flooding velocity in the separator will always limit the capacity of a system as shown in Fig. 4-5. Reactor input, therefore, cannot exceed a fixed maximum flow of gas per hour, but for any given flow rate including this maximum, the so-called recycle ratio,* $R/(R + F)$, can be independently varied from zero to unity. In this expression, R and F are, respectively, the moles of recycle and the moles of feed, based on some arbitrary time period. When the recycle ratio is zero, then only fresh feed but no

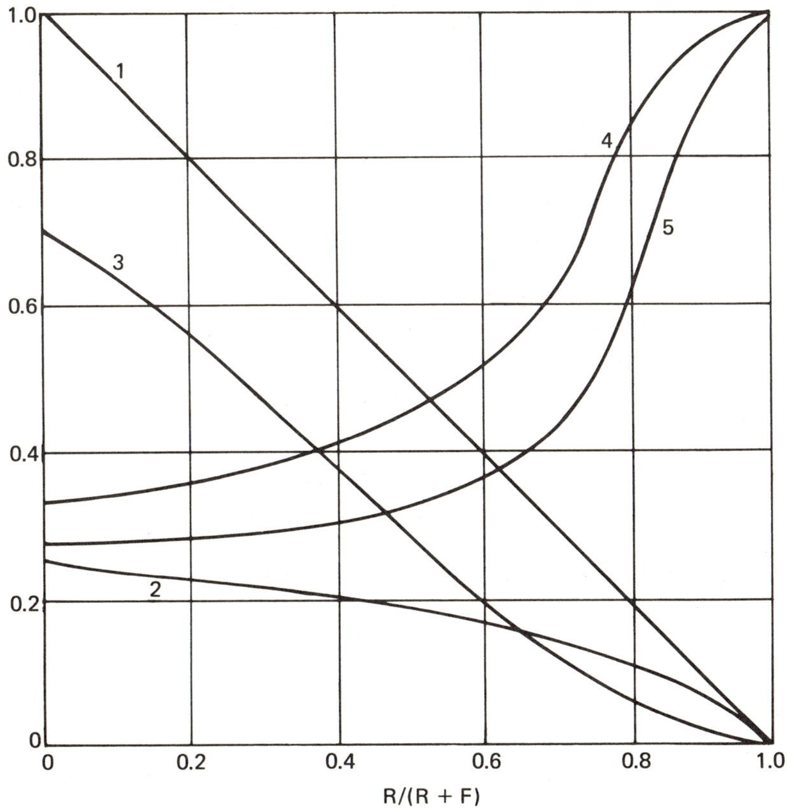

Fig. 4-9. The effect of varying the recycle ratio, namely $R/(R + F)$, upon the performance of the reactor in Ex. 4-2. The feed here is 40% A, 40% B, and 20% N_2 as inert gases to the reactor. The following reaction, whose equilibrium constant is unity, occurs:

$$A + B = 2M$$

Curve 1: $F/(F + R)$
Curve 2: $M/(F + R)$
Curve 3: $P/(F + R)$
Curve 4: $M/0.8\ F$
Curve 5: $0.2\ F/P$

*Like so many other terms in this field, the definition of "recycle ratio" varies from industry to industry. In some, recycle ratio means R/F, in others it is $R/(R + F)$, in still others, the reciprocal F/R is used, etc. Unless otherwise indicated, the recycle ratio is defined in this book as $R/(R + F)$.

recycle enters the reactor, while when the recycle ratio is unity, then no fresh gas but only recycle enters the reactor.

The effect on performance of varying the recycle ratio can again be illustrated with the flow sheet of Fig. 4-5 when carrying out reaction (4-8). As in example 4-2, the composition of the fresh feed will be 20% inert substances, 40% A, and 40% B, and equilibrium is assumed to be always attained in the reactor off-gases. The effect of any given value of $R/(R + F)$ upon process performance can be calculated exactly as in example 4-2. The results are shown graphically in Fig. 4-9, and are typical of equilibrium-controlled recycle situations. Inspection of this figure shows that, for a fixed reactor feed quantity, as $R/(R + F)$ increases:

1. Both moles of fresh feed (curve 1) and moles of product (curve 2) diminish; thus, the greater the recycle ratio, the smaller the output of desired product per unit time.
2. Moles of purge P also diminish (curve 3). The purge always carries off all of the inert substances in the feed, and so will decrease as the feed rate decreases.
3. The material efficiency increases (curve 4). Material efficiency is defined here as

$$\frac{100 \text{ (Moles pure } M \text{ produced)}}{\text{Moles } (A + B) \text{ in fresh feed}}$$

4. The concentration of inert material in the purge increases (curve 5), corresponding to the increasing efficiency. The concentration of inert substances in the recycle loop equals that in the purge, and so also increases.

Obviously, at total recycle, feed, purge, and product rates fall to zero, and the recycle gas contains only inert substances.

It is clear that the proper operation of this flow sheet for maximum profits calls for a careful economic balance. That is, at low recycle ratios, a relatively large amount of product is produced but material efficiency is poor. Under these conditions, capital, labor, and maintenance charges (which total to a fixed sum per unit time) are low per unit of product produced, but raw material costs are high. The reverse exists at high recycle ratios, for which production rates are low but material efficiencies are high.

Management of Heat Energy 5

There are three general types of problems involved in the management of heat in chemical plants:

1. Sensible heat. Chemical reactions are usually carried out at temperatures above that of the surroundings, either to gain the benefit of higher reaction rates, or because equilibrium becomes more favorable, or both. Sensible heat must, therefore, be supplied in heating the reactants to reaction temperature, and also removed from the products, if these (as is usually the case) are cooled down to the temperature of the surroundings before leaving the plant.

Occasionally, reactors are operated at below the temperature of the surroundings, as in certain butadiene polymerizations. Obviously, the reactants here must be somehow cooled to reactor temperature, and the products heated to atmospheric temperature. Only in the relatively rare case of reactor operation at the temperature of the surroundings is there no problem of sensible heat supply.

2. Heat of reaction. Most reactions are more or less vigorously exothermal or endothermal, so that after the reactants are brought to reactor temperature, provision must be made for removing or supplying the heat of reaction.

3. Energy effects during product recovery. The desired product of a chemical operation often leaves the reactor mixed with by-products, unconverted raw materials, etc. Separation and purification usually requires energy, generally in the form of heat.

The supply of sensible heat is closely related to, and often inseparable from, the supply (or removal) of the heat of reaction and of product recovery. These three areas are discussed further after initial consideration of the costs of heat energy.

Energy costs. Since heat is almost invariably produced or consumed by chemical reactions, it is helpful to obtain some perspective on its costs. The three important fuels used for heating are natural gas, heavy (residual) fuel oil, and coals. Electrical energy is also used in electrothermal operations for heating, but to a much more limited extent. The relative costs of heat energy vary widely from place to place because of differences in transportation cost, reflecting varying distances to sources of supply. Thus, natural gas is available for 20 cents per 1000 cu ft in the Southwest on a long-term contract basis, but costs perhaps five times as much in the northeastern states. Residual fuel oil is worth $1.60 per barrel in Venezuela and $2.00 per barrel in Boston. Bituminous coal can be bought for about $3.50 per ton at the mine mouth in the Ohio River Valley on a long-term contract, but costs $8.00 per ton in Boston. Electric energy can be had for under four mils (0.4¢/kw-hr) in the T.V.A. region, at Niagara, in the Northwest, etc., but costs increase rapidly as fuel costs and as the distance from the site of power generation increases. Costs of electricity from nuclear power plants are being reduced, and are already competitive with plants operating on fossil fuels in many parts of this country.

TABLE 5-1. TYPICAL FUEL COSTS OF HEAT (BOSTON)

Fuel	Unit Fuel Costs	Fuel Costs per Million Btu of Heat
Natural gas*	$0.65/1000 cu ft	$0.65
Coal	$8.00/ton	$0.32
Bunker C fuel oil	$2.00/bbl	$0.32
Electricity	0.8 cents/kw-hr	$2.70

*These costs are estimated for firm supplies. For interruptible gas, costs are closer to oil costs per million Btu.

A glance at Table 5-1 shows why electrical energy is not used industrially for heating unless the need for it is critical. That is, heat from electrical energy is generally far more expensive than from the usual combustible fuels. Electrical heating is used only when its unique advantages justify its costs, namely, when it is necessary to keep products of combustion out of the furnace system, to develop very high temperature, to attain a high degree of control, etc.

Sensible heat management. Sensible heat is often recycled within a chemical process, in that the heat required to raise the temperature of one process stream is obtained by cooling a second process stream. Such energy re-use is accomplished by use of heat exchangers. Before considering the location of exchangers in a process, a brief review of the relevant characteristics of these familiar devices is presented in the next few pages.

Heat exchangers take many forms, but gas-gas or liquid-liquid exchangers typically involve two adjacent ducts separated by a common wall through which heat flows, as shown in Fig. 5-1(a) and (b). Such an exchanger can be operated countercurrently or cocurrently, depending on whether the streams are made to flow through the device in opposite directions as in Fig. 5-1(a), or in the same direction as in Fig. 5-1(b). The two streams of a heat exchanger need to be kept separated by a wall only when they react or dissolve in each other. For inert solid-gas systems, the exchanger may take the form of a cylindrical shaft down which solids travel counter to a rising gas stream. Other modifications of heat exchangers exist, depending upon the nature of the streams involved.

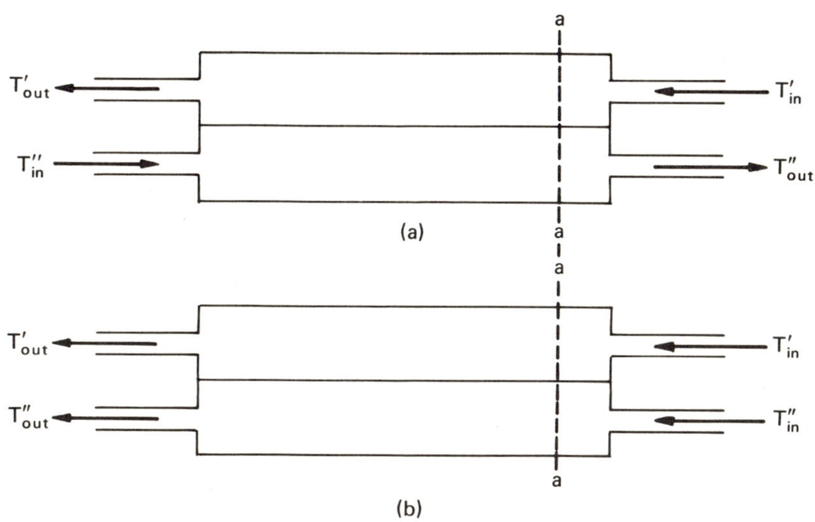

Fig. 5-1. (a) A heat exchanger operating countercurrently. (b) A heat exchanger operating cocurrently.

Exchanger performance. The exterior of a heat exchanger is generally well insulated to minimize heat loss to the surroundings, since the purpose of this device is to conserve energy, allowing heat to flow only between the two process streams involved. An idealized exchanger is therefore adiabatic, and when operating in steady-state, the loss in enthalpy of hot stream exactly equals the gain in enthalpy of the cold stream. The magnitude of this heat transferred between streams per unit time is designated as q, usually expressed as Btu/hr.

Exchangers for sensible heat are ordinarily so located in a process that no chemical reactions or phase changes, such as vaporization or crystallization, occur within them, and so only sensible heat exchange is involved. When (as is normally a good first approximation) the heat capacity of each stream is constant and independent of temperature, then for each of the exchangers shown in Fig. 5-1(a) and (b),

$$q = \Sigma n' MC'_p(T'_{\text{in}} - T'_{\text{out}}) = \Sigma n'' MC''_p(T''_{\text{out}} - T''_{\text{in}}) \tag{5-1}$$

where the single primes refer to the hot stream, the double primes to the cold stream, and the subscripts "in" and "out" designate the entering and leaving streams, The symbol ΣnMC_p represents the total heat capacity per unit time of the flowing streams, while T is temperature. Equation (5-1) usually requires no modification to allow for pressure drop of the streams flowing through the exchanger, since heat capacities of gases and liquids are generally little affected by pressure.

It will, of course, be remembered that Eq. (5-1) is an approximation, since in it, heat capacities are taken as constant. It is often desirable to replace Eq. (5-1) by the more accurate relation

$$q = \int_{T_{\text{out}}}^{T_{\text{in}}} \Sigma n' MC'_p dT' = \int_{T_{\text{in}}}^{T_{\text{out}}} \Sigma n'' MC''_p dT'' \tag{5-2}$$

where the molal heat capacities are now temperature functions.

A moment's reflection shows that with both countercurrent and cocurrent exchangers, a "hot" end and a "cold" end can always be identified, the hot end being that at which the stream having the higher temperature enters the exchanger. Again, the temperature difference between two streams at any section through the exchanger, such as section (a-a) of Fig. 5-1, can be defined as

$$\Delta_a = (T'_a - T''_a)$$

Obviously, in countercurrent operation:

at the hot end: $\quad \Delta_H = (T'_{\text{in}} - T''_{\text{out}})$

at the cold end: $\quad \Delta_C = (T'_{\text{out}} - T''_{\text{in}})$

In cocurrent operation:

at the hot end: $\quad \Delta_H = (T'_{\text{in}} - T''_{\text{in}})$

at the cold end: $\quad \Delta_C = (T'_{\text{out}} - T''_{\text{out}})$

The heat transferred in either of these exchangers, when operating in a steady-state, can now be related to A, the exchange area, and to the overall heat transfer coefficient, as follows:

$$q = UA\Delta_{LM} \tag{5-3}$$

where Δ_{LM} is the so-called log mean temperature driving force which is valid when U is constant. By definition,

$$\Delta_{LM} = \frac{\Delta_{\text{hot}} - \Delta_{\text{cold}}}{\ln \Delta_{\text{hot}}/\Delta_{\text{cold}}} \tag{5-4}$$

Derivations of the foregoing two familiar equations are presented in elementary texts on heat transfer, and so are not further discussed here. It will, however, be remembered that Δ_{LM} values are within 4% or less of the arithmetic mean of Δ_{cold} and Δ_{hot}, provided that one value of Δ is not more than twice the other.

Countercurrent exchanger advantages. As illustrated in example 5-1 below for a fixed overall heat transfer coefficient, the exchanger area required to transfer a given amount of heat between two streams in countercurrent flow is smaller than in cocurrent flow. Moreover, in cocurrent operations, the emergent stream temperatures are obviously limited to that value attainable when the streams are directly mixed together. The temperature of the cold stream can under no circumstances exceed this mixing temperature, nor can the hot stream temperature fall below it. In countercurrent operations, of course, no such limitations exist.

EXAMPLE 5-1

A hot stream of liquid is to be cooled from 200–145°F by heat exchange with water available at 60°F. Flow quantities and heat capacities are as follows:

	Hot Stream	Water
Flow rate, lb/sec	180	100
Heat capacity, Btu/(lb) (°F)	0.835	1.0

(a) Contrast the area requirements of a countercurrent and a cocurrent exchanger for this task, assuming the same overall coefficient of heat transfer in each device. (b) Repeat, when cooling the hot stream from 200–110°F.

Answer: (a) The heat lost by the hot stream in either exchanger, by Eq. (5-1), is as follows:

$$q = 180 \times 0.835\,(200 - 145) = 8250 \text{ Btu/sec}$$

The temperature of the leaving water stream is

$$T_{\text{water}} = 8250/100 + 60 = 142.5°\text{F}$$

Temperature driving forces are, therefore, as follows:

	Countercurrent	Parallel Flow
Δ_{hot}	$= 200 - 142.5 = 57.5°\text{F}$	$200 - 60 = 140°\text{F}$
Δ_{cold}	$= 145 - 60 = 85°\text{F}$	$145 - 142.5 = 2.5$
Δ_{LM}	$= \dfrac{85 - 57.5}{2.3 \log (85/57.5)} = 70°\text{F}$	$\dfrac{140 - 2.5}{(2.3 \log 145/2.5)} = 34.2$

Obviously,

$$\frac{\text{Area for countercurrent operation}}{\text{Area for cocurrent operation}} = \frac{34.2}{70} = 0.49$$

(b) In countercurrent operations, the leaving water temperature is (200 − 110)(0.835)(180)/100 + 60 or 195°F. The exchanger area required can be calculated from Eq. (5-3) if U is known, and is obviously finite. It is also clear that it is impossible to attain an exit hot stream temperature of 110°F in a cocurrent exchanger.

For the reasons just discussed and illustrated, commercial exchangers are almost invariably countercurrent. Cocurrent operation is encountered only when there are special factors requiring its use, such as the need to protect the cold stream from overheating, or to prevent the development of excessive viscosities in the hot stream upon cooling. The remaining discussion, therefore, deals with countercurrent operation.

Type 1: Balanced countercurrent heat exchangers. Inspection of Eq. (5-1) shows that, when $\Sigma n'MC'_p$ is equal to $\Sigma n''MC''_p$, then the temperature change for the hot stream equals that for the cold. It follows that the difference in temperature between the two streams at any arbitrarily chosen section *a-a* [see Fig. 5-1(a)] equals that at the exchanger's ends, or

$$(T'_{\text{out}} - T''_{\text{in}}) = (T'_{\text{in}} - T''_{\text{out}}) = (T'_a - T''_a) \tag{5-5}$$

Obviously, as the heat exchange area is made larger for a given set of entering stream temperatures, namely T'_{in} and T''_{in}, the difference in temperature at all points within the exchanger between the two streams $(T'_a - T''_a)$ becomes smaller. When the heat exchange area (or the heat transfer coefficient) becomes infinite, then this temperature difference diminishes to zero.

Exchangers in which the value of ΣnMC_p for the two countercurrently flowing streams are equal are here called "balanced" heat exchangers. When they also have an infinite area or an infinite coefficient, then they are called "balanced infinite" heat exchangers. It follows from Eq. (5-1) that in all balanced exchangers, whether infinite or not, the rise in temperature of the cold stream exactly equals the fall in temperature of the hot stream, or

$$T''_{\text{out}} - T''_{\text{in}} = T'_{\text{in}} - T'_{\text{out}} \tag{5-6}$$

Type 2: Unbalanced countercurrent heat exchangers. In the general case, the ΣnMC_p values for the two countercurrently flowing streams are not equal. Such heat exchangers are here called "unbalanced" heat exchangers. When, in addition, the area or the heat exchange coefficient is infinite, then the device is called an "unbalanced infinite" exchanger.

Equation (5-2) applies to an unbalanced exchanger as well as to a balanced exchanger. For an unbalanced exchanger, however, the change in temperature of the hot stream, namely $(T'_{\text{in}} - T'_{\text{out}})$, is obviously not equal to $(T''_{\text{out}} - T''_{\text{in}})$, the rise in temperature of the cold stream, and so the temperature difference $(T'_a - T''_a)$ is no longer constant throughout the exchanger. This latter temperature difference is smallest at one end of the exchanger, namely, that end at which the stream with the larger value of ΣnMC_p enters. This is a consequence of the fact that the stream with the larger ΣnMC_p value undergoes a smaller change in temperature in passing through the exchanger than the other stream.

Obviously, for a given set of entrance temperatures, namely, T'_{in} and T''_{in} in Fig. 5-1, the quantity $(T'_a - T''_a)$ for any given temperature T'_a becomes smaller as the exchanger area increases. The smallest attainable value of this temperature difference, namely, zero, is encountered in an infinite unbalanced exchanger at that end at which the stream with the greater heat capacity enters. Note that even with infinite area (or an infinite coefficient) the temperature difference in an unbalanced exchanger is zero only at this one point, but is finite everywhere else.

Exchanger efficiency. It is obvious that complete heat exchange can be attained in a balanced infinite countercurrent exchanger. That is, the hot stream can transfer all its heat to the cold stream, while being cooled to the temperature at which the cold stream enters. The cold stream, in turn, is heated all the way to the temperature at which the hot stream enters the exchanger. Since $(T'_a - T''_a)$ is zero at all sections, heat always flows isothermally between the two streams, and so from the viewpoint of the second law of thermodynamics, this exchanger is completely reversible.

The unbalanced infinite countercurrent exchanger has a poorer efficiency than a balanced exchanger. That is, heat flows isothermally between the two streams at only one point, namely, at that end of the exchanger where the stream having greater heat capacity enters. Note that even with infinite area (or an infinite coefficient) the quantity $(T'_a - T''_a)$ is zero at only one end of the exchanger, but is finite at all other sections. As a consequence, the temperature change of the stream having the greater value of ΣnMC_p will be smaller than that of the other stream. Since most of the heat flows with a finite temperature difference, this infinite exchanger is not reversible in the thermodynamic sense, and the greater this temperature difference, the less efficient the exchanger becomes.

System planning. Consider again a generalized chemical reaction (1-1), for which it will be assumed that $\Delta\Sigma nMC_p$ is zero. This means that, when operating in a flow reactor, the total heat capacity of the feed stream equals that of the product stream. Under these conditions, regardless of reactor temperature, it is always possible to find an arrangement of balanced infinite heat exchangers whereby the total sensible heat requirement for heating the reactants to the reactor temperature equals the sensible heat of cooling the products to the temperature of the surroundings. In developing a flow sheet, it is, therefore, convenient to assume at the outset that balanced heat exchangers are used, thus simplifying the early planning stages by avoiding the task of predicting a heat transfer coefficient, and of integrating complex heat capacity equations. After orientation is obtained in this manner on the general characteristics of the flow sheet, then the calculations may be repeated using an approximate heat transfer coefficient, recognizing that the smallest value for the temperature difference "across" an exchanger is often around 20–40°F. In later calculations, the coefficient is carefully calculated and the exchanger is sized in accordance with the usual considerations of economic balance.

EXAMPLE 5-2

The following endothermal gas reaction is to be carried out catalytically and isothermally in a cooled reactor at 600°F:

$$A + B = M + N \qquad \Delta H^\circ = +20 \text{ kcal}$$

Reactants A and B are available at 25°C, and can be mixed without reaction occurring until the catalyst is reached, whereupon reaction may be assumed to go to completion.

Two of many possible arrangements of heat exchangers for supplying sensible heat to reactants A and B are presented in Figs. 5-2 and 5-3. The objective, of course, is to use the heat of the product stream to preheat A and B to as close as possible to the reactor temperature of 600°F. Which of the arrangements of Figs. 5-2 and 5-3 is the better? Are further improvements possible? Molal heat capacities of A, B, M, and N are 7, 23, 15, and 15 cal/(g

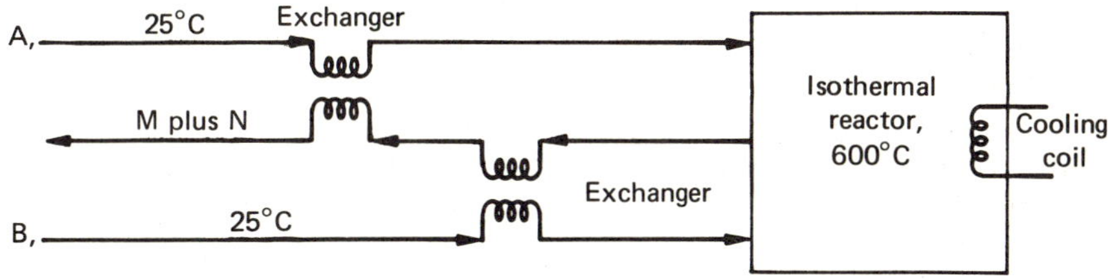

Fig. 5-2. The flow sheet for Ex. 5-2, using infinite unbalanced exchangers.

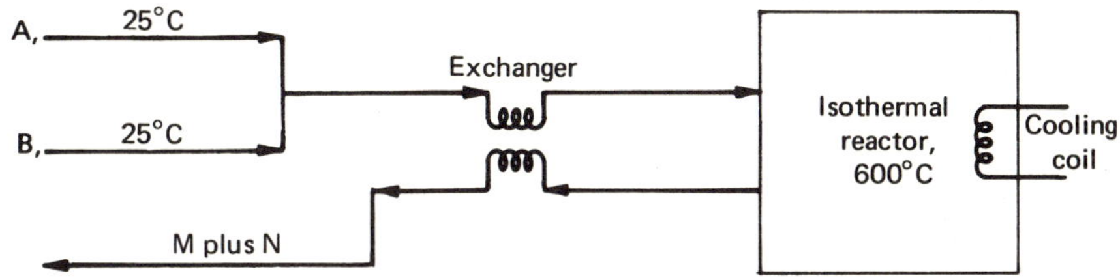

Fig. 5-3. The flowsheet for Ex. 5-3, using infinite balanced exchangers.

mole) (°K), respectively, resulting in a value for $\Delta\Sigma nMC_p$ of zero for this reaction.

Answer: Inspection shows that the exchangers of Fig. 5-2 are not balanced, in that the heat capacities of the mixed hot product stream is greater than that of the cold stream in each exchanger. The exchanger of Fig. 5-3, on the other hand, is balanced. Because the exchangers of Fig. 5-2 are not balanced, it is obvious that stream *A* cannot be preheated to the reactor temperature, while both *A* and *B* can be so preheated in Fig. 5-3. It follows that scheme 5-3 is very much the better of the two, and indeed cannot be improved upon, since it involves only balanced infinite exchangers.

HEAT SUPPLY—ENDOTHERMAL REACTIONS

The heat of reaction required by an endothermal reaction must not only be supplied in proper quantity but also at the correct temperature level. Thus melting a metal such as copper may be viewed as a simple example of an endothermal operation, in which only heat supplied above the melting temperature of copper at 1063°C is effective. If heating is with hot combustion gases and these gases are below 1063°C, then no melting occurs regardless of how much of these gases pass through the furnace. The existence of minimum allowable operating temperatures, dictated by the temperature of melting, vaporization, decomposition, etc., is typical of monovariant systems.

Thus, calcium carbonate develops a dissociation pressure of 1 atm at 898°C, and only heat provided above this temperature will cause decomposition to proceed in a reactor having a pressure for CO_2 of 1 atm.

Reactor temperatures for endothermal reactions are, of course, often determined by considerations of kinetics as well as by thermodynamics. As mentioned earlier, there is a kindling temperature for many reactions below which rates are negligible despite a favorable equilibrium. Thus, the rate of dehydrogenation of butane to butenes becomes very small below about 400°C over the commonly used catalysts, even though equilibrium is favorable at these lower temperatures. In such cases, heat of reaction must be supplied at above the kindling temperatures of the reactions.

In addition to a minimum operating temperature, a maximum temperature often exists above which reagents undergoing an endothermal reaction should not be heated. This maximum temperature may be determined by decomposition or cracking of one of the reactants present, by deterioration of the catalyst, etc. As a consequence, it becomes necessary to operate within a restricted temperature interval. Problems of reactor design become more difficult with a narrowing of this interval and with increases in the amounts of heat to be supplied or removed.

Four general methods for supplying a reactor with the heat required by an endothermal reaction are now presented. In at least three of these, both the enthalpy of reaction and the sensible heat requirements for preheating the feed streams may be simultaneously supplied within the reactor. In such cases, the advantages of using the sensible heat of the product stream to preheat the feed are usually deliberately lost. Discussion focuses particularly upon the supply of chemical heat rather than the supply of sensible heat in the reactor, but it should be remembered that it is sometimes advantageous to stress both simultaneously.

Indirect heating. Heat may be added indirectly to an endothermal system through heat exchange surfaces in the reactor itself, as when reagents undergoing endothermal reaction flow through an externally fired pipe coil furnace. Such furnaces are common in the petroleum industry. High temperature heat can also be supplied to a reaction system by use of a regenerator, such as hot checkerbrick, as when cracking natural gas to make "thermal" carbon black. Similarly, the catalyst carrier itself can serve to store heat in the manner of a regenerator, as in one process for dehydrogenating butanes to butadiene.

Heat is sometimes also supplied to a reactor by recycling part of the reacting mass through an external heater and back to the reactor. Thus, the endothermal heat requirements in one form of catalytic cracking are supplied by cycling the hot catalyst between the reaction vessel and an external heater.

Electric heating. Indirect electrical heating, in which the reacting mass is separated from electrical resistors by a wall, presents the disadvantage of high energy costs. Such costs usually more than offset one important advantage of electrical heating, namely, the case and simplicity of controlling energy input to a system. Various direct methods of generating heat within the reacting system by use of electrical energy also exist which, however, are rather specialized, and applicable to few situations. Thus induction furnaces can be used only if the charge itself is a conductor of electricity, as with a mass of molten metal. Direct electrical resistance heating is likewise possible only if the mass within the reactor is an electrical conductor, as in the case of a bed of coke, or a plasma. Direct arc heating has the disadvantage of generating within the reactor a small zone of extremely high temperature which few materials can tolerate. In all these cases, the high cost of electrical heat offsets the advantage that no products of combustion form within the reactor to contaminate the product.

Reaction combination. When more than one reaction occurs in a reacting mass, then the net enthalpy change equals the algebraic sum of the enthalpy changes of the individual reactions. Both exothermal and endothermal reactions can be made to occur in some systems, and when their proportion can be varied, then the net heat requirements per unit of product can be changed accordingly. Thus, the reduction of iron oxides by hydrogen is endothermal while that by carbon monoxide is exothermal. By use of a H_2–CO mixture of the proper proportions, gaseous reduction of the iron oxides occurs with a value for the overall ΔH of zero. In the blast furnace, air is blown through the ore–coke charge, and the heat evolved in burning part of the coke to CO supplies the heat needed both for the endothermal reduction of ore with coke, and for preheating the raw materials. Again, in the adiabatic production of HCN by the strongly endothermal reaction of CH_4 with NH_3, air is added to the reaction mixture to generate the necessary heat by partial combustion of the methane. Obviously, in cases like these, an exothermal reaction can be combined with an endothermal reaction in such proportions that the reacting system's enthalpy change can be varied over a wide range.

In practice, this technique of combining exo- with endothermal reactions is limited for several reasons. The reagents added may react disadvantageously with the product of the primary reaction, recovery of the desired product may become difficult because of dilution with the new reagents, the catalyst used may be ruined in the altered system, etc. For reasons like these, other techniques for heat supply or removal are more generally used.

Sensible heat. Reactants are sometimes heated well above the kindling temperature of a reaction before entering the reactor catalyst bed or be fore mixing. The heat required by the endothermal reaction is then supplied by the sensible heat released during cooling to the quench temperature. Again, such a procedure is successful only when the reagents and products do not undergo undesired side reactions or decomposition at the high temperature levels of the preheat.

HEAT REMOVAL—EXOTHERMAL REACTIONS

When no heat is required in raw material preparation or in product recovery, and when all raw materials enter and products leave at the temperature of the surroundings, then there is a net production of heat in any plant in which an exothermal reaction is run. Under these conditions, the heat of reaction usually leaves the plant as sensible heat in cooling water or in hot stack gases. Three general flow sheet designs for exothermal reactions, involving either adiabatic reactors followed by product coolers or else internally cooled reactors followed by product coolers, are now discussed.

Adiabatic reactors. The temperature of a stream of reactants undergoing an exothermal reaction inevitably rises in passing through an adiabatic reactor. As a consequence, the higher the temperature of the feed stream, the higher the product stream temperature. With exothermal as well as endothermal reactions it is often necessary to preheat the feed stream to higher temperatures to "kindle" the reaction. The following example illustrates preheating feed to reaction temperature by direct mixing into the reacting system within the reactor. Success is attained because rates are high despite the low reactant concentrations resulting from such mixing, and because no complicating side reactions occur. Operating in this fashion, the product stream obviously attains the "adiabatic reaction temperature."

EXAMPLE 5-3

Zinc pyrites (ZnS) are oxidized in a continuous adiabatic fluid bed operation.* The solids are fluidized with 150% of the theoretical amount of air (21% O_2, 79% N_2) required by the following reaction:

$$ZnS + 1\tfrac{1}{2}\,O_2 = ZnO + SO_2 \qquad \Delta H^\circ_{25^\circ C} = -\,105.5\ \text{kcal}$$

As is typical of good fluidized operation,* excellent mixing of solids makes the bed isothermal at the adiabatic reaction temperature T_r. The equilibrium constant at this elevated temperature is large, rates are high, and the solid product to be withdrawn is estimated at 95 mole% ZnO, the rest is ZnS. Pyrites and air enter the bed at 70°F, while the solid product and off-gases

**Note:* Fluidization of solids is described in Chap. 9.

leave at T_r. The hot off-gase sare used to generate steam in a waste heat boiler. Using constant average heat capacities [MC_p values over this temperature range are 12.8, 12.2, 7.6, 8.1, and 12.6 cal/(g mole) (°C), respectively, for ZnS, ZnO, N_2, O_2, and SO_2], calculate T_r.

Answer: Choose 1 mole of ZnO in the product as basis, which contains $\frac{5}{95}$ moles of unconverted ZnS. The oxygen fed as air is then $1.5 \times \frac{150}{100}$, or 2.25 moles, while the nitrogen associated with this oxygen is $\frac{79}{21} \times 1.5 \times \frac{150}{100}$, or 8.43 moles. The oxygen consumed in the reaction is obviously 1.5 moles/mole of ZnO product.

Product Gas		Moles	MC_p	$(nMC_p)_{gas}$
N_2		8.43	7.6	63.8
O_2: 2.25 − 1.5	=	0.75	8.1	6.08
SO_2		1.0	12.6	12.6
				82.5

$$(\Sigma nMC_p)_{\text{solids}} = 12.8 \times \tfrac{0.05}{0.95} + 12.2 = 12.9$$

$$(82.5 + 12.9)(T_r - 70) = (105{,}500) \times 1.8 \qquad T_r = 2060°\text{F}$$

The pyrites burner of example 5-3 is unusual in the sense that the adiabatic reaction temperature of 2060°F with 150% excess air is high enough for good reaction rates at the reagent concentrations existing in the reactor, yet low enough to avoid severe problems with materials of construction. In other cases, there is often insufficient or excess reaction heat over the sensible heat requirements needed for the feed streams to reach the desired reaction temperature, and so supplementary heat exchange facilities for cooling or heating are required.

Equilibrium reactions in adiabatic systems. As already pointed out, feed must be preheated for reactions having an elevated kindling temperature. When preheating feed directly in a reactor by mixing it into the reacting mass, as just illustrated, then both reaction rates and by-product formation may become disadvantageously affected. To avoid such problems, preheating is often done outside the reactor, after which the feed is passed through the reactor in such a way that back mixing is minimized.

Preheating the feed outside the reactor obviously results in increased reactor temperature levels. Conversion is consequently reduced in reactions whose equilibrium constants are small at such elevated temperature, as in the air oxidation of SO_2 to SO_3, which kindles at 400°C over the catalysts used today in industry. Being vigorously exothermal, the equilibrium constant of this reaction diminishes rapidly with increasing temperature, and is small enough at the typical reactor exit temperatures so that conversion is seriously limited. One way of dealing with such an equilibrium limited reaction is to send the gas through several adiabatic converters in series, with interstage cooling. The benefits obtained from such an operation are indicated in example 5-4. Also shown are the calculational procedures involved when dealing with an exothermal equilibrium reaction in adiabatic systems, wherein reaction ceases when the temperature rises to a value at which the composition of the system satisfies the equilibrium constant.

EXAMPLE 5-4

In modern sulfuric acid manufacture, a gas mixture of SO_2, O_2, and N_2 is made by burning elemental sulfur with an excess of dry air (21% O_2, the rest N_2) as follows:

$$S + O_2 = SO_2 \tag{5-7}$$

After cooling in a waste heat boiler, this burner gas is made to flow through a series of three or four adiabatic converters in series, with interstage cooling between each pair of reactors. Each converter contains vanadium pentoxide, which catalyzes the following reaction:

$$SO_2 + \tfrac{1}{2}O_2 = SO_3 \qquad \Delta H^\circ = -22 \text{ kcal} \tag{5-8}$$

The catalyst is typically in the form of a fixed bed 3 ft deep of $\frac{1}{4} \times \frac{1}{4}$ in. porous ceramic cylinders impregnated with vanadium pentoxide. This catalyst, which causes reaction (5-8) to kindle at about 400°C, is so effective that the off-gases from each converter bed are substantially at equilibrium.

The equilibrium constant for reaction (5-8) is as follows for selected temperatures:

T, °C	$\log K_p$	T, °C	$\log K_p$
371	3.039	704	0.371
593	1.025	816	−0.144

(a) In a typical sulfur burner, elemental sulfur is converted to SO_2 by combustion with excess air by reaction (5-7). Some SO_3 may also form simultaneously by reaction (5-8), but only in minor amounts because the equilibrium constant of this reaction is small at the elevated temperature of the combustion process. When charging m atoms of sulfur per 100 moles of dry air, then the burner gas contains a total of m moles of SO_2 plus SO_3, together with 79 moles of N_2, and unconsumed oxygen. If the fraction of the sulfur converted to SO_3 is f, then for 79 moles of N_2 the off-gas contains SO_2, SO_3, O_2, and N_2 in quantities and with partial pressures as listed below:

For 79 Moles of N_2

	Moles	Partial pressures, atm
SO_2	$m(1-f)$	$\pi m(1-f)/(100-mf/2)$
SO_3	mf	$\pi mf/(100-mf/2)$
O_2	$21-m-mf/2$	$\pi(21-m-mf/2)/(100-mf/2)$
N_2	79	$\pi(79)/(100-mf/2)$
Total	$100-mf/2$	π

The foregoing table is based upon a total pressure of π atm.

Prepare a chart whose ordinate is f as just defined, and whose abscissa is temperature in °C. Now locate lines showing the variation at equilibrium of f versus temperature for each of the following constant values of m: 0, 6, 10, 12, 14, 16, and 17, all at a constant total pressure of 1 atm. (b) A burner gas analyzing 79% N_2, 9% O_2, 12% SO_2, and zero SO_3 (for this gas, m is 12 and f is zero) is fed at 400°C to the first of three adiabatic converters operated in series at atmospheric pressure. The product gases leaving each converter are at equilibrium, and are cooled to 400°C before entering the next converter. What is the temperature and composition of the gas leaving the first converter? Entering the second? Leaving the second converter? For convenience, make the following approximation: for O_2, N_2, SO_2, and SO_3, MC_p values are respectively 7, 7, 12, and 15.5 Btu/(lb moles) (°F).

Answer: (a) The equilibrium constant K_p for reaction (5-8), which has the units $(\text{atm})^{-1/2}$, is written as follows:

$$K_p = \frac{p_{SO_3}}{p_{SO_2} p_{O_2}^{1/2}} \tag{5-9}$$

Substituting from the table just presented gives:

$$K_p = \frac{mf(100 - mf/2)^{1/2}}{m(1-f)(21 - m - mf/2)^{1/2}\pi^{1/2}} \tag{5-10}$$

For an m value of 12, K_p can be evaluated at various arbitrarily selected values of f from Eq. (5-10), after which the corresponding equilibrium temperature can be determined by interpolating between the values of log K_p versus T presented above. Thus, when m is 12 and f is arbitrarily chosen as 0.625, then K_p at 1 atm pressure is 6.1 from Eq. (5-10), which corresponds to a temperature of 650°C. Additional equilibrium values of f versus T can be determined for an m value of 12 in this fashion, and these results are shown graphically in Fig. 5-4. The same procedure is used to locate the other lines in this figure. (b) Reaction occurs adiabatically in the gas as it flows through the converter, with temperature increasing as this exothermal reaction proceeds. The attainable equilibrium conversion diminishes, however, with increasing temperature. The reacting gas temperature, therefore, rises as reaction proceeds until the corresponding equilibrium constant is satisfied, after which no further change in composition or temperature can occur.

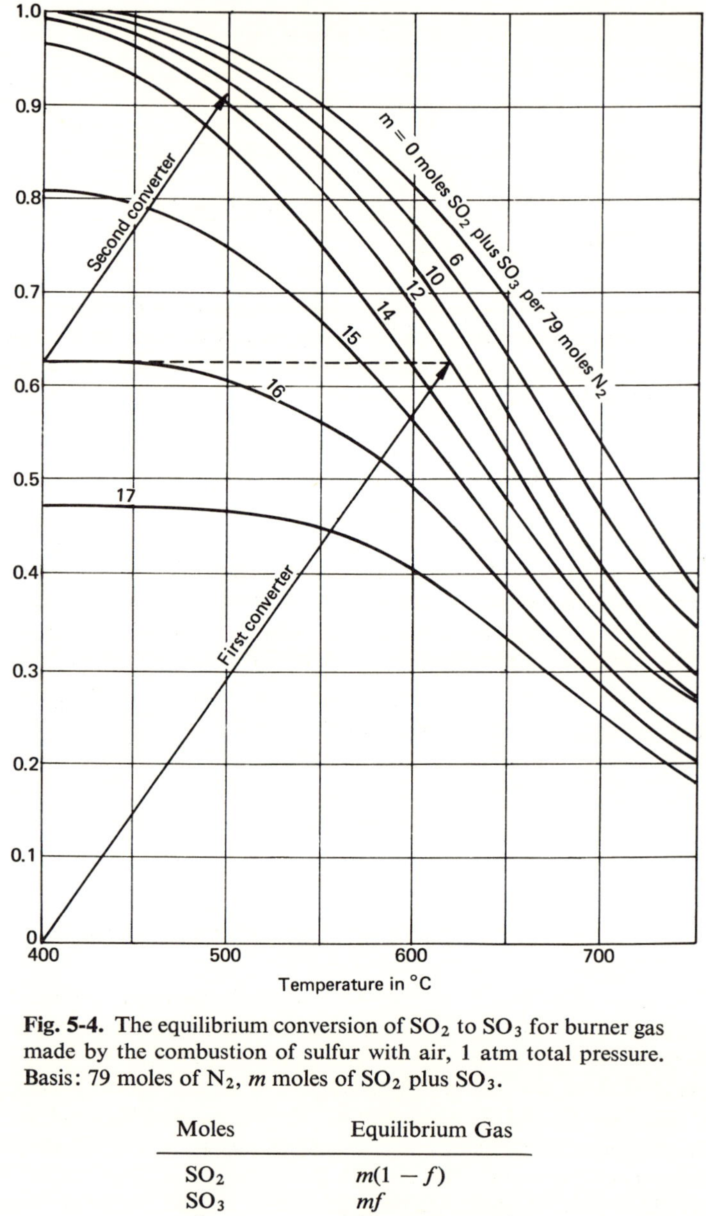

Fig. 5-4. The equilibrium conversion of SO_2 to SO_3 for burner gas made by the combustion of sulfur with air, 1 atm total pressure. Basis: 79 moles of N_2, m moles of SO_2 plus SO_3.

Moles	Equilibrium Gas
SO_2	$m(1-f)$
SO_3	mf
O_2	$21 - m - \frac{1}{2}mf$
N_2	79
	$100 - \frac{1}{2}mf$

Sulfur exists in the gas phase only as SO_2 plus SO_3, with f representing the fraction of this sulfur present as SO_3. The lines on this diagram marked "First converter" and "Second converter" represent performance in Ex. 5-4.

This adiabatic converter operation is readily formulated in quantitative terms. For 79 moles N_2, the moles of this burner feed gas (for which f is zero and m is 12) are presented in column 1 of the following table, with the moles of product gas at any arbitrary conversion f shown in column 2. The heat capacity of the gas at conversion f is as in column 3. Inspection shows that $\Delta\Sigma nMC_p$ for reaction (5-8) is zero, consequently the heat capacity of the reacting mass is independent of f, and is constant at 760 Btu/(°F) (79 moles N_2).

	(1) Fresh Feed, Moles	(2) Gas at Conversion f, Moles	(3) Heat Capacity at Conversion f
N_2	79	79	7(79)
O_2	9	$9 - 12f/2$	$7(9 - 12f/2)$
SO_2	12	$(12)(1 - f)$	$(12(12)(1 - f)$
SO_3	0	12	$(15.5)(12f)$
	100	$100 - 12f/2$	760

The change of composition is related to temperature by an energy balance around the first adiabatic converter, as follows:

$$760(T - 400) - (22{,}000)(12)f = 0 \tag{5-11}$$

This equation must be solved simultaneously with the graphical equation in Fig. 5-4 for the equilibrium conversion f versus temperature using the line in Fig. 5-4 at which m equals 12. The answer is conveniently attained by plotting Eq. (5-11) in Fig. 5-4, yielding the dotted line labeled "1st Converter." Inspection shows that the equilibrium gas leaving this converter has a value for f of 0.62 at a temperature of 620°C. The mole fraction of SO_3 in this gas is then $(12f)/(100 - 12f/2)$, or 0.077. Similarly, the mole fractions of SO_2, O_2, and N_2 are 0.047, 0.055, and 0.82, respectively.

The gas leaving the first converter is now cooled to 400°C before entering the second converter. No reaction occurs in the cooler since no catalyst is present, consequently the composition of the gas leaving the first converter equals that of the gas entering the second one. This cooling step is represented graphically in Fig. 5-4 by the dotted horizontal line labeled "First Cooler." This cooled gas, which is obviously no longer at equilibrium, undergoes further exothermal reaction (5-8) under adiabatic condition in the second converter, where the temperature again rises. For 79 moles of nitrogen, the composition and heat capacity of the gas leaving the second reactor are shown in the two last columns of the foregoing table, hence,

$$760(T - 400) + (-22{,}000)(12f_2 - 12f_1) = 0 \tag{5-12}$$

where f_1 and f_2 are fractional conversions in the first and second converters, and f_1 is known to be 0.62.

The exit composition and temperature from the second converter is found by solving Eq. (5-12) simultaneously with the line in Fig. 5-4 for a value of m equalling 12. This is again conveniently done graphically by plotting Eq. (5-12) in Fig. 5-4, which results in a straight line intersecting the equilibrium line at a value for f_2 of 0.90, and a temperature of 500°C. As expected, Eqs. (5-11) and (5-12) plot as parallel straight lines.

This procedure can be extended to the third and fourth converters, as required.

Cooled reactors. For many exothermal reactions such as organic oxidations, adiabatic reactor operations are not allowable because the temperature levels reached lead to cracking and other undesired side reactions. In these cases, the heat of reaction must be somehow removed from within the reactor to avoid development of excessive temperatures. The cooling means used must be such that the reacting system is nowhere chilled to below its kindling temperature. Three general methods for reactor cooling are now presented.

In indirectly cooled reactors, heat is removed from the reacting mass through heat transfer surfaces installed directly in the reactor. These surfaces often take the form of tubes through which a cooling liquid circulates, or within which a coolant boils. Cooling is often also accomplished by cycling part of the reacting mass through a cooler external to the heater. With liquid systems, the reacting liquid itself is often allowed to boil as heat of reaction is evolved, and the resultant vapors are condensed and refluxed into the reactor. In all these cases, the heat removed generally ends up in the water circulating through coolers or condensers. Occasionally, atmospheric air is used as a cooling fluid.

Cooling by reaction combination. As discussed earlier, by combination of an endothermal reaction with an exothermal reaction, large changes can be made in net heat requirements of a reacting system. Thus, when a bed of a coal is burned in an updraft of air, the temperatures developed in the burning zone due to the vigorously exothermal combustion reactions occurring are very high and may result in ash melting. Such melting is disadvantageous in some types of furnaces and is avoided by mixing steam with the combustion air, whereupon the following endothermal reaction proceeds simultaneously with the combustion reaction:

$$C + H_2O = CO + H_2 \qquad \Delta H^\circ = +32 \text{ kcal}$$

As a result, bed temperature is reduced. The H_2 and CO formed by this reaction can be burned in the combustion space above the fuel bed by suitable addition of more air, and the endothermal heat of the foregoing reaction is recovered in this way. Again, much formaldehyde is produced by passing a methanol-air mixture over platinum, the methanol being in part oxidized exothermally to formaldehyde, in part dehydrogenated endothermally to formaldehyde. By suitably adjusting the ratio of air to methanol in the feed, these two reactions proceed in such proportions that the net isothermal enthalpy change is either positive, negative, or zero.

HEATS OF PRODUCT RECOVERY

In recovery and purification of a product from a chemical reaction, considerable quantities of heat are often required which are usually obtained from process steam. Simultaneously, large amounts of lower temperature heat may be rejected from this recovery system, generally as sensible heat in cooling water.

Simple condensation. In processes in which the desired product appears as a vapor component in an off-gas stream, recovery is sometimes by simple cooling and condensation. In these situations, no process steam is required and the heat of condensation is rejected to a waste heat boiler or to cooling water. Occasionally, vapor-liquid equilibria are favorable enough to allow the desired product to condense in a pure condition. For example, 1, 2-dichlorethane is made by the addition of Cl_2 to ethylene as catalyzed by $CuCl_2$, $FeCl_3$, or $SbCl_5$. To minimize the formation of more highly chlorinated by-products, a large excess of ethylene is used. The dichlorethane (b.p. 83.5°C) produced is condensed directly from the product gas with minor contamination from more highly chlorinated ethanes. For some uses, redistillation of this chlorinated product is required. Another example of selective product recovery by simple condensation is in the production of zinc in which the liquid metal (m.p. 419°C, b.p. 907°C) is condensed out of the retort off-gases, analyzing about 50% zinc-vapor, the remainder being primarily CO. A water cooled condenser is used, with temperatures maintained high enough to avoid metal solidification. In this case, the condenser can be used as a boiler to generate steam under pressure. There are, of course, other operations in which the desired product leaves a reactor as vapor relatively free from impurities, as in the production of tetraethyl lead or in carbon disulfide manufacture. In all these cases, redistillation may be necessary to comply with rigid purity specifications.

Work energy is sometimes required to make possible condensations of the type just discussed, such as when refrigeration is used to reach low condensation temperatures at atmospheric pressure operations. As an alternative, it is sometimes cheaper to operate the system at elevated pressures, thus increasing the product's partial pressure to levels where effective recovery using cooling water at ordinary temperatures becomes possible. In all these recovery methods, heat leaves the plant as sensible heat in the cooling water.

In some cases, product recovery itself involves strongly exothermal or endothermal reactions. Thus, the off-gases from the last converter of a sulfuric acid plant pass to an absorption tower for recovery of SO_3 in 98% sulfuric acid. Large amounts of heat are generated as SO_3 reacts with water in the absorbing acid to form more H_2SO_4. This heat is generated at the temperature of the absorption, usually below 150°C, and either leaves the plant as sensible heat in the cooling water, or is used to preheat boiler feed water.

Quenching. Quenching involves chilling a reacting mass very rapidly to below the kindling temperature of the reaction which is to be arrested. Quenching is typically necessary when (a) slower cooling would result in reversion in endothermal reactions for which equilibrium constants become less favorable with decreasing temperature, and (b) when the further occurrence of a sequential reaction must be prevented. Rapid cooling for quench purposes is generally done by injecting cold gases, liquids, or solids into the reactor product stream. The cooling agent selected is preferably inert to avoid undesirable reaction with the products.

Product recovery and purification. Vapor-liquid relations are generally such that the product recovered from a reactor off-gas stream is impure, being mixed with other reaction partners. Thus, all four chlorinated methanes, namely CH_3Cl, CH_2Cl_2, $CHCl_3$, and CCl_4 form when 1 mole of methane reacts with, for example, 2 moles of chlorine, and condensates from the product gases are mixtures of these materials. Separation of pure compounds requires careful fractionation, with unavoidable consumption of steam and cooling water.

Products are often recovered from gases by absorption rather than by condensation. Thus, formaldehyde is made by the partial oxidation of methanol with air, and product recovery is by water scrubbing of the reactor off-gases. The amount of water used is gauged to the formaldehyde present to produce the 30% commercial grade. Higher strength formaldehyde is produced from 30% formaldehyde by distillation.

In liquid phase reactions, similar problems are encountered. The liquid product stream from the reactor usually contains several materials in solution, and separation of a pure product requires distillation or other heat-consuming operations. Liquid-liquid extraction procedures may be used, but ultimate product recovery is generally by distillation.

Flow sheet planning. With exothermal reactions, it is evident that the heat of reaction can, in theory, be utilized to supply at least part of any heat consumed in product recovery and separation. When the latter heat needs are relatively small, or when heat is evolved in product recovery, then this plant is a net heat producer, and a flowsheet can always be found whereby the plant supplies all of its own heat requirements. Whether such a flow sheet is adopted, of course, depends upon it economic attractiveness relative to alternative schemes.

Situations are often encountered in which, even though steam is purchased for use in preheaters, stills, etc., larger amounts of low temperature heat are rejected by this plant to cooling water. Reasons for such operations are as follows:

1. Low temperature reactor heat. Many organic reactions must be operated at 150°C or below. For example, vinyl chloride is made from acetylene and HCl at temperatures of about 80°C, the heat of this exothermal reaction being transferred to cooling water below the normal boiling temperature of vinyl chloride. This low temperature heat could also have been used in the subsequent distillation of vinyl chloride to produce a pure product, but such a distillation would have to be carried out at subatmospheric pressure. It is usually cheaper to buy steam for atmospheric pressure distillation than to work at reduced pressure.

2. Process control problems. Different stages of a process often have very different operating schedules and control requirements, making heat exchange between them difficult. For example, in some batch polymerizations, heat evolution rates vary with time, consequently cooling must be very carefully matched to the reaction being carried out. It is difficult to utilize the heat so evolved for subsequent distillation operations without sacrificing the flexibility required for good control.

3. Maintenance difficulties. A chemical plant is always complex, presenting continuing puzzles of operation and maintenance. Heat recovery equipment represents added complication to a plant, and incremental costs of its operation may more than offset the value of the heat recovered.

It should, however, be noted that processes do exist in which heat of reaction is extensively recovered. The possibility of generating process steam is always considered when a reactor can be cooled by boiling water at pressures of 50 psi or higher. A waste heat boiler is considered for a hot gas stream at temperatures above perhaps 700°F. Of course, the amount of heat made available in this way must be large enough to justify the effort, and use for the recovered heat must be at hand. If no outlet for this heat exists within the plant (a fairly uncommon situation), then it may be argued that this heat could be used to generate electrical energy for sale to local public utilities. Apart from the legal problems involved in such an operation, electrical power can be produced competitively from waste heat only, if the latter is available at high temperature, in large quantity, and in localities reasonably accessible to power distribution facilities and markets. The value of heat for power production falls rapidly with lowered temperature, and excess heat from most chemical plants, unfortunately, is usually generated at low temperatures relative to power plant practice. Moreover, for power production to show low costs, relatively large installations are operated in which the heat quantities required are much greater than are commonly available from chemical plants. For reasons such as these, by-product heat from chemical processes is often rejected to the surroundings, even though it is "free."

6 Homogeneous Reaction Rates

The conditions under which chemical reactions attain commercially interesting rates are of importance to the industrial chemist, since these have a profound influence upon reactor selection and flow sheet design. For example, reactor type and dimensions, residence times, operating conditions, and mixing patterns obviously determine product stream composition and the type of recovery equipment installed. The occurrence of side reactions, and of poisoning if a catalyst is used, help fix the purity requirements of the materials to be charged. For reasons such as these, the rates of homogeneous reactions are examined presently, with later chapters devoted to rates in heterogeneous systems.

When more than one reagent is involved in a homogeneous reaction, as in the liquid-phase esterification of alcohols and acids or in olefin chlorination in the gas phase, then the participating molecules must first be made to collide before they can combine. In practice, this requires mixing of the reagents, either before or after heating to the proper reactor feed temperature, with heat of reaction supplied or removed as needed. The conversion attained in a given reactor is therefore determined by the slower of three steps, namely mixing, heat transfer, or kinetics of the reaction itself.

MIXING CONTROLLED RATES

The relative change with temperature of the rates of mixing and chemical combination was illustrated earlier by the behavior of hydrogen and oxygen. It was pointed out that these gases, when hot, combine so rapidly that product formation per unit time is determined by the rate at which they are brought together, hence mixing is said to control the process. Because of very slow reaction rates when cold, however, these gases can be mixed and stored for months at room temperature without detectable combination, hence chemical reaction is now rate-controlling. Reaction rates are not always slow at room temperature, of course, as with strong aqueous acids and bases, which combine so rapidly even at 0°C that their reaction rates have never been accurately measured. At the other extreme are reagent mixtures which, even at elevated temperature, require the help of a catalyst to attain reasonable combination rates, as in the gas phase formation of SO_3 from SO_2 and oxygen at atmospheric pressure below 700°C.

It is evident that when reaction rates are slow, reagent mixing can be completed long before the reaction has proceeded very far, and so reactor size and design are determined by kinetics. On the other hand, when reaction rates are great, then mixing becomes the rate-limiting step which fixes reactor design. Information on kinetics is not now required, since combination occurs so rapidly after mixing that the additional reactor space requirements for the reaction itself are minute. Mixing can also be made to occur rapidly, and so reactor size in a mixing-controlled situation may be quite small. Reduced reactor size means that little space is available for heat transfer surfaces, consequently these are reactors in which mixing controls are often adiabatic.

Sometimes, mixing must be made rapidly for reasons other than reduced reactor size. Thus, rapid mixing may avoid the appearance of undesirable products which form over certain ranges of reactant concentration ratios. For example, the chlorination of propylene to allyl chloride proceeds at 600°C (50), with good yields attained at molal ratios of chlorine to propylene of $\frac{1}{2}$ and below, but undesired di- and trichlorides appear at higher ratios. Since rates of formation of these more extensively chlorinated materials are great, mixing must be rapid to minimize exposure of propylene to regions of higher chlorine concentrations.

Mechanism of mixing. The mixing of gases or liquids is always more or less time-consuming, depending upon the system and the energy expended. Thus, in the early stages of the batch mixing of fluids in a stirred vessel, concentrations at any fixed point are found to fluctuate. Depending on the vigor of stirring, on fluid characteristics like viscosity, and on the system's dimensions, these fluctuations diminish with time and finally become undetectable. Similarly, when mixing in the flow systems of Fig. 6-1, wide concentration fluctuations are always found at points a short distance downstream from where the streams are first brought together. These fluctuations diminish with increasing distance downstream from the mixing point, and finally disappear.

Molecular diffusion, namely the random movement of individual molecules past each other in either gaseous or liquid solutions, is always involved in all of these mixing processes. In a closed vessel without agitation or stirring, mixing is solely by natural convection and molecular diffusion. It will be remembered that molecular diffusion tends to be slow, becoming relatively rapid only when concentration gradients are steep, as across thin films. Final completion of any mixing process always involves diffusion, but mixing can be greatly speeded up by turbulence. Turbulence is the relative movement and intermingling of large packets of fluid, and develops when energy is expended in a fluid by a paddle, propellor, or similar device to develop regions

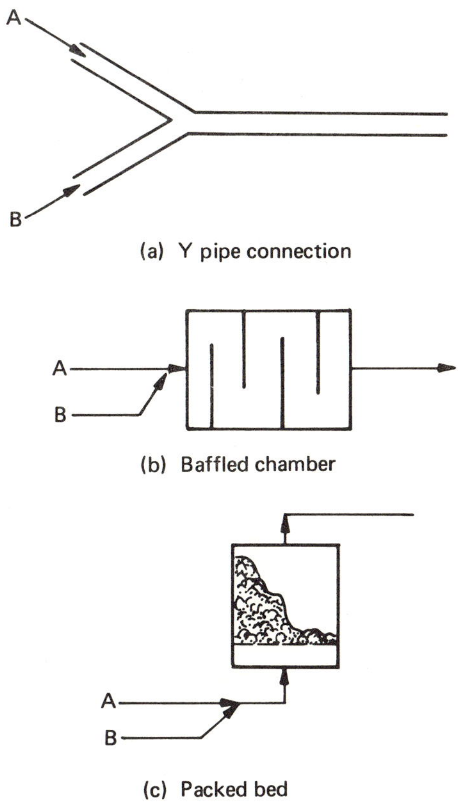

Fig. 6-1. Mixing devices.

of higher velocity. Mixing time is greatly reduced by turbulence because of the reduction in the distances over which molecular diffusion must take place.

Mixing in equipment. Turbulent mixing can be made to occur in many ways. In Fig. 6-1(a), fluids moving in two ducts join in a "*Y*" as shown, and when flow is rapid enough to be turbulent, mixing is extensively completed within 50 pipe diameters downstream from the mixing point. Mixing in a baffled chamber is shown in Fig. 6-1(b), and in a chamber filled with broken solids in Fig. 6-1(c). Both baffles and packing promote turbulence and mixing, and may be used either in liquid or in gaseous systems.

Liquids are often mixed and caused to react while being passed continuously through a stirred vessel, as in Fig. 6-2. Such stirred vessels are also

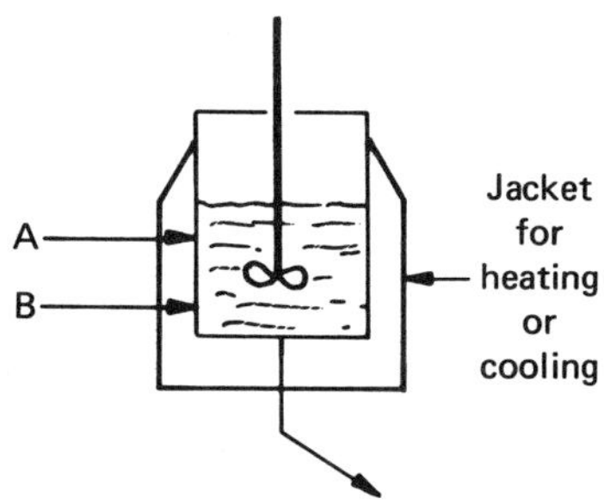

Fig. 6-2. A stirred vessel. This vessel can be operated either on a "batch" basis, or continuously as shown.

used for batch mixing and batch reaction. Perfect mixing, which can be approached by use of a good agitator, means that the liquid composition within the vessel is everywhere the same, and equal to that of the exit stream. Really rapid mixing of gases or liquids requires the expenditure of considerable quantities of energy to generate vigorous turbulence, as in passage through pumps and blowers, or by causing high velocity jets of the fluids to be mixed to impinge upon each other. Several of the possible arrangements for carrying out such mixing continuously are shown in Fig. 6-3. In each case, the extensive turbulence and development of high shear upon impact brings about rapid mixing. Given sufficiently high flow velocities, the arrangement of Fig. 6-1(a) can also be used.

Another type of mixing situation is as shown in Fig. 6-4, where one fluid moves through an orifice port into a large space where it impinges upon a relatively stagnant atmosphere of a second fluid. The resultant mixing patterns have been extensively studied, since this case is commonly encountered in furnaces in which a fuel gas issues into the combustion space from a burner port located in the furnace wall. When gas velocity is low, the

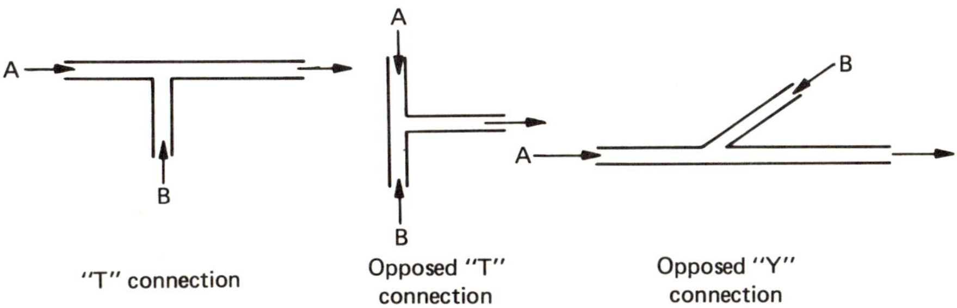

Fig. 6-3. Connections for jet mixing.

flames which form are laminar, having a smooth steady shape in quiet air. Such flames become turbulent at higher gas velocities, are unsteady and without definite shape, forming and reforming constantly. These two flame types, shown in Fig. 6-5, are observable on any Bunsen burner operated with closed air ports.

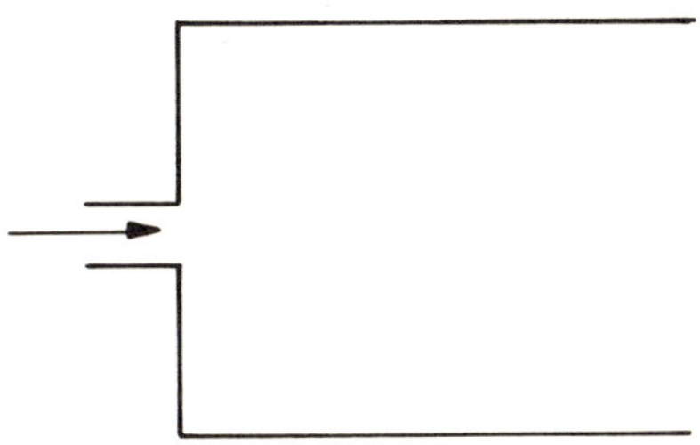

Fig. 6-4. A gas port into a large chamber.

Fig. 6-5. Laminar and turbulent flames reflecting low and high Reynolds numbers.

The dimensions and time-averaged compositions of a jet of fluid issuing as in Fig. 6-4 through a circular port into a relatively stagnant atmosphere can be generalized (58, 59) by use of the reduced parameters of Fig. 6-6. Reduced distances are presented in terms of port diameters, namely L/D, while the reduced concentration C_r is the time-averaged mole fraction of the nozzle fluid in the total fluid at the sampling point. The magnitudes of the fluctuations from the averaged concentrations of Fig. 6-6 depend upon operating conditions. The contours of Fig. 6-6 apply, regardless of whether reaction is occurring or not, as long as the flow of the jet fluid is turbulent, with Reynold's numbers in the port of over 3000. For a given fluid system and a given port, inspection shows that the distance out from the wall needed to attain a given degree of mixing is independent of jet velocity, consequently mixing time must vary inversely with jet velocity. Obviously, mixing time

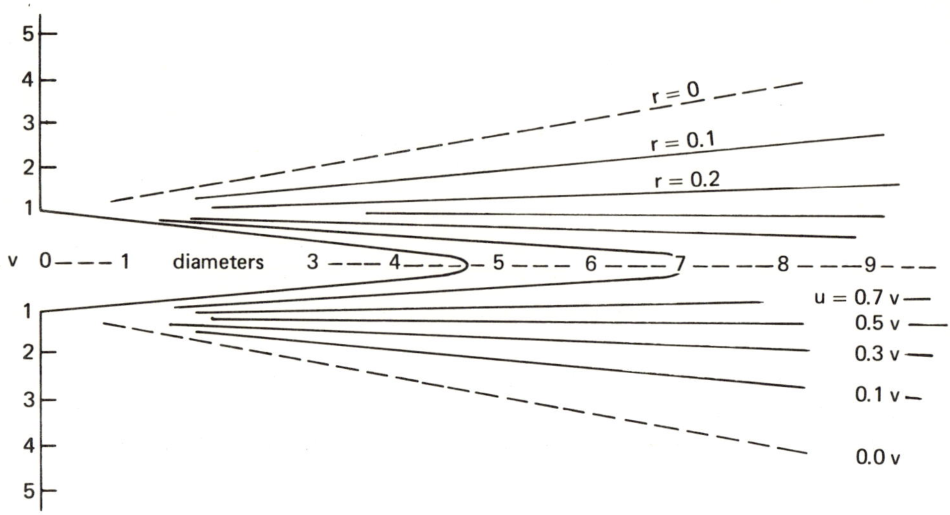

Fig. 6-6. A round jet issuing into a stagnant ambient medium of the same density. The solid lines are for constant time-averaged velocity, where v is the velocity in the port. These are also the lines for which r, the ratio (moles jet fluid per moles ambient fluid) remains constant (59).

for a fixed flow volume could also be decreased by increasing the number of ports, but keeping the total port area constant.

HOMOGENEOUS REACTION KINETICS

After mixing, miscible reagents form a homogeneous gaseous or liquid solution, in which reaction then occurs. The rate of a homogeneous reaction in either the mixed gaseous or liquid-phase is conveniently expressed in terms of the appearance or disappearance of an arbitrarily chosen key component. For reaction (1-1), the rate of production r_M of the key component M in moles per unit time and per unit volume is expressed as follows:

$$r_M = \frac{dn_M}{V d\theta} \tag{6-1}$$

where V is the volume of the phase in which reaction occurs, n_M represents the moles of reactant M in this phase, and θ is time. Any reaction partner of reaction (1-1) may be chosen as key, with rates of appearance and disappearance of these partners related by the reaction stoichiometry:

$$-r_A = -\frac{a}{b}r_B = \frac{a}{m}r_M = \frac{a}{n}r_N \tag{6-2}$$

Some indication of the influence of reagent concentration upon the rate of a chemical reaction as expressed in Eq. (6-1) can be obtained from the Kinetic Theory. That is, molecules of the reagents involved in a reaction must collide for a molecule of product to form. Simplifying reaction (1-1) by assuming coefficients a, b, m, and n all to be unity, then a molecule of M (and of N) forms by this reaction only upon occurrence of a binary collision between molecules of A and B. Reaction rate r_M will depend upon this collision frequency, which is in turn proportional to the product of the concentration of the reactants expressed as moles per unit volume, namely $C_A \cdot C_B$. This simple collision model requires modification, however, since the number of molecules reacting per unit time is usually far smaller than the number of molecular collisions in this same time period. Moreover, according to the Kinetic Theory, collision frequency increases with the square root of the absolute temperature, whereas temperature is usually found to have a far greater effect than this upon reaction rate. These somewhat contradictory findings are explained in modern rate theory by assuming that only a small fraction of the collisions between molecules in a reacting system result in combination. For reaction to occur, a collision must involve an energy of translation, vibration, etc., in excess of the quantity E, called the activation energy. The fraction of total collisions per unit time having energies in excess of E may be represented as $e^{-E/RT}$, so that the greater E becomes, the smaller the fraction of collisions which result in reaction at a given temperature. The exponential form of the term $e^{-E/RT}$ accounts for the unexpectedly large effect of temperature upon chemical reaction rates, and is usually in agreement with experimental findings. Thus, the rate of formation of M becomes

$$r_M = \alpha T^{1/2} e^{-E/RT} \cdot c_A c_B \tag{6-3}$$

where α is a constant, the term $T^{1/2}$ allows for the variation of collision frequency between A and B with temperature, $e^{-E/RT}$ represents the fraction of the collisions having that minimum violence required for combination, while product $c_A c_B$ is proportional to the collision frequency. The term $e^{-E/RT}$ is generally so much more important than the term $T^{1/2}$ that the latter is often omitted. The full significance of constant α is considered by modern reaction rate theory, but lies beyond the scope of this book. For purposes of industrial application, Eq. (6-3) is rewritten as follows:

$$r_M = k\, c_A c_B \tag{6-4}$$

where the experimentally determined rate coefficient k is constant at any given temperature, varying with $e^{-E/RT}$. The activation energy E is a constant for each case, but is different from one reaction to the next.

In the foregoing discussion, it is recognized that almost all collisions occurring between molecules in a gas are binary, with collisions involving three molecules being very rare and collisions between more than three molecules being practically nonexistent. Most reaction steps are, therefore, probably bimolecular, that is, involve reaction between two colliding molecules. However, it is relatively unusual to encounter the simple case discussed in the preceding paragraphs, whereby binary collisions lead directly to the formation of the final product. Instead, the material formed by reaction between two colliding molecules is an intermediate which itself may undergo further collisions and reaction, with several such steps involved before the final product appears. These steps are often so rapid that the intermediates never build up to a high concentration. The simple rate Eqs. (6-3) and (6-4) obviously do not apply to such a complex reaction mechanism. Fortunately, regardless of mechanism, the rate of a great many homogeneous reactions can be represented by the following empirical expression:

$$r_M = kc_A^{\alpha}c_B^{\beta}c_C^{\gamma} \tag{6-5}$$

Experimental studies must be made to establish k, which is again constant at any given temperature, and proportional to $e^{-E/RT}$. Exponents α, β, and γ determined from these rate data, may be positive, negative, or zero, need not be integers, and bear no necessary relation to the stoichiometry of the overall reaction (1-1). The activation energy E is determined from the slope of the straight line obtained when ln k is plotted against reciprocal temperature.

EXAMPLE 6-1

Assume that the formation of M by reaction (1–1) at temperature T_1 may be represented by Eq. (6–5). If the rate doubles every ψ°C, (a) develop an expression for r_{T_2}, namely the rate at T_2, in terms of the rate constant k_{T_1} at T_1; (b) if the rate doubles between 373–383°K, what is the magnitude of the activation energy E?; and (c) if the rate doubles between 1000–1010°C, what is E?

Answer: (a) At any given concentration of reactants, if the rate doubles every ψ°C, then it follows directly that

$$\frac{r_{T_2}}{r_{T_1}} = 2^{(T_2-T_1)/\psi} \tag{6-6}$$

But r_{T_1} equals $k_1c_A^{\alpha}c_B^{\beta}$ by Eq. (6–5). It also follows from this equation that k_{T_2}/k_{T_1} equals r_{T_2}/r_{T_1}. Combining,

$$r_{T_2} = 2^{(T_2-T_1)/\psi}k_{T_1}c_A^{\alpha}c_B^{\beta} \tag{6-7}$$

(b) From Eq. (6–4), since k varies with $e^{-E/RT}$,

$$\frac{r_{T_2}}{r_{T_1}} = \exp\frac{E}{R}\left(\frac{1}{T_1} - \frac{1}{T_2}\right) \tag{6-8}$$

Combining Eqs. (6–6) and (6–8) and taking logarithms,

$$E = \frac{RT_1 \cdot T_2}{\psi}\ln 2 \tag{6-9}$$

Recognizing that R is 1.99 cal/(g mole)(°K), T_1 is 373°K; T_2 is 383°K, and that ψ is 10°, then from Eq. (6–9), E is 20,000 cal/g mole.

(c) If rate doubles between 1000–1010°K, then from Eq. (6–9), the corresponding activation energy is 140,000 cal/g mole. As shown in this example, therefore, the assumption that the rate doubles at all temperature levels within a given temperature interval, such as 10°C, is incorrect, since this is not consistent with a constant activation energy. The assumption of a doubling in some temperature interval is nevertheless often made for convenience in calculation, providing the range of temperature covered is small.

Further consideration of Eq. (6–5) shows that it refers to a unit volume of system, since r is defined as $dn/Vd\theta$. This equation applies to systems of any volume, as is made evident by multiplying Eq. (6–5) by V:

$$+\frac{dn_M}{d\theta} = k\left(\frac{n_A^\alpha}{V^{\alpha-1}}\right)c_B^\beta c_C^\gamma$$

All the rate expressions discussed so far apply only to reactions which go substantially to completion at equilibrium, that is, to reactions for which the equilibrium, constant K is large. For reactions which do not go to completion at equilibrium, the net rate expression sometimes takes the following form:

$$r_M = k_F c_A^\alpha c_B^\beta c_C^\gamma - k_R c_M^\mu c_N^\eta \tag{6-10}$$

where k_F and k_R are the isothermal rate constants for the forward and the reverse reactions. At equilibrium, where r_M is zero, Eq. (6–10) should simplify to the equilibrium constant expression for reaction (1–1). Thus, when α, β, γ, μ, and η, respectively, equal coefficients a, b, c, m, and n of Eq. (1-1), and when r is zero, then it is easy to show that Eq. (6–10) simplifies to the equilibrium constant expression, Eq. (1–6). Such simplification is not always possible, since equations like Eq. (6–10) are empirical, and may not apply all the way to equilibrium.

Determination of reaction rates. It follows from the foregoing discussion that, while the rate theory provides a valuable basis for interpreting results, it cannot be used to predict even the form of the rate expression for any given homogeneous chemical reaction. Moreover, rates are often very sensitive to the presence of impurities introduced in the feed or by the materials of construction of the apparatus used, which may cause changes in both the form of the rate expression and in the magnitude of the rate itself. Because of these difficulties, the industrial chemist must determine reaction rates in his systems by direct experimentation, duplicating plant conditions as nearly as possible to avoid misleading results.

A commonly used method for the experimental determination of the rate of a homogeneous gas-phase or liquid-phase reaction involves adding known quantities of the reactants to a batch reactor arranged to operate isothermally at constant volume, mixing thoroughly, and then following the change in concentration of the key component with time. This concentration is determined by any suitable method, such as light absorption, electrical conductivity, etc., or by taking samples at appropriate intervals for direct analysis. When necessary, the total volume and pressure of the reacting mass is also recorded, making due allowance for any samples withdrawn. The two curves of Fig. 6-7 show the typical variation with time of the concentration of the key component A when A and B react isothermally.

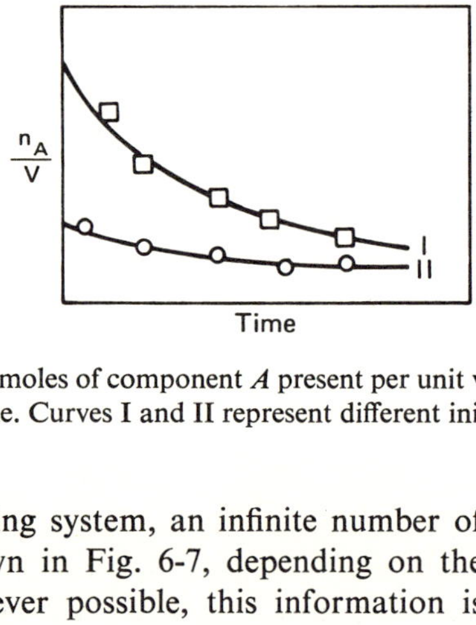

Fig. 6-7. The moles of component A present per unit volume of the system vs. time. Curves I and II represent different initial compositions.

For a given reacting system, an infinite number of different isothermal curves could be drawn in Fig. 6-7, depending on the composition of the initial charge. Whenever possible, this information is summarized in the form of the differential Eq. (6-5). When such an equation can be found, then its convenience lies in the ease with which it can be integrated for various conditions of feed ratios, pressures, and temperatures.

Batch reactions. As indicated by Eqs. (6-4), (6-5), and (6-10), the rate of a homogeneous reaction at any given moment, namely $dn_M/Vd\theta$, is uniquely determined by the system temperature and the concentrations of the reaction partners. This is true regardless of the history of the system, and of whether operations are at constant or changing volume or pressure, of whether the system is closed or being supplied with fresh reagents, etc.

Calculations of the extent to which a reaction proceeds in any system over a given time period requires information on the initial system composition and on the relevant rate expressions. By applying material balance on the key component, input (both by material flow into the system and by chemical reaction) minus output equals accumulation. In a batch system, materials flow in and out of the reactor is zero, hence, formation of the key component by chemical reaction equals accumulation, or,

$$Vrd\theta = dn \tag{6-11}$$

where n is the moles of the key component formed, V is volume of the reacting mass, and r is rate. Application of this equation to batch systems, both for constant volume and for changing volume is demonstrated in example 6-2.

EXAMPLE 6-2

A homogeneous gas phase reaction and its rate expression are as follows:

$$2A = M \qquad r_A = -kc_A^2$$

where c_A is the concentration of A in moles per unit volume and r is defined by Eq. (6–1). At 500°F, the equilibrium constant of this reaction is very large, while the rate constant k equals 20 cu ft/(sec) (mole).

(a) A batch reactor of fixed volume is charged at 500°F and 1 atm pressure with an equimolal mixture of reagent A and an inert substance. When keeping the temperature fixed at 500°F, how long does it take for 75% of the A initially present to react? Note that the total pressure falls in this case while reaction occurs, because of a decrease in the moles of the system.

(b) Repeat part (a) when operating in a reactor comprising a cylinder fitted with a movable piston, wherein constant pressure is maintained by changing the reactor volume. The initial volume of this system is to equal that of the reactor in part (a).

Answer: (a) Choosing as basis 1 mole of charge, with Y mole of key component A present at any time θ:

	Mole	
	Initial	At any conversion
A	0.5	Y
M	0	$0.5(0.5 - Y)$
Inert substance	0.5	0.5
S, total	1.0	$0.75 + 0.5Y$

Since the reactor volume is constant at the volume of 1 mole of a perfect gas at 500°F and 1 atm pressure, namely 700 cu ft, then

$$r_A = -\frac{kY_A^2}{(700)^2}$$

Combining with Eq. (6–11) and recognizing that n is identical with Y on the basis chosen,

$$\theta = -\frac{700}{k}\int \frac{dY}{Y^2}$$

Integating Y between the limits of 0.5 and 0.125,

$$\theta = \frac{700}{20}\left(\frac{1}{Y_2} - \frac{1}{Y_1}\right) = \frac{700}{20}\left(\frac{1}{0.125} - \frac{1}{0.50}\right) = 210 \text{ sec}$$

(b) The mole table of part (a) again applies. The volume of the system however, diminishes as reaction occurs at constant pressure and equals $700(0.75 + 0.5Y)$, consequently,

$$r_M = \frac{kY^2}{700^2(0.75 + 0.5Y)^2}$$

Substituting into Eq. (6–11) and rearranging:

$$\theta = -\frac{700}{k}\int \frac{(0.75 + 0.5Y)\,dY}{Y^2}$$

or

$$\theta = \frac{700}{k}\left[0.75\left(\frac{1}{Y_2} - \frac{1}{Y_1}\right) + 0.5\ln\frac{Y_1}{Y_2}\right]$$

Between the limits for Y of 0.5 and 0.125 as before, θ is found to be 183 sec for this case.

Flow reactors. Reactors are often operated under steady flow conditions, rather than batchwise, especially in larger plants. This means that raw materials are fed and products withdrawn continuously, no accumulation occurs within the reactor, and although temperature, pressure, and concentration may vary from one point to another in the reactor, they are constant with time at any fixed point.

Two contrasting types of flow reactors are the "pipe" reactor and the stirred vessel. In the first, reactants enter at one end, travel through the duct while undergoing reaction, and leave at the other end. In the idealized "piston flow" or "plug flow" case, it is assumed that no axial mixing occurs, and that all parcels of fluid have an equal velocity and an equal residence time in the reactor. Thus, concentration and temperature may both change in the direction on flow, but not at right angles to this direction. On the other hand, in the idealized case of the "perfectly stirred" vessel, the feed streams are so rapidly incorporated into the tank contents that concentrations are the same in all parts of the vessel, and the composition of the leaving product stream is the same as that of the reacting mass within the tank. A simple form of the stirred vessel shown in Fig. 6-2 is a tank equipped with an agitator. By rapid mixing, temperature as well as concentration is made uniform throughout the tank, regardless of the feed stream temperature and composition.

While both are idealizations, some reactors approximate piston flow while others approximate perfect mixing. The mathematics of these two limiting cases have received much attention, are relatively simple, and are discussed in the following paragraphs.

(a) Piston flow. Consider the pipe reactor of length L and cross section A, shown schematically in Fig. 6-8, in which the generalized reaction (1-1) is being carried out either in the gas or the liquid phase. The feed enters this pipe reactor at a constant rate of F mole per unit time and each mole of this feed contains Y_0 moles of the key component M. As reaction proceeds in the fluid moving through the reactor, there is a change in Y, namely in the moles of key component present in a parcel of fluid which initially contained 1 mole of fresh feed. It should be noted that Y is not a mole fraction or a concentration, but a ratio. Similarly, F is a constant regardless of whether there is a change in the moles flowing due to chemical reaction. Obviously, for a reactor in a steady-state, FY moles of key component per unit time passes through any imaginary reactor plane located at right angles to the flow direction.

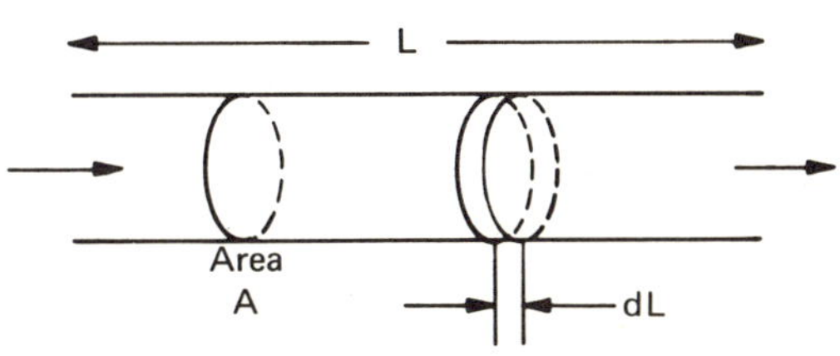

Fig. 6-8. A flow reactor showing a section of differential length dL.

The volume of the reactor section of differential length shown in Fig. 6-8 is AdL. The formation of the key component per unit time in this section by chemical reaction is $rAdL$, where r is evaluated at the average concentration of the key component within this section. As the reactants pass through this section, the ratio of key component changes from Y to $(Y + dY)$, consequently by using material balance on this component for unit time

$$FY + r_M A dL = F(Y + dY) \tag{6-12}$$

Rearranging,

$$\frac{AL}{F} = \int_{Y_1}^{Y_2} \frac{dY}{r_M} \tag{6-13}$$

Integration of Eq. (6-13) can be performed when r_M is known as a graphical or analytical function of Y, as will be illustrated in example 6-3. It should be noted that product (AL) in Eq. (6-13) equals the volume of this piston flow reactor, namely V_{PF}.

(b) Perfectly mixed reactor. A relation analogous to Eq. (6-13) can be developed for the perfectly mixed reactor of Fig. 6-2, into which the feed enters at a constant rate of F mole per unit time. By using material balance per unit time on the key component M around this stirred tank of volume V_{ST} one obtains

$$FY_{\text{out}} = FY_{\text{in}} + r_M V_{ST} \tag{6-14}$$

It is again to be remembered here that F is the constant feed rate entering the tank, Y is the moles of key at any time/mole of this entering feed, while r_M is the rate evaluated at the tank composition.

EXAMPLE 6-3

The gas phase reaction of example 6–2 together with its rate expression is again presented below:

$$2A = M$$
$$r_A = -kc_A^2$$

As before, the equilibrium constant of this reaction is very large at 500°F, while the rate constant k again equals 20 cu ft/(sec) (mole).

(a) Conversion is to be carried out isothermally at 500°F and at a substantially constant pressure of 1 atm in a pipe reactor operating in piston flow. What is the length of pipe required to convert 75% of A in the feed to M, if the feed analyses 50 mole % A, the rest being inert substance? The feed rate, namely, F/A, is 0.005 lb mole/(sq ft) (sec).

(b) A stirred reactor is to be substituted for the flow reactor just described to operate on the same feed at the same temperature and pressure, and attain the same degree of conversion. What is the ratio of the volume of the stirred reactor to that of the piston flow reactor, namely, V_{ST}/V_{PF}? [Industrial gas reactions are not often operated in reactors approximating a well-stirred vessel. Part (b) of this example is, nevertheless, presented here to illustrate calculational methods.]

Answer: (a) the specific volume of an ideal gas at 500°F and 1 atm, namely v, is 700 cu ft/lb mole. Choosing A as the key component, then the following mole table again applies per mole of feed:

		lb Mole	
Constituents	Feed	At any conversion	At 75% conversion
A	0.5	Y	0.125
M	0	$0.5(0.5 - Y)$	0.1875
Inert substances	0.5	0.5	0.5
S, total	1.0	$0.75 + 0.5Y$	0.8125

Substituting from this table into the rate expression, for constant pressure operation:

$$r_A = -\frac{kY^2}{(v^2)(0.75 + 0.5Y)^2}$$

Combining with Eq. (6-13), and remembering that A is now the key component:

$$L = -\frac{Fv^2}{kA}\int \frac{(0.75 + 0.5Y)^2\,dY}{Y^2}$$

$$= \frac{Fv^2}{kA}\left\{0.56\left(\frac{1}{Y_2} - \frac{1}{Y_1}\right) + 0.75\ln\frac{Y_1}{Y_2} + 0.25(Y_1 - Y_2)\right\}$$

Since Y_1 is 0.5 and Y_2 is 0.125, then

$$L = \frac{4.3Fv^2}{Ak}$$

But v is 700 cu ft, F/A is 0.005 lb moles/(sq ft) (sec), and k is 20 in these units, hence L equals 530 ft.

(b) From Eq. (6-14)

$$V_{ST} = \frac{F(Y_{\text{out}} - Y_{\text{in}})}{r_A} = \frac{F(0.125 - 0.5)}{r_A}$$

But from the rate expression in part (a) above, at a value for Y of 0.125

$$r_A = -\frac{k(0.125)^2}{v^2(0.75 + 0.5 \times 0.125)^2} = -\frac{0.0236k}{v^2}$$

Combining the two foregoing relations:

$$V_{ST} = \frac{15.8Fv^2}{k}$$

But V_{PF} was shown in part (a) of this example to be $(4.6\,Fv^2/k)$; consequently,

$$\frac{V_{ST}}{V_{PF}} = \frac{15.8}{4.3} = 3.7$$

Inspection of these relations shows also that the ratio (V_{ST}/V_{PF}) increases as Y decreases, indicating the advantages of piston flow as far as reactor size is concerned.

Residence times for piston flow. It is sometimes desirable to calculate residence times as well as conversions in piston-flow reactor systems. When the value of δ in Eq. (1-7) is other than zero, then as reaction proceeds, there is a change in S, the total moles of the system per mole of feed. Thus the value of δ for the reaction of example 6-3 is minus unity and for any given value of Y, S equals $(0.75 + 0.5Y)$, as shown in the mole table of example 6-3. The volume of fluid flowing past any cross section in a pipe reactor per unit time is FSv. The residence time $d\theta$ of the fluid in a differential section of pipe of length dL and volume AdL is then

$$d\theta = \frac{AdL}{FSv}$$

Combining with Eq. (6-12) and integrating over the whole pipe:

$$\theta = \int_{Y_{\text{in}}}^{Y_{\text{out}}} \frac{dY}{Svr_M} \tag{6-15}$$

The time θ in Eq. (6-15) is that required for a slug of fluid of differential thickness to travel from one end of the pipe reactor to the other. Since piston flow is assumed, this slug travels through the tube with no mixing into the fluid ahead or behind.

EXAMPLE 6-4

The gas phase reaction of examples 6-2 and 6-3 in this chapter is shown again below, along with its rate expression:

$$2A = M$$

$$r_A = -kc_A^2$$

As before, the equilibrium constant for this reaction is very large, and k at 500°C is 20 cu ft/(sec) (mole). The gas feed contains 50 mole % A and 50 mole % inert substances. For a 75% conversion of A to M, what is the residence time in a piston flow reactor?

Answer: Choosing A as the key component, the mole table per mole of feed is

	lb Mole	
Reagents	Feed	At any conversion
A	0.5	Y
M	0	$0.5(0.5 - Y)$
Inert substances	0.5	0.5
S total	1.0	$0.75 + 0.5Y$

The rate expression is as before:

$$r_A = -\frac{kY^2}{v^2(0.75 + 0.5Y)^2}$$

Equation (6-15) now becomes

$$\theta = -\int_{Y_{in}}^{Y_{out}} \frac{v^2(0.75 + 0.5Y)^2\,dY}{kY^2Sv}$$

Substituting $(0.75 + 0.5Y)$ for S, 700 for v, and integrating Y between the limits of 0.5 and 0.125:

$$\theta = \frac{700}{k}\left\{0.75\left(\frac{1}{Y_2} - \frac{1}{Y_1}\right) + 0.5\ln\frac{Y_1}{Y_2}\right\} = 182\text{ sec}$$

Residence times, stirred reactor. It is easily shown experimentally that there is a wide variation in the residence times of different particles of a liquid stream flowing through a well-stirred vessel. Thus, assume two inert but miscible liquids A and B are available, which have the same density. The stirred vessel of Fig. 6-2, always containing V' lb of liquid, is filled with liquid A, which it is receiving and discharging at a steady rate of F' lb per unit time. Without changing flow rates and assuming perfect mixing, liquid B is suddenly substituted for liquid A in the feed stream, consequently B is instantly mixed into the contents of the vessel and starts to appear in the product stream. The experimentally determined weight fraction of B in the kettle and in the product stream is plotted against the "reduced time," namely $(F'\theta/V')$, as shown by curve (1) in Fig. 6-9. Points on this curve obviously represent the fraction of the contents of the vessel which have been present for $F'\theta/V'$ time units or less, where $F'\theta/V'$ is zero at the moment of changeover from A to B. For example, when $(F'\theta/V')$ is 1.38, then the liquid in the tank is found by chemical analysis to be 75%

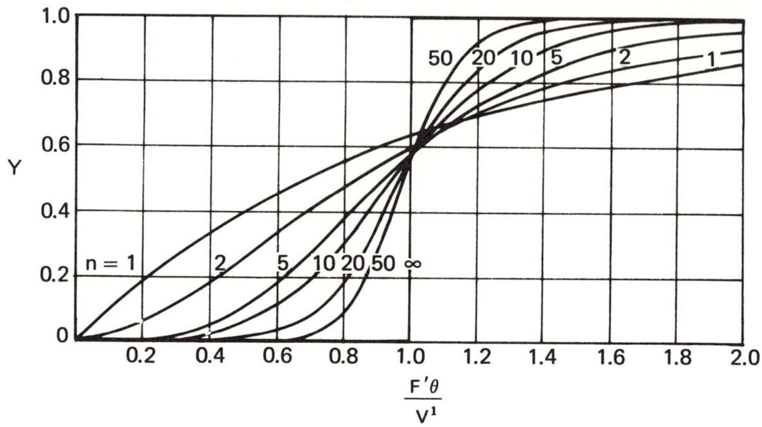

Fig. 6-9. The weight fraction of component B in the exit streams of Ex. 6-5. For each curve the number of identical well-stirred vessels is n, with the sum of the volume of these n vessels equaling V' for each curve (79). The reduced time $F'\theta/V'$ is zero at the moment of feed change-over from A to B.

B, indicating that 75% of the contents have been present in the tank for 1.38 reduced time units, or less.

Curve (1) in Fig. 6-9 can be determined by calculation as well as by experiment. Thus, if Y is the weight fraction of component B in the kettle (and in the product stream), then over a differential time period $d\theta$, since input minus output of B to the kettle equals its accumulation,

$$F'd\theta - F'Yd\theta = V'dY \tag{6-16}$$

Rearranging and integrating, with zero time being the instant of changeover from A to B:

$$Y = 1 - e^{-F'\theta/V'} \tag{6-17}$$

This equation fits curve (1) in Fig. 6-9 exactly.

EXAMPLE 6-5

The single well-stirred vessel of volume V' discussed above is replaced by two identical well-stirred vessels I and II, each having a volume $V'/2$, namely half that of the original vessel. At the outset, vessels I and II are both filled with pure liquid A, but are connected in series, so that pure feed enters vessel I, flows from I to II, with overflow from II comprising the system product. Operations are continuous at the same constant feed rate as before, namely F' lb per unit time. Liquid B is now again suddenly substituted for liquid A in the feed. Develop a relation between Y_2, the weight fraction of B in the product stream of vessel II, and the reduced time $F'\theta/V'$.

Note: Liquids A and B are again inert, miscible, and of equal density, and perfect mixing is assumed in each vessel.

Answer: Reasoning as in the derivation of Eq. (6-17), then for vessel I, whose volume is $V'/2$,

$$Y_1 = 1 - e^{-2F'\theta/V'}$$

where Y_1 is the weight fraction B in the effluent. For vessel II, whose volume is again $V'/2$,

$$F'(Y_1 - Y_2)d\theta = \tfrac{1}{2}\, V'dY_2$$

Rearranging and combining the two foregoing equations:

$$\frac{dY_2}{d\theta} + \frac{2F'Y_2}{V'} = \frac{2F'}{V'}(1 - e^{-2F'\theta/V'})$$

This equation is linear in Y_2, hence, the integrating factor is $e^{-2F'\theta/V'}$, and so

$$Y_2 e^{2F'\theta/V'} = \int \frac{2F'}{V'}(1 - e^{-2F'\theta/V'})e^{2F'\theta/V'}\, d\theta + \text{constant}$$

or

$$Y_2 e^{2F'\theta/V'} = e^{2F'\theta/V'} - \frac{2F'\theta}{V'} + \text{constant}$$

Since Y_2 is zero when θ is zero, the constant is minus unity. Substituting and rearranging:

$$Y_2 = 1 - \left(\frac{2F'\theta}{V'} + 1\right)e^{-2F'\theta/V'}$$

The foregoing relation is plotted as curve (2) in Fig. 6-9. In similar fashion, equations can be developed for series operation of n identical tanks, whose volumes again sum up to V' units of mass. Inspection of the typical curves in Fig. 6-9 shows that piston flow is approached as n increases.

Reaction rate expressions. It has been assumed in the discussion presented in this chapter that algebraic rate expressions like the one of example 6-5 are available for the reactions under consideration. In many industrial situations, rate data exist only in graphical form. The calculational procedures outlined in the preceeding paragraphs, of course, can be used equally well with graphical rate data as with analytical rate expressions like Eq. (6-5).

Reaction rate expressions developed in batch reactors of constant volume are occasionally written as follows:

$$\frac{dc_A}{d\theta} = -kc_Ac_B \tag{6-16a}$$

or

$$\frac{dp_A}{d\theta} = -k'p_Ap_B \tag{6-16b}$$

The student is at times confused by these equations, since they appear at first sight not to be applicable to a steady-state operation, such as a stirred tank reactor, in which neither partial pressures nor concentrations at any given point change with time as the reaction proceeds. For such reactors, both equations must first be converted into the form of Eq. (6-5). Thus, since c equals (n/V) by definition and since $(dc/d\theta)$ equals $(dn/Vd\theta)$ at constant volume, then Eq. (6-16a) becomes

$$\frac{dn_A}{Vd\theta} = -k\left(\frac{n_A}{V}\right)\left(\frac{n_B}{V}\right) \tag{6-16c}$$

Similarly, p equals (nRT/v), and differentiating at constant volume, dp equals $(RTdn/V)$, hence Eq. (6-16b) becomes

$$\frac{dn_A}{Vd\theta} = -k'RT^2\left(\frac{n_A}{V}\right)\left(\frac{n_B}{V}\right) \tag{6-16d}$$

Obviously, constant k of Eq. (6-5) equals k of Eq. (6-16c) and also $(k'RT^2)$ of Eq. (6-16d). Transformed in this way, Eqs. (6-16c) and (6-16d) can now be used for stirred reactors in flow systems, as illustrated earlier in example 6-4.

Reactors for homogeneous reactions. Batch reactions involving liquids are common, and are often carried out in stirred vessels. Depending on the circumstances, tanks, vats, and autoclaves are likewise used. In some cases, a batch reaction is carried out in a pipe loop through which the reacting liquid is circulated.

As pointed out in earlier paragraphs, stirred vessels and pipe reactors are commonly used for continuous operations of liquid phase reactions. Stirred reactors are compact but suffer from the disadvantage common to all mixed systems, namely, that reactant concentrations are low, and reaction rates are reduced accordingly. Pipe reactors approximate piston flow, but are relatively costly because of the high ratio of wall area to volume for long tubes, requiring the use of much metal. A cheaper vessel for flow operations would be a cylinder having a volume equal to the pipe reactor, but with a height per diameter ratio near unity, for which the ratio of wall area to volume is small. Unfortunately, however, cooling or heating such vessels is obviously less effective, and much lower linear flow velocities must be used to attain residence times equal to those of the pipe reactors. Backmixing and stagnant pockets of liquid tend to develop at such low velocities, so that piston flow is no longer even approximated. Such undesirable flow patterns are in some degree to be offset by filling these vessels with a suitable solid packing.

Unlike liquid phase reactions, industrial gas phase reactions are usually operated on a continuous flow basis rather than batchwise, and are rarely carried out in stirred reactors. Reactors in which piston flow is approximated are common, and may be pipe reactors or catalyst packed vessels of various types. When carrying out a homogeneous reaction having high rates, then jet mixing type reactors such as gas burners are also used.

The internal combustion reciprocating and turbine-type engines are special types of reactors for the burning of fuels, and can be used both for homogeneous gas phase, for gas-liquid, and gas-solid reactions. In theory, these reactors are suitable for any exothermal reaction, with an advantage over other reactors that at least part of the enthalpy change involved appears as work rather than as heat. To date, however, it has not proved practical to carry out reactions other than the usual fuel combustion reactions in these devices. Research interest continues in oxidations like those of ammonia to nitric oxide, and of methanol to formaldehyde.

Simultaneous reactions. Both simultaneous and sequential reactions are frequently encountered in reacting systems, especially in organic operations. Commonly cited examples of simultaneous reactions are the chlorinations and nitrations of benzene, where substitutions occur partially in the meta- and partially in the para-positions. A somewhat more complex example is the sulfonation of naphthalene to yield the α- and β-monosulfonic acids simultaneously:

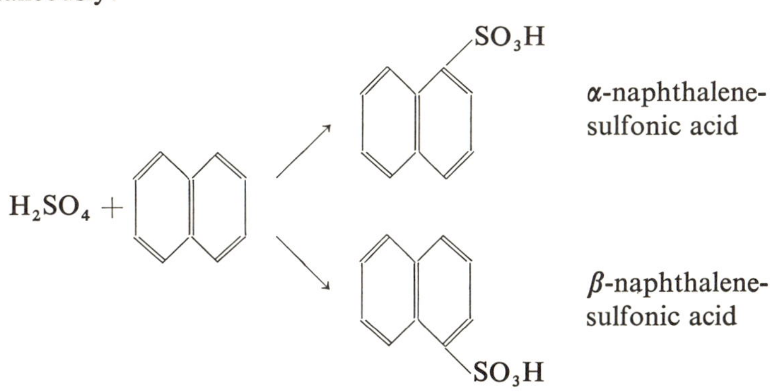

Higher temperatures favor the β-form over the α one, but under no conditions is either of them the sole product. If, for example, the object is to make the α-isomer, reducing the temperature increases attainable yields and so reduces raw material cost, but cooling costs increase. Fixed charges likewise increase, since production from a given reactor falls due to lower reaction rates. Selection of proper design is a problem in optimization.

Sequential reactions often occur along with simultaneous reactions. Thus, the two napthalenemonosulfonic acids formed simultaneously, as just described, tend to be sulfonated further to disulfonic acids. Another common sequence is the formation of a desired organic product in one reaction step, followed by decomposition or "cracking" of this product. Thus, ethane is converted to ethylene in a high temperature, homogeneous dehydrogenation reaction in the gas phase, followed by cracking of the ethylene to hydrogen and carbon. As an indication of the complicated situations which can develop if care is not exercised, a considerable fraction of the ethylene so formed polymerizes in a further sequential reaction to higher olefins, which also crack, etc.

Selectivity. In any operation involving simultaneous and sequential reactions, whether homogeneous or heterogeneous, the question of selectivity is important. The word selectivity refers to the yield of the desired product, and is generally defined as the moles of desired product formed divided by the moles of a given reagent which disappears. Selectivity is obviously not the same as conversion per pass, defined as moles of a reagent disappearing per mole of that reagent fed.

The selectivity of a reaction may be adversely affected by backmixing when sequential and simultaneous reactions occur as illustrated in the following paragraphs by some simple isothermal systems for which algebraic rate expressions are assumed to be available. It is especially important to approximate plug flow in these cases, with careful control of residence times.

Kinetics of simultaneous reactions. Given the necessary rate equations, outputs of simultaneous reactions are usually not difficult to calculate. For example, assume the rates of the two sulfonation reactions for naphthalene discussed in the preceding paragraphs to be of the form

$$\frac{dn_{\text{alpha}}}{V d\theta} = k_\alpha c^a_{\text{acid}} c^b_{\text{naphthalene}}$$

$$\frac{dn_{\text{beta}}}{V d\theta} = k_\beta c^{a'}_{\text{acid}} c^{b'}_{\text{naphthalene}}$$

When a and a' and also b and b' are equal, then by combining and integrating

$$\frac{n_{\text{alpha}}}{n_{\text{beta}}} = \frac{k_\alpha}{k_\beta}$$

In this case, the ratio of two isomers remains constant as reaction proceeds. When the exponents are not equal, then the isomer ratio is a function of composition.

Kinetics of sequential reactions. As with simultaneous reactions, many different situations involving sequential reactions are encountered and have been extensively discussed in the literature (84). Perhaps the simplest case is the isothermal conversion in batch reactors of A to B and of B to C by two first order homogeneous gas phase or liquid phase reactions:

$$A \xrightarrow{k_1} B \tag{6-18}$$

$$B \xrightarrow{k_2} C \tag{6-19}$$

If A, B, and C also represent the mole fractions of these materials at any time θ, and if the initial system contains only inert substances plus A at a starting concentration of A_0, then

$$A_0 = A + B + C \tag{6-20}$$

The loss of A in time $d\theta$ is

$$dA = -k_1 A d\theta \tag{6-21}$$

which integrates directly to

$$\frac{A}{A_0} = e^{-k_1\theta} \tag{6-22}$$

The net formation of B is the result of input from reaction (6-18) and output from reaction (6-19), or

$$dB = k_1 A d\theta - k_2 B d\theta \tag{6-23}$$

Combining with Eq. (6-22) and rearranging:

$$\frac{dB}{d\theta} + k_2 B = A_0 k_1 e^{-k_1\theta} \tag{6-24}$$

This is a linear equation of the first degree in B, hence its integrating factor is $e^{k_2\theta}$, yielding

$$Be^{k_2\theta} = \int_0^\theta A_0 k_1 e^{-k_1\theta} e^{k_2\theta}\, d\theta + \text{constant} \tag{6-25}$$

Completing the integration and evaluating the constant term by recognizing that B is zero when θ is zero:

$$\frac{B}{A_0} = \frac{k_1}{k_1 - k_2}(e^{-k_2\theta} - e^{-k_1\theta}) \tag{6-26}$$

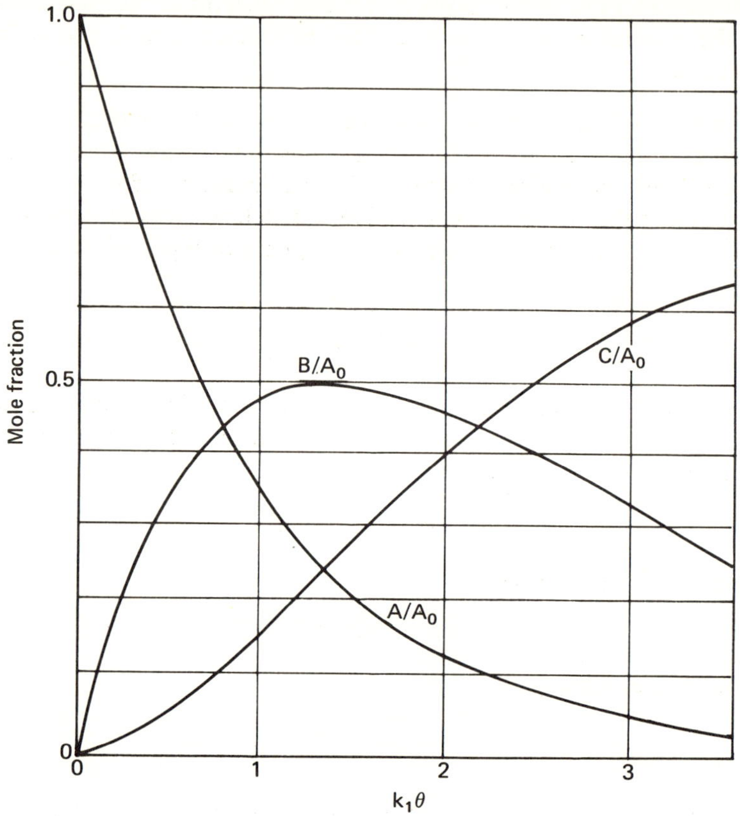

Fig. 6-10. The effect of time on reactant concentration in slug flow reactors for the first-order reactions (6-18) and (6-19). For the case plotted, k_1/k_2 is chosen as 2.0, with values calculated by Eq. (6-22) and (6-26).

A graphical solution of Eqs. (6-22), (6-26), and (6-20) is presented in Fig. 6-10 for the typical case in which k_1/k_2 is 2.0. Inspection shows that B goes through a maximum, as is characteristic of all intermediate products formed in sequential reactions regardless of the relative magnitude of k_1 and k_2. For the case shown in Fig. 6-10, the maximum attainable mole fraction of B is 0.5, equivalent to 50% conversion of A to B.

The maximum value of B, and the time θ at which it is attained, change with the ratio k_1/k_2. Thus, B is a maximum when dB is zero, hence, from Eq. (6-23):

$$\frac{B}{A} = \frac{k_1}{k_2} \tag{6-27}$$

Dividing Eq. (6-26) by Eq. (6-22), and substituting Eq. (6-27) one gets

$$k_1\theta_{max} = \frac{\ln(k_1/k_2)}{1-(k_2/k_1)} \tag{6-28}$$

Since from Eq. (6-28)

$$e^{k_1\theta_{max}} = \left(\frac{k_1}{k_2}\right)^{1/(1-k_2/k_1)}$$

then upon substituting into Eq. (6-26):

$$\frac{B_{max}}{A_0} = \left(\frac{k_1}{k_2}\right)^{k_2/(k_2-k_1)} \tag{6-29}$$

The variation of B_{max}/A_0 with k_1/k_2 is shown graphically as the upper line in Fig. 6-11. It is interesting that as k_1/k_2 becomes smaller, B_{max} decreases, since this means that the rate of the decomposition reaction (6-19) increases relative to the formation reaction (6-18). Inspection shows further that k_1/k_2 must be quite large in order for high yields of B to be attainable.

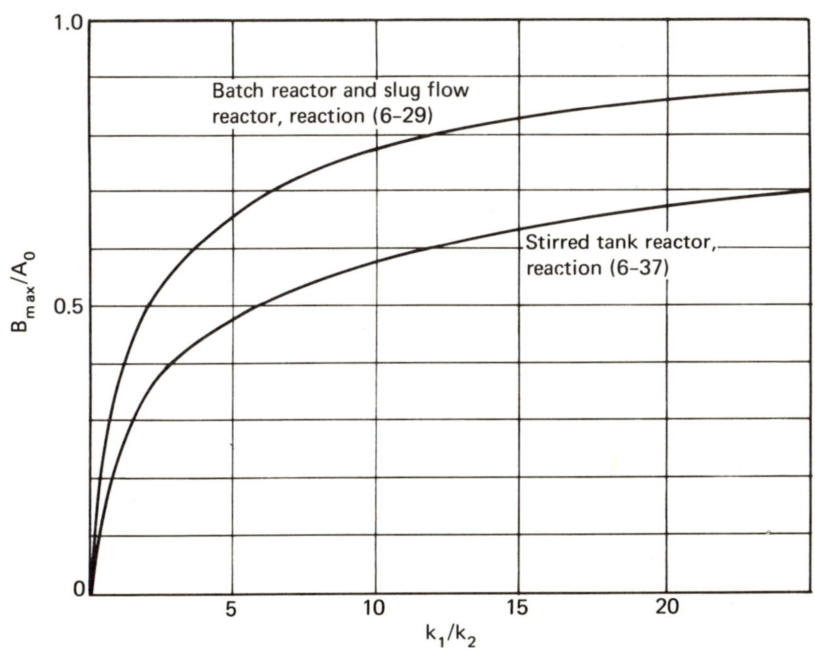

Fig. 6-11. The maximum yields of B obtainable by the first-order reactions (6-18) and (6-19), as a function of k_1/k_2.

Sequential reactions in flow systems. Obviously the foregoing equations, developed for isothermal batch systems, also apply without modification to steady-state flow systems, provided piston flow exists. In this case, each differential slug of material may be viewed as a batch system, and the residence time in the reactor equals the time required by the flowing liquid or gas to traverse the reactor. Provided reactions (6-18) and (6-19) occur, the foregoing equations apply directly to each slug.

When operating continuously and in a steady-state in a stirred reactor such as illustrated in Fig. 6-2, then the applicable equations differ somewhat from those just developed for plug flow. Assuming perfect mixing, then by using material balance on A and B around the vessel of volume V for the constant feed rate F one gets

$$FA_0 = FA + k_1VA \tag{6-30}$$

$$k_1AV = FB + k_2VB \tag{6-31}$$

Rearranging and combining:

$$\frac{A}{A_0} = \frac{F}{F + k_1V} \tag{6-32}$$

$$\frac{B}{A_0} = \frac{Fk_1V}{(F + k_2V)(F + k_1V)} \tag{6-33}$$

It is convenient to define the quantity V/F as τ, which obviously has the units of time. When the density of the entering and the leaving streams are equal, and also for gas systems where there is no change in the number of moles as is very nearly true with liquid systems, then τ equals the average residence time within the reactor. Introducing τ into the foregoing equations:

$$\frac{A}{A_0} = \frac{1}{1 + k_1\tau} \tag{6-34}$$

$$\frac{B}{A_0} = \frac{k_1\tau}{(1 + k_2\tau)(1 + k_1\tau)} \tag{6-35}$$

The variation of B/A with τ for a value of k_1/k_1 arbitrarily chosen as $\frac{1}{2}$ is shown in Fig. 6-12.

Inspection again shows the typical development of a maximum in the concentration of the intermediate B. The flow rate of feed, which will make possible this maximum production of B in any given apparatus of fixed volume, is determined by differentiating Eq. (6-33) with respect to F. The same result is attained by differentiating Eq. (6-35) with respect to τ:

$$\tau_{\max} = \frac{1}{(k_1k_2)^{1/2}} \tag{6-36}$$

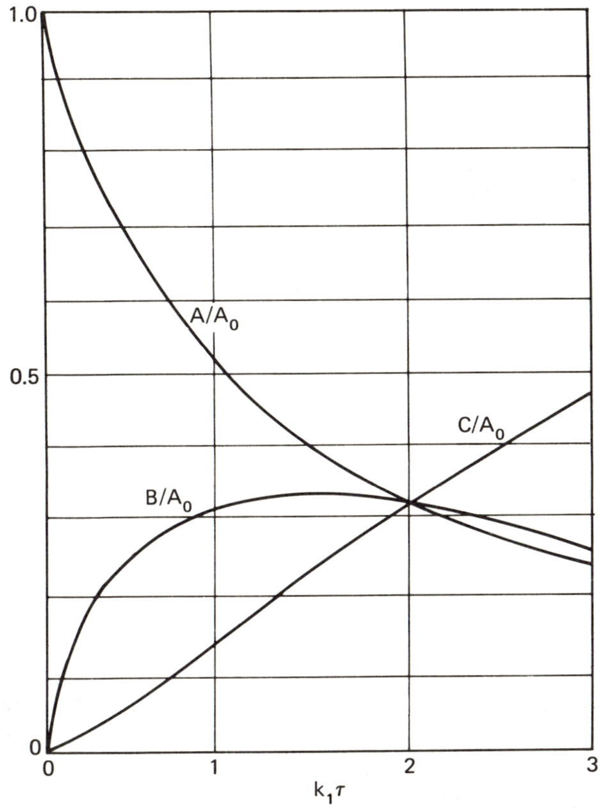

Fig. 6-12. Concentration vs. time in a stirred reactor operating in steady-state flow, for the first-order reactions (6-18) and (6-19). For the case plotted, k_1/k_2 equals 2 and A_0 is unity, calculated from Eq. (6-34) and (6-35).

By combining Eqs. (6-36) and (6-35):

$$\frac{B_{max}}{A_0} = \frac{\dfrac{k_1}{k_2}}{\left\{1 + \left(\dfrac{k_1}{k_2}\right)^{1/2}\right\}^2} \tag{6-37}$$

Values of B_{max} as calculated from the foregoing equation are presented as the lower curve in Fig. 6-11. Inspection shows that maximum conversion to B in this stirred reactor is always less than in the slug flow case. The disadvantageous effect of mixing upon selectivity is clearly shown in this figure.

Catalytic Reactions and Reactors 7

A material which when present in a reacting system causes the reaction rate to change without itself being consumed is called a catalyst. Catalysis is said to be homogeneous when the catalyst and the reaction partners all exist in the same phase, as in the liquid-phase esterification of ethanol with acetic acid, catalyzed by hydrogen ions. Given the relevant rate expressions, calculations of reactor performance in such systems are similar to the calculations for homogeneous reactions discussed in the preceding chapter. Catalysis is heterogeneous when the catalyst is present in a phase different from that carrying the reagents, as in the vanadium pentoxide catalyzed oxidation of gaseous SO_2 to SO_3, and the quantitative treatment of such systems is considered in this chapter.

Heterogeneous catalysis. Heterogeneous catalysis of a gas reaction by solids generally involves chemisorption of one or more of the reactants on the catalyst surface, followed by chemical reaction, and desorption of the products. A catalytically active solid surface is a complex system, on which most or all of the reaction occurs at certain locations called active centers. These active centers apparently differ from neighboring surface sites in the configuration of the local force fields, causing the reagents adsorbed here to become vulnerable to reaction. Active centers may involve crystal edges, active atoms projecting far above the surrounding surface, etc. Additives incorporated into the catalyst, called promoters, appear to increase the number of such active centers, and may also improve the system's stability.

A catalytic material is usually highly specific, in the sense that a good catalyst for one reaction is not necessarily suitable for another one. Thus, the platinum metal which catalyzes the oxidation of ammonia to nitric oxide cannot replace the clay catalysts used in hydrocarbon cracking, and vice versa. This is true even when both catalysts are active adsorbents for the reagents involved, since only if the adsorbed molecules become properly vulnerable to attack does a good catalyst result.

Up to very recently, formulation of catalysts has been more an art than a science, with only a few generalizations available: hydrogenations are often catalyzed by nickel and platinum, organic oxidations by silver, acetylene reactions by mercury salts, etc. Trial and error has been in large part resorted to in the discovery of a catalyst, especially for new reactions. Important advances and new applications of catalysis are expected to result from present intensive studies in the field.

Catalyst preparation. Materials used in heterogeneous catalysis are generally complex solids having high internal pore area, whose effectiveness is determined by the physical and chemical structure of their surfaces. Catalytic activity of a material is determined by the history of its preparation and, unless great care is taken to duplicate procedures, materials showing wide variation in performance are produced. Typical materials and methods of manufacture are as outlined:

1. Natural catalyst materials. Examples here are the clays used in the cracking of hydrocarbons. These clays are first pretreated with sulfuric acid, washed, dried, and pelletized before use.
2. Skeleton catalysts. These catalysts can be made by selectively leaching one or more components out of alloys, leaving the remaining metal as a skeleton of high surface area. Thus, Raney nickel catalyst used in hydrogenation is made by leaching aluminum-nickel alloys with sodium hydroxide. Again, the low temperature reduction of pellets of iron oxide with hydrogen yields iron sponge of very high pore area, effective in certain cracking reactions. The final product pellets have the same size as the pellets before processing.
3. Decomposition catalysts are made by decomposing granules of thermally unstable compounds, such as nitrates, formates, ammonium tungstates, molybdates, etc. The resultant oxides may be used directly or converted at low temperature to porous metal with a suitable reductant like hydrogen. Sometimes these salts are first impregnated into a porous carrier, after which decomposition and reduction are undertaken.
4. Precipitation catalysts are formed by precipitation of gels, followed by

formation of pellets of the desired size and thermal decomposition of the gel substance.

Catalysts are generally made either as fine powders or as pellets, all to controlled sizes. Final activation may require exposure to certain atmospheres under stipulated conditions of temperature, etc. Traces of poisons or promoters may be deliberately added to improve selectivity.

Mass transport. Materials transport is always involved in heterogeneous catalysis (132). Reagents must find their way from the bulk of the reactant stream to the vicinity of the catalyst surface, diffuse through the stagnant film around the particle, diffuse into the pores (if any), adsorb on the surface, and react. The same series of steps must be traversed in reverse by the products of the reaction. Any one of these steps may be rate limiting, if it is slow as compared to the others. The slowest step is generally either the chemical reaction itself or pore diffusion.

Catalyst pellets for fixed bed operations are usually spherical or cylindrical in the size range of $\frac{1}{16}$–$\frac{1}{2}$ in. The lower size limit is determined by the bed pressure drop, which, of course, increases at a fixed feed rate for a bed of given overall dimensions as pellet size decreases. The upper size limit is imposed by diffusional limitations within the pellet pores. That is, the reagents undergo reaction as they penetrate into the pores, and the concentration of these reagents diminishes accordingly. With the relatively long pores of larger pellets, reagent concentration may fall to zero before the center region of the pellet is reached, and so the catalyzed reaction is restricted to the exterior layers of the pellet. Since catalysts generally represent a considerable investment and also require reactor space, it is desirable to maintain as high a level of activity throughout a catalyst pellet as possible.

Effectiveness factor. A measure of the level of activity of a pelletized catalyst in which this problem of pore diffusion exists is the so-called "effectiveness factor," namely η. The effectiveness factor is defined as the ratio of the amount of reaction actually occurring per unit time in a pellet to that which would occur if the pellets were pulverized, exposing all of their internal area to the reagent concentration existing at the pellet surface. The effectiveness factor is easily formulated for the following simple case: assume an isothermal catalyst pellet shaped like a plate as shown in Fig. 7-1, in which all the pores are perpendicular to the face. These pores are to be cylindrical with radius R, extend entirely through the pellet, and so have a length equal to the pellet thickness. Assume the reaction occurring to be of the form $A = B$, and to have a very large equilibrium constant. Moreover, this reaction is of zero order, consequently the number of moles of A reacting per unit time per unit pore area is constant at α regardless of the concentration of A, until this reagent has completely disappeared. It follows that a section of

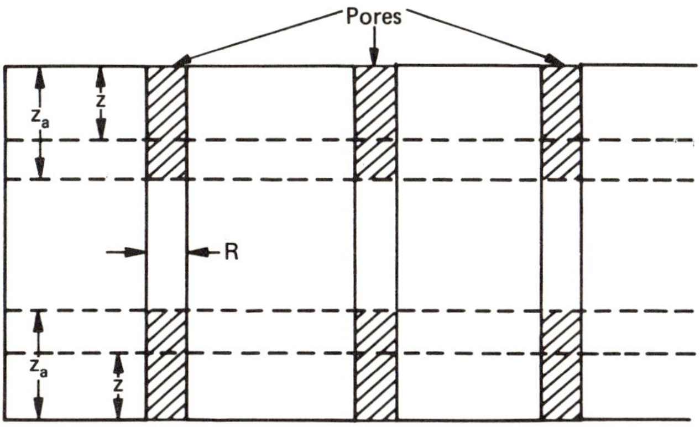

Fig. 7-1. An idealized plate-like catalyst pellet with cylindrical pores. The cross-hatched regions represent the active pore lengths.

pore wall at both ends of every pore must be actively supporting catalysis, each active section having the length Z_a. No reaction occurs in the pore between these two active sections because the reagent has disappeared from the gas phase before penetrating into this region.

Now let Z be the distance measured inward from either pellet surface. In isothermal operation, the rate of disappearance of A by reaction on the pore walls between Z_a and Z (see Fig. 7-1) must equal the amount entering this section of the pore by diffusion at Z. If the coefficient for diffusion is D, then*

$$-D\pi R^2\left(\frac{dc}{dz}\right) = 2\pi R(Z_a - Z)\alpha \tag{7-1}$$

If c_0 is the concentration of the reacting component at either entrance to the pore, where Z is of course zero, then by simplifying and rearranging:

$$-\int_0^{c_0} dc = \int_{Z_a}^{0} \frac{2\alpha}{DR}(Z_a - Z)\,dZ \tag{7-2}$$

Integrating one gets

$$c_0 = \frac{\alpha Z_a^2}{RD}$$

or

$$Z_a = \sqrt{\frac{c_0 DR}{\alpha}}$$

A moment's reflection shows that η, the effectiveness factor for this case,

*It is assumed in Eq. (7-1) that reagent concentration does not vary radially in a pore.

must equal the ratio of the active pore length to the half-length of the pore, namely Z_T, or

$$\eta = \frac{Z_a}{Z_T} \tag{7-4}$$

Substituting from above:

$$\eta = \frac{1}{Z_T}\sqrt{\frac{c_0 DR}{\alpha}}^{*} \tag{7-5}$$

Very similar expressions are obtained (148) for pellets in which first and second order reactions occur, for pellets having small pores radiating from the large pores, for spherical or cylindrical pellets, etc.

Inspection of the foregoing equation shows that η varies with the square root of reactant concentration around the particle, inversely with increasing pore length (pellet size), and inversely with the square root of α. It is further evident from Eq. (7-5) that η is influenced by temperature chiefly to the extent that the reaction rate term α itself is changed.

An obvious consequence of the foregoing is that for a given bed weight, the amount of reaction occurring (other things being equal) is independent of pellet size when the pellets are small, but diminishes with pellet size as the pellets are made larger.

Quantitative relations, fixed bed. The moles of product formed per unit time and per unit weight of a catalyst bed is determined by the area of the accessible surfaces in this bed. For a given catalyst of any stated pellet size, the accessible area is proportional to the bed weight. Since accessible area is not easily measured, the rate of a catalyzed reaction is conveniently expressed as moles of material produced (or consumed) per unit time and per unit weight of the particular catalyst used. For the familiar reaction

$$aA + bB = mM + nN \tag{7-6}$$

the reaction rate r is again expressed in terms of an arbitrarily selected key component when a catalyst is used:

$$r_M = \frac{dn_M}{W d\theta} \tag{7-7}$$

The term n_M represents the moles of M present, and W is the catalyst weight.

*The right side of Eq. (7-5) is the dimensionless Thiele modulus as defined by Eq. (3-5) of reference (148).

The rates of appearance and disappearance of the various reaction partners are again related:

$$-r_A = -\frac{a}{b}r_B = \frac{a}{m}r_M = \frac{a}{n}r_N \tag{7-8}$$

where a, b, m, and n are the stoichiometric coefficients of Eq. (7-6). The rate itself is usually some function of reagent concentrations:

$$r_M = kc_A^{\alpha}c_B^{\beta} \tag{7-9}$$

Now consider a catalyst bed in a reactor of cross section A, volume V, and packed density ρ. The reacting mass moved through this reactor in piston flow, and operations are in a steady-state. The feed to the bed is constant at F moles per hr. If there are Y moles of the key component M present per mole of feed at any bed cross section located a distance L from the entrance, then by using material balance on component M over a differential length dL of this catalyst bed,

$$FdY = r_M \rho A dL \tag{7-10}$$

where the term on the right equals rdW. If Y_0 is the moles of key component per mole of feed, which enters the reactor when L is zero, then

$$L = \frac{F}{A\rho}\int_{Y_0}^{Y}\frac{dY}{r_M} \tag{7-11}$$

Note the similarity of the foregoing relation to Eq. (6-13).

EXAMPLE 7-1

The kinetics of the catalyzed oxidation of SO_2 to SO_3 have been extensively discussed in the literature (35a). One of the better-known rate equations is that of Eklund (29), applying to a bed of $\frac{1}{4}$ in. commercial vanadium catalyst pellets weighing 30 lb/cu ft:

$$r = \sqrt{\frac{p_{SO_2}}{p_{SO_3}}}\left\{p_{O_2} - \left(\frac{p_{SO_3}}{Kp_{SO_2}}\right)^2\right\}k \tag{7-12}$$

where K is the equilibrium constant for the reaction

$$SO_2 + \tfrac{1}{2}O_2 = SO_3 \qquad \Delta H = -23{,}000 \text{ cal/g mole}$$

with all pressures in atm, while r is in (g moles SO_3)/(sec) (g catalyst). The rate constant k has the units (g moles of SO_3)/(g catalyst)(sec)(atm). Numerical values are:

T, °C	437	455	476	525	554	600	660
$k \times 10^6$	10.2	41.8	84.7	183	224	(307)*	(415)*
K, atm$^{-1/2}$ (29)	195	131	85	33.4	20.4	10.0	4.1

*Extrapolated.

(a) A feed gas is to contain 10% SO_2, 11% O_2, and 79% N_2. On a graph of r versus T, locate lines of constant per cent conversion of SO_2 to SO_3 for an atmospheric pressure operation. Heat capacities for these gases are as in example 5-4.

(b) This feed is to enter an adiabatic fixed bed converter at 437°C and a superficial approach velocity of 5 ft/sec. Operations are in a steady-state at atmospheric pressure, with piston flow of the gas. How deep must the bed be to attain 95% of the equilibrium conversion?

Answer: (a) As in example 5-4, let f be the mole ratio $SO_3/(SO_2 + SO_3)$, while m is moles $(SO_2 + SO_3)$ per 79 moles N_2 in the gas. For this case, m obviously equals 10 moles. The following mole table applies per mole of feed gas:

	Moles Present			
	Feed	Product	at 50% Conversion	nMC_p
SO_2	0.10	$0.1(1-f)$	0.05	$12(0.1)(0.1-f)$
SO_3	0	$0.1f$	0.05	$(15.5)(0.1f)$
O_2	0.11	$0.11 - 0.1f/2$	0.085	$(7)(0.11 - 01f/2)$
N_2	0.79	0.79	0.79	$(7)(0.79)$
Total	1.00	$1.00 - 0.1f/2$	0.975	7.5

(a) The partial pressure of SO_2 at 50% conversion is 0.05/0.975 atm, of O_2 is 0.085/0.975, etc. Introducing these partial pressures into Eq. (7-12), then, at 50% conversion,

$$r = k\sqrt{1}\left\{\frac{0.085}{0.975} - \left(\frac{1}{K}\right)^2\right\} \tag{7-13}$$

Substituting values of k and K at various temperatures, the 50% line in Fig. 7-2 can be drawn. The other lines on this chart are prepared similarly. Note that for each constant per cent conversion, r rises from a low value to a maximum one as temperature increases, but then falls to zero again as equilibrium is approached. The choice of semilog coordinates for Fig. 7-2 was arbitrarily made to keep the chart to manageable size. (b) With this gas entering at 437°C, calculation as in example 4-5 shows that the off-gases from an adiabatic converter would be at equilibrium at 635°C, with a 63% conversion of the initial SO_2 to SO_3. At 95% of equilibrium, conversion is 0.95×63 or 60%.

The heat capacity per mole of feed gas at any value of f is presented in the last column of the foregoing table, and is shown to be 7.5 cal/(g mole)(°C),

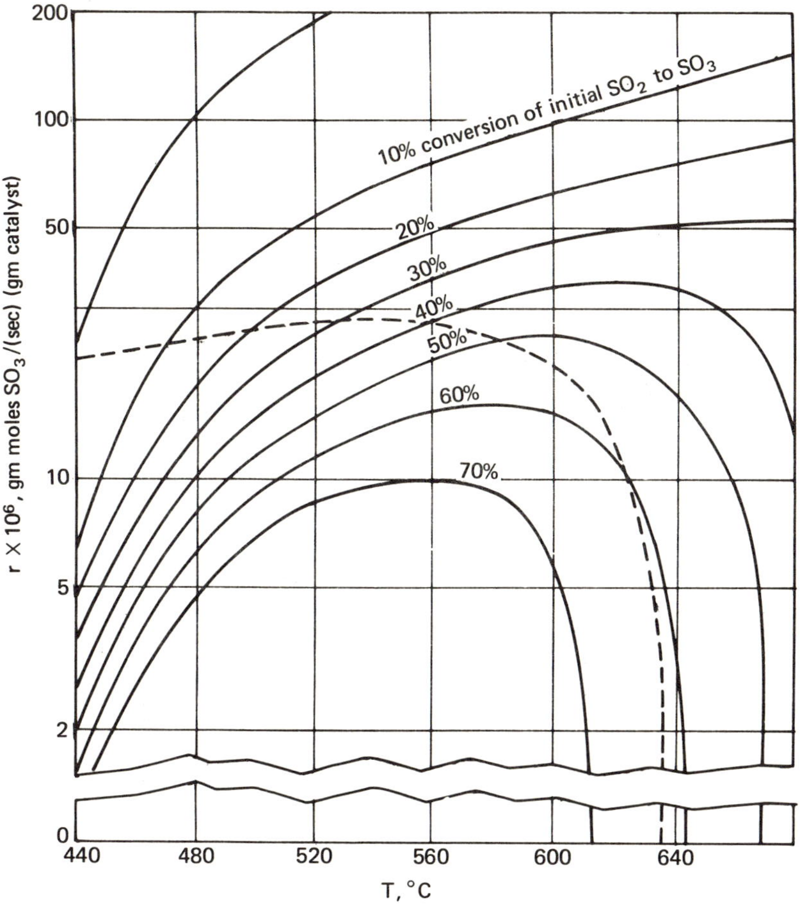

Fig. 7-2. Rate vs. T by the Eklund equation (example 7-1) for various conversions of SO_2 to SO_3. Initial gas composition is 10% SO_2, 11% O_2, and 79% N_2. The dotted line represents the temperature-composition history through the adiabatic converter under consideration.

independent of the conversion attained. For a mole of feed, the energy balance for this reactor becomes

$$7.5(T - 437) - \frac{23000mf}{100} = 0$$

Remembering that m equals 10 for this case, temperatures for the various arbitrarily selected values of f tabulated below in column (1), are now calculated from this equation and listed in column (3):

(1) f	(2) Conversion, %	(3) T°, C	(4) $r \times 10^6$	(5) $(1/r) \times 10^{-6}$
0.01	1	440	22	0.0455
0.1	10	468	23	0.0435
0.2	20	499	26	0.0385
0.3	30	529	28	0.036
0.4	40	559	26	0.0385
0.5	50	590	23	0.0435
0.6	60	621	11	0.091
0.63	63	635	0	(equilibrium)

The per cent conversions in column (2) above, of course, equal $100f$, and are plotted versus temperatures in Fig. 7-2 as the dotted line. The instantaneous values of the rate r corresponding to the various temperatures and compositions through the converter are read from this chart and tabulated in column (4). By definition, the moles of SO_3 formed at any point per mole of feed gas,

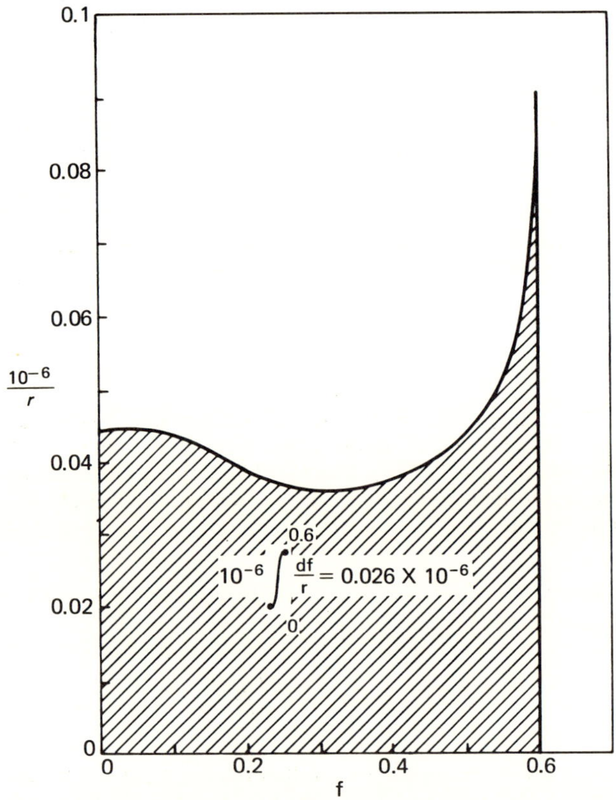

Fig. 7-3. Graphical interpretation for example 7-1.

namely Y, equals mf. The length of this adiabatic converter is then as represented by Eq. (7-11), in which the integral $\int dY/r$ can be replaced by

$$\frac{m}{100}\int\frac{df}{r}$$

The latter term is evaluated in Fig. 7-3 by plotting $1/r$ [column (5) of the preceding table] against f, and determining the area under the curve between values for f of 0.0 and 0.6. This integral is found here to equal 0.026×10^6, hence $\int dY/r$ equals 2,600. It should be noted in passing that r is numerically the same when expressed as (lb mole)/(sec)(lb catalyst), as when expressed as (g moles)/(sec)(g catalyst).

A is unity per sq ft of cross section and F, the reactor feed, at a 5 ft/sec superficial entrance velocity becomes

$$F = \frac{5}{359} \times \frac{273}{(273 + 437)} = 0.0054 \text{ lb moles/(sec)(sq ft)}$$

Since the catalyst density ρ is 30 lb/cu ft, then from Eq. (7-11), L becomes

$$L = \frac{0.0054}{30} \times 2600 = 0.47 \text{ ft}$$

Inspection of Fig. 7-3 shows that increasing the desired per cent conversion from 95–97% of equilibrium would require (a) a much longer bed, and (b) accurate rate data near equilibrium for the calculation of the needed depth of bed.

Reactor stability. Regardless of whether or not catalysts are used, adiabatic reactors are usually simpler to design and cheaper to construct than reactors which contain heat transfer surfaces. The design procedure for piston flow, as just illustrated in example 7-1, is obviously applicable either to exothermal or to endothermal reactions. When endothermal, the feed stream is suitably preheated before entering the reactor, in which temperature then diminishes with bed depth as reaction proceeds. Given sufficiently deep beds, temperature continues to fall until rates become negligible or equilibria become unfavorable. In exothermal reactions, the reactants are preheated to the kindling temperature, and then increase in temperature in passing through the catalyst bed as reaction proceeds. In a bed of indefinite length, temperature increases until reaction ceases because of attainment of equilibrium or the exhaustion of reactants. In actual practice, beds must often be kept relatively shallow to limit the extent of reaction, thus avoiding those high temperatures at which reagent decomposition or catalyst damage occurs.

In theory, then, any reaction can be carried out adiabatically, but in practice the allowable conversion per pass becomes smaller as the interval between the permissible lower and higher temperature extremes in the reactor diminishes. Of course, conversion between these temperature limits can be raised by increasing the heat capacity of the reagent stream, by adding an inert substance. There are obviously limits to such additions, however, due to such factors as increased costs of product recovery at greater dilution.

The advantages of heat transfer surfaces within a reactor are clear. By maintaining an optimum temperature through proper cooling or heating, product output rate from a given reactor is maximized. Design must be such, however, that reactor operations are stable, which means that after any brief accidental departure from a given steady-state condition of operation, the system will return spontaneously to the original conditions. For example, a small accidental increase in reactor operating temperature may result from any one of a number of causes, such as a fluctuation in the feed temperature. If the reactor returns spontaneously to its original temperature when the disturbance ceases, then operating conditions are stable. On the other hand, if such a small disturbance triggers a large and uncontrollable change in the reactor temperature, then these operating conditions are unstable. Unstable operations are obviously to be avoided, especially with those exothermal reactions for which it is vital to operate below a given temperature level in the interests of safety. In reading the following discussion, it should be noted that stability problems are encountered both in catalyzed and uncatalyzed situations.

Flow reaction. When a reaction occurs in a stirred reactor using heat transfer surfaces for heating or cooling, then for a given feed condition, there are usually two or more possible operating temperatures, with one of these being unstable. For simplicity, this will now be demonstrated by use of heat balance on an exothermal homogeneous reaction occurring in a stirred reactor of volume V operating continuously and in a steady-state at a constant temperature t. Three separate heat rates can be identified:

1. Chemical heat rate, q_{chem}, is the rate at which heat is generated by the chemical reaction in the kettle. For reaction (1-1), the standard enthalpy change per mole of the key component M is $\Delta H°/m$. The kettle contents are of uniform composition because of excellent mixing, hence,

$$q_{\text{chem}} = -rV\frac{\Delta H°}{m} \tag{7-14}$$

   Note that q_{chem} is positive for exothermal operations and r is in moles/(volume) (time).
2. Sensible heat rate, q_{sens}, is the sensible heat of the feed. When the feed is at a temperature T_F and enters the kettle at F lb mole per unit time; then q_{sens} becomes

$$q_{\text{sens}} = MC_p \cdot F(t - T_F) \tag{7-15}$$

where MC_p is the molal heat capacity of the feed and t is steady-state kettle temperature.

3. Cooling rate, q_{cool}, is the heat transferred to the cooling fluid. Because of excellent stirring, the kettle contents are at the uniform temperature t. For this reactor, the cooling area is A, the coolant temperature is assumed constant at T_w, and the overall heat transfer coefficient is U. The cooling rate is

$$q_{\text{cool}} = UA(t - T_w) \tag{7-16}$$

It is obvious that the three foregoing heat rates are related as follows in a steady-state operation:

$$q_{\text{chem}} = q_{\text{sens}} + q_{\text{cool}} \tag{7-17}$$

Substituting Eqs. (7-14), (7-15), and (7-16) into Eq. (7-17):

$$-\frac{\Delta H^\circ}{m} \cdot V \cdot r_M = UA(t - T_w) + MC_pF(t - T_F) \tag{7-18}$$

Since r_M is an exponential function of the kettle operating temperature, it is not surprising to find that, other variables being held constant, Eq. (7-18) is satisfied by two different values for t. This is illustrated for a simple zero order reaction in example 7-2.

EXAMPLE 7-2

The following liquid phase reaction is carried out continuously in the cooled, well stirred kettle of Fig. 6-2:

$$A + B = 2M \qquad \Delta H^\circ = -80{,}000 \text{ Btu/lb mole}$$

The rate of this reaction, assumed to be of zero order and to double every 20°F, is as follows:

$$r_M = 0.05(2)^{t-80/20} \text{ lb moles } M/(\text{cu ft})(\text{hr})$$

where t is in °F.

Reagents A and B, each with a heat capacity of 125 Btu/(lb mole)(°F), are each fed at 80°F to the kettle at a constant rate of 2 lb moles/hr. The specific volumes of A, B, and M are each 1 cu ft/lb mole. The liquid level is adjusted so that the kettle contains 4 cu ft of liquid at all times. The kettle is equipped with 5 sq ft of heat transfer area, the cooling water is at 80°F, and the overall coefficient of heat transfer is 100 Btu/(sq ft)(hr)(°F).

What are the possible values of t, the steady-state operating temperatures for this system? How many moles of M are produced per hour at each of these temperatures?

Answer: Substituting into Eq. (7-18):

$$(80{,}000/2)(4)(0.05)2^{(t-80)/2} = 100 \times 5(t - 80) + 2 \times 2 \times 125(t - 80)$$

where t is the steady-state kettle temperature. The left side of the foregoing equation is, of course, q_{chem}, while the right side is $(q_{\text{cool}} + q_{\text{sens}})$. Solving, two possible values for t are found, namely 92.5° and 136°F.

The variations of q_{chem} and also of $(q_{\text{cool}} + q_{\text{sens}})$ with kettle temperature t are shown graphically in Fig. 7-4. As is to be expected from the form of Eqs. (7-15) and (7-16), the quantity $(q_{\text{cool}} + q_{\text{sens}})$ plots as a straight line against t, intersecting the abscissa at 80°F. This straight line is in turn cut by the curve for q_{chem} at two points S and U, which represent, respectively, the possible steady-state operating points at 92.5° and 136°F.

The production of M in lb moles per hour obviously equals Vr_M. Substituting for r_M from Eq. (7-19) and remembering that V is 4 cu ft, then 0.31 moles of M is produced per hour at 92.5°F., with a mole fraction in the product stream of 0.31/4 or 0.078. At 136°F, production of M equals 1.4 moles/hr, with a mole fraction in the product stream of 1.4/4 or 0.35.

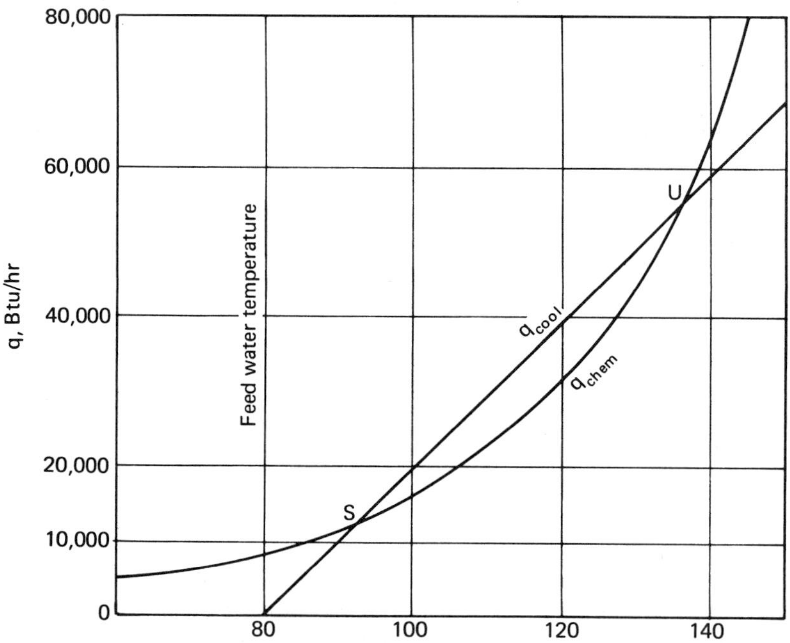

Fig. 7-4. Variation of q_{chem} and $(q_{\text{cool}} + q_{\text{sens}})$ with temperature for example 7-2. Points S and U represent possible operating conditions.

Unstable operation. Operation of the kettle of example 7-2 at 92.5°F, namely at point S, is shown to be stable by the following reasoning. Imagine the occurrence of a small disturbance, such as a momentary increase in the cooling water temperature or the feed temperature, causing the kettle temperature to rise a little above 92.5°C. As a result, by inspection of Fig. 7-4, both q_{chem} and $(q_{\text{cool}} + q_{\text{sens}})$ increase. Inspection also shows that $(q_{\text{cool}} + q_{\text{sens}})$ increases more than does q_{chem}, consequently, the kettle contents must cool again to 92.5°C when the disturbance is terminated and the original feed and cooling water temperatures are reestablished. Similarly, if the kettle temperature falls a trifle below 92.5°C, then after termination of this disturbance, this temperature is restored because the term $(q_{\text{sens}} + q_{\text{cool}})$ is now smaller than q_{chem}. Thus, point S is a stable operating point.

By similar reasoning, point U is unstable. If the temperature of the kettle contents accidentally and briefly rises above 136°F at point U, then this temperature continues to rise and at an accelerating rate, since q_{chem} is now always greater than $(q_{\text{cool}} + q_{\text{sens}})$, causing heat to accumulate in the system. Alternatively, if the temperature of the kettle contents falls briefly below 136°F, then it will continue to fall, since the term $(q_{\text{cool}} + q_{\text{sens}})$ is now greater than q_{chem}, until point S is reached. Thus, point U is unstable, since any slight variation in kettle temperature causes a permanent departure from this point. It is obvious that, if the temperature is displaced above point U, the results may not only be undesirable but even catastrophic, depending on the system's characteristics.

The foregoing can be generalized as follows: For a stable operation any increased amount of heat generated per unit time due to a momentarily increased reactor temperature must be easily removable without changing the temperature of the coolant or of the feed. Alternatively,

$$\frac{dq_{\text{chem}}}{dt} \leq \frac{d(q_{\text{cool}} + q_{\text{sens}})}{dt} \tag{7-19}$$

Note that this criterion is satisfied at point S, not at point U.

Criterion for stability. There are values for the coolant temperature T_w and the feed temperature T_F for which the stable operation of the kettle just considered is not possible. To simplify discussion, assume for the present that T_F and T_w are always equal, and have the common value T_{Fw}. For this situation, Eq. (7-18) becomes

$$-\frac{\Delta H^\circ}{m} Vr_M = (UA + MC_pF)(t - T_{Fw}) \tag{7-20}$$

It will be remembered that the left-hand side of this equation is q_{chem}, while the right hand side is the term $(q_{\text{cool}} + q_{\text{sens}})$. This term $(q_{\text{cool}} + q_{\text{sens}})$ formed a straight line when plotted in Fig. 7-4. Obviously, a change in T_{Fw} will move this line to the left or right in Fig. 7-4, without, however, changing its slope as long as U, A, MC_p, and F are constant. Lines of $(q_{\text{cool}} + q_{\text{sens}})$ versus t for example 7-4 are plotted in Fig. 7-5 for various values of T_{Fw}, line A being for 80°F, line B for 60°F, line C for 89°F, and line D for 100°F. Also presented in Fig. 7-5 is the line q_{chem} from Fig. 7-4. Obviously, line D represents an unstable operation, since it does not intersect q_{chem} at any point. Line C, deliberately drawn to be tangent to curve q_{chem}, defines the greatest value for T_{Fw} for which stable operations are possible.

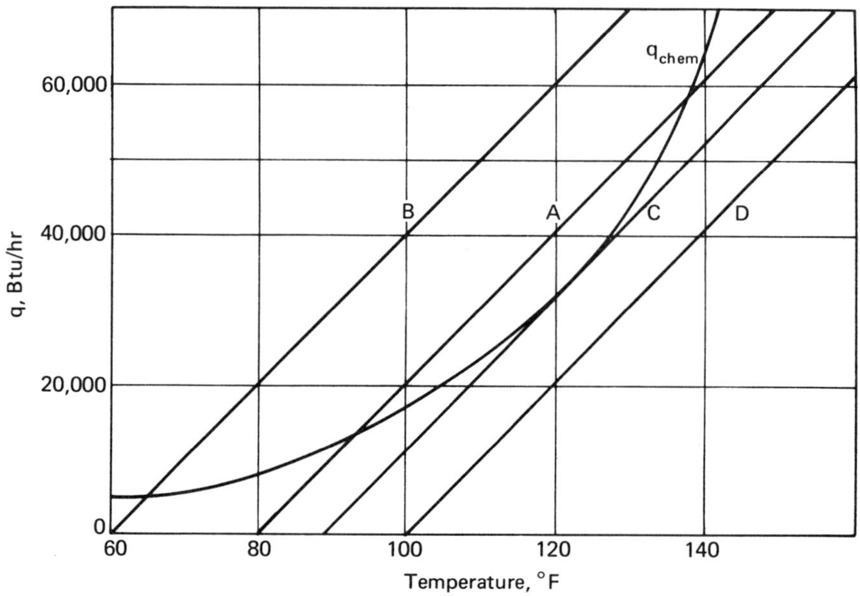

Fig. 7-5. The effect of changed cooling water temperature on kettle performance.

It is now possible to calculate the generalized temperature differences which characterize tangent line C in Fig. 7-5 for zero order reactions (12). Thus, reaction rate r is of the familiar approximate form (see example 6-1)

$$\frac{r}{r_0} = 2^{(t-T_0)/\psi} \tag{7-21}$$

where ψ is the temperature interval over which the rate of this zero order reaction doubles, while r_0 is the rate at T_0. Introducing Eq. (7-21) into Eq. (7-18),

$$-r_0\frac{V\Delta H^\circ}{m}\cdot 2^{(t-T_0)/\psi} = UA(t - T_w) + MC_pF(t - T_F) \tag{7-22}$$

At the point of tangency of line C with q_{reaction},

$$\frac{d(q_{\text{cool}} + q_{\text{sens}})}{dt} = \frac{dq_{\text{chem}}}{dt} \tag{7-23}$$

Consequently, differentiating Eq. (7-22),

$$-r_0 \frac{V\Delta H^\circ}{m} \cdot 2^{(t-T_0)/\psi} \cdot \frac{\ln 2}{\psi} = UA + MC_pF \tag{7-24}$$

Dividing Eq. (7-22) by Eq. (7-24),

$$\frac{\psi}{\ln 2} = \frac{UA(t - T_w) + MC_pF(t - T_F)}{UA + MC_pF} \tag{7-25}$$

when T_w equals T_F, and again designating this temperature as T_{Fw},

$$t - T_{Fw} = \frac{\psi}{\ln 2}$$

or

$$t - T_{Fw} = 1.45\psi \tag{7-26}$$

This relation for stirred batch reactors is independent of magnitudes of V, A, ΔH, T_0, ψ, U, T_{Fw}, etc., and so in this sense is perfectly general. That is, as is evident from inspection of Fig. 7-5, the largest permissible temperature difference between the reacting mass temperature and the coolant temperature for stable operation is 1.45 times the temperature interval needed to double the reaction rate. A greater temperature difference than this will obviously result in a runaway reaction.

Modification of this criteria for cases in which T_F and T_w are not equal, or in which only one of these temperatures is changed follows directly from Eq. (7-25).

EXAMPLE 7-3

Reaction time is considered excessive when operations of example 7-2 are carried out at point S of Fig. 7-4, namely, at 92.5°F. It is now desired to operate safely at the temperature of 136°F, equivalent to point U of Fig. 7-4, which, however, was shown in example 7-2 to be unstable. How could stable operation at 136°F be attained?

Answer: If T_w and T_F are to be kept equal, then their lowest allowable temperature for stable kettle operation at 136°F would involve satisfying Eq. (7-19), that is, operations would require that $(q_{\text{cool}} + q_{\text{sens}})$ be tangent to q_{chem} at the kettle temperature. Under these conditions, by Eq. (7-26), (t

$- T_{Fw}$) is again 29°F when ψ is 20°F. Thus, a line for ($q_{\text{cool}} + q_{\text{sens}}$) tangent to q_{chem} at 136°F in Fig. 7-4 and replotted in Fig. 7-6, will intersect the x-axis at (136 − 29), or 107°F. To obtain the proper slope for the line ($q_{\text{cool}} + q_{\text{sens}}$) in Fig. 7-6, the proper amount of heat transfer area must be installed. Substituting into Eq. (7-19), A is found to be 14.3 sq ft. It is worth noting in passing that it would be difficult to install this much cooling area in a kettle of 4 cu ft volume.

The tangent point just calculated is not a good choice for stable operations, since it allows no margin of safety. A somewhat steeper slope for ($q_{\text{cool}} + q_{\text{sens}}$) as shown by the dotted line, requiring a larger cooling surface and a higher cooling water temperature, would be preferable for kettle operations at 136°F.

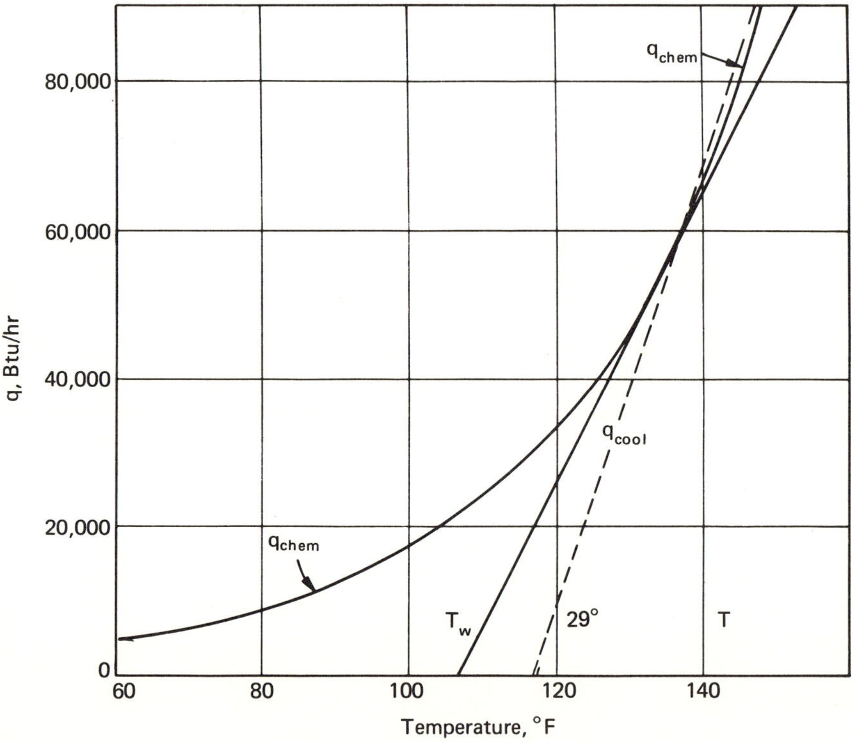

Fig. 7-6. Operations at 139.2°F, as in example 7-3.

Reactor control. Stable operating conditions in cooled stirred reactors are often found to be uneconomical in that large costly heat transfer areas are required because of the small temperature differences between kettle contents and the cooling water. Operations at an unstable point such as point U in Fig. 7-4 may then be undertaken, with controls installed to correct for deviations from this condition. Such controls could involve continuously changing the cooling water temperature, reducing the temperature at point U as T_{Fw} rises, and increasing this temperature as T_F falls. Only when the mass of the system is large relative to the rate of heat generation are temperature changes slow enough so that such control becomes possible. In addition, for safety, automatic quenching of some sort must be provided when reactor temperature climbs to above the predetermined safe maximum level. Control can also be attained by means other than external cooling. Thus, in a kettle operating either batchwise or in continuous flow, evaporation of a volatile reaction partner or diluent from the reacting mass itself and its return as condensate will usually carry off any quantity of heat generated and so avoids the instability problem entirely. Again, use of a nonvolatile diluent which increases the sensible heat of the feed diminishes the maximum temperature rise, but also diminishes equipment capacity.

Stability in stirred reactors, general. Both reaction rate and q_{chem} are functions of concentration in other than zero-order reactions. Since q_{chem} diminishes as the reaction proceeds, a higher order reaction is somewhat less prone to a runaway performance than a zero-order reaction. Nevertheless, a conservative approach to design would consider the higher order reaction equivalent to a zero-order reaction whose rate equals the initial rate of the higher order reaction.

The criterion for stability in an isothermal kettle reactor operating batchwise rather than on a flow basis can be developed in the same way as just illustrated. For isothermal operation, the batch kettle must be initially charged with reagents at the steady-state operating temperature, making q_{sens} zero for this case. Since q_{chem} now equals q_{cool}, Eq. (7-18) must apply, except that the second term on the right-hand side of this equation is zero. The maximum value for $(t - T_w)$ is again given by Eq. (7-26), with T_{Fw} equalling T_w.

The criterion for stability in isothermal operation of a piston flow system, such as a pipe reactor, is the same as for the batch system just discussed. Radial uniformity of temperature and composition through the piston flow system is of course assumed. Piston flow is then equivalent to the movement of a succession of small slugs through the system, each slug being a batch system at constant temperature.

Stability in catalyst-packed tubes. The discussion of reactor instability just presented has been limited to vessels in which the temperature of the reacting mass is uniform throughout because of vigorous mixing. Instability can also exist in systems which are not isothermal, as when an exothermal reaction occurs in a solid in which a temperature gradient develops as heat flows outward by conduction to the cooled surfaces of the solid. Similarly, a radial temperature gradient develops within cooled tubes packed with catalyst granules on which an exothermal reaction occurs as reagents flow through the tube. That is, the heat generated within the mass by reaction must flow outward to the cooling water under the action of a temperature gradient.

The criterion for stability of a system of this type is as before: a system is unstable, if a small temporary temperature increase anywhere within the mass results in such a large increase in internal heat generation per unit time that this heat cannot be conducted out radially unless the cooling water temperature is lowered. On the other hand, no runaway situation develops if the increased heat generated per unit time as a result of a temporary increase in catalyst temperature can be easily conducted away without requiring a decrease in the cooling water temperature (53).

EXAMPLE 7-4

A 1-in. I.D. tube is packed with $\frac{1}{8}$-in. of catalyst spheres, on which an exothermal reaction occurs, generating 4000 Btu/cu ft of bed at 500°F. Reaction rate and therefore heat generation rate doubles every 30°F. The catalyst bed has an equivalent conductivity of 0.3 Btu/(hr)(sq ft)(°F)/(ft). The overall coefficient of heat transfer through the wall to the coolant, namely h, is 16 Btu/(hr)(sq ft)(°F). (a) What is the temperature of the catalyst at the tube wall if the tube's center temperature is 500°F? Repeat for a center temperature of 505°F. (b) With a center temperature of 500°F, what is the cooling water temperature? Is this operation stable? (c) If the wall heat transfer coefficient had been 1.0 Btu/(hr)(sq ft)(°F) instead of 16, what would the cooling water temperature have been for a center temperature of 500°C? Would this operation have been stable? (d) How could the accuracy of these calculations be improved?

Answer: (a) While the differential equation for radial heat flow is easily formulated for this case, analytical solution is rather difficult. An approximate method of calculation is therefore presented here which, though less elegant, is mathematically simple and can be made as exact as desired. This analysis is based on a unit length of tube, for which L is 1-ft. Divide the tube into five concentric cylinders by drawing the concentric circles with radii r_1 of $\frac{1}{10}$ in., r_2 of $\frac{2}{10}$ in., r_3 of $\frac{3}{10}$ in., and r_4 of $\frac{4}{10}$ in. Assume the temperature of the center cylinder "a" of radius r_1 to be uniform at T_1, of the cylindrical shell "b" having inside and outside radii of r_1 and r_2 to be T_2, of shell "c" to be T_3, etc. The heat generated in the center cylinder per hour, which is at 500°F, is q_1, where

$$q_1 = (\pi r_1^2 L) \times 4000$$

Since L is unity and r_1 is $\frac{1}{10}$ in., then q_1 is 0.87 Btu/(hr)(ft of tube length). Similarly, heat generation in the cylindrical shells is

$$q_2 = \pi(r_2^2 - r_1^2)L \times 4000 \times 2^{(T_2-500)/30}$$
$$q_3 = \pi(r_3^2 - r_2^2)L \times 4000 \times 2^{(T_3-500)/30}$$

etc.

The temperature gradient between two adjacent cylinders will be taken as the temperature difference between these two cylinders divided by the difference of their radii. Thus the gradient between cylinders a and b is $(T_1 - T_2)/(r_2 - r_1)$, the one between cylinders 2 and 3 is $(T_2 - T_3)/(r_3 - r_2)$, and that between 3 and 4 is $(T_3 - T_4)/(r_4 - r_3)$, etc. Similarly, the area for heat flow between cylinders a and b is $2\,\pi r_1 L$, the one between b and c is $2\,\pi r_2 L$, etc., where L is unity. The heat conduction equation is now used to calculate T_2 by equating the heat generated in cylinder a, namely q_1, to the heat conducted away to cylinder b:

$$q_1 = 2k\pi r_1\left(\frac{T_1 - T_2}{r_2 - r_1}\right)$$

or

$$0.87 = 2\pi\cdot 0.3 \times \frac{0.1}{12}\left(\frac{500 - T_2}{0.2/12 - 0.1/12}\right)$$

hence $T_2 = 499.5°\text{F}$. Temperature T_3 is next calculated from a heat balance on the hollow cylinder b. Knowing T_2 to be 499.5°F, the heat generated within this cylinder can be calculated from the foregoing expression for q_2, and is found to equal 2.6 Btu/hr. This quantity of heat, plus q_1, must equal the heat conducted out through the outside face area of cylinder b, namely $2\pi r_2 L$. The expression to be solved for T_3 is then

$$0.87 + 2.60 = 0.3 \times 2\pi r_2 \times \frac{(T_2 - T_3)}{r_3 - r_2}$$

All terms are known in this expression except T_3, which upon solving is found to be 498.6°F. Similarly, the heat flowing into cylinder c is $(q_1 + q_2)$, that flowing into cylinder d is $(q_1 + q_2 + q_3)$, etc. The temperature of each hollow cylinder can therefore be calculated exactly as was done for cylinder b above, with the following results:

	Center T, 500°F		Center T, 505°F	
Cylinder, No.	T, °F	Heat Generated in each Shell, Btu/hr	T, °F	Heat Generated in each Shell, Btu/hr
1	500	0.87	505	0.98
2	499.53	2.6	504.5	2.9
3	498.6	4.2	503.4	4.7
4	497.3	5.75	501.9	6.4
5	495.5	7.25	499.9	7.9
Total		20.7		22.9

Thus the average temperature of the outer cylinder is 495.5°F for a 500°F center temperature, and 499.9°F for a 505°F center temperature. (b) For a

500°F center temperature, the heat to be conducted away as calculated in part (a) is 20.7 Btu/hr. The wall area for a 1-ft length of this 1-in. tube is $\pi/12$ sq ft. With an overall wall coefficient of 16, the temperature T_w for the cooling water is

$$20.7 = \frac{\pi}{12} \times 16(495.5 - T_w)$$

or T_w equals 490.5°F. For this operation to be stable, the heat generated when the center temperature is raised from 500 to 505°F should be successfully conducted away without requiring an increase T_w. As shown in the above table, the temperature at the wall is 499.9°F, consequently the heat conducted away becomes $\pi/12 \times 16(499.9 - 490.5)$ or 39.3 Btu/hr for a unit length of tube. This heat exceeds that generated, namely 22.9 Btu/hr, consequently the tube will cool down again, and the operation is stable. (c) For a 500°F center temperature, the cooling water temperature becomes

$$20.7 = 1.0 \times \frac{\pi}{12}(495.5 - T_w)$$

hence T_w equals 416.5°F. For this cooling water temperature and a 505°F center temperature, the heat conducted away is $3 \times \pi/12(499.9 - 416.5)$ or 21.8 Btu/hr. Since this value is smaller than the heat generated in this case, namely 22.9 Btu/hr, the system is now unstable. (d) These calculations can be made more accurate by choosing smaller intervals between r_1, r_2, and r_3.

Reactions Between Gases and Solids 8

The complex problems of heterogeneous reaction kinetics are explored here for three typical situations: (a) decomposition reactions wherein a solid is converted into a gas, or a solid and a gas, (b) gasification reactions in which a solid combines with a gas to form a gas or a solid plus a gas, and (c) deposition reactions in which a solid forms from gaseous reactants. Systems involving a single sphere of reacting solid are first considered, after which multiple particle systems are briefly discussed.

Decomposition: solid = gas + solid. The equilibrium behavior of monovariant decomposition reactions was illustrated in Chap. 2 by reaction (2-9) involving the conversion of calcium carbonate to calcium oxide and CO_2. The rates of such decomposition reactions always increase with temperature. Even in vacuum, these rates are usually undetectably slow below those temperatures at which equilibrium pressures are less than 10^{-4} atm but as already mentioned in Chap. 2, they increase as temperatures climb, and are usually fairly rapid when temperatures are such that decomposition pressures exceed 0.1 atm. This can be demonstrated by heating a lump of calcium carbonate in a closed, evacuated vessel having a volume 100-fold greater than that of the solid. At 590°C, attainment of the equilibrium dissociation pressure of 1 mm of mercury is found to require hours, while only a few minutes are required at 800°C, where the dissociation pressure has increased to 0.23 atm. When heating this same calcium carbonate at 800°C in the open air at atmospheric pressure, a relatively slow decomposition rate is again observed. This slow rate relative to vacuum reflects the fact that the CO_2 produced at 0.23 atm must diffuse away from the lump's external and internal pore surfaces through the surrounding air to make room for additional CO_2 to be formed. This diffusion is slow, consequently mass transfer controls. Finally, when heating at 910°C where the decomposition pressure is somewhat above 1 atm, decomposition proceeds very rapidly and at an equal rate in air or in a vacuum. No mass transfer limitations exist now since the CO_2 tends to be generated at above atmospheric pressure, and so sweeps away the surrounding air in mass flow.

Depending on circumstances, therefore, decomposition of limestone is typically controlled either by kinetics, diffusion, or heat transfer. The finding that kinetics control at low temperature but not at high is to be expected, since reaction rates usually increase with temperature far more rapidly than, for example, diffusion rates. In industry, of course, operations are preferable under conditions of high rate and small reactor space, such as are attainable at high temperature in limestone decomposition.

Mathematical formulation. The system next discussed involves a solid reacting to form a solid plus a gas, as in a thermal decomposition. The solid specimen under study is to be a single nonporous sphere of initial radius R_0. As reaction proceeds and gas escapes, a growing shell of porous solid product forms around a core of unused solid reactant of diminishing size, as shown in Fig. 8-1. When the specimen is in an isothermal environment so that heat is supplied uniformly from all sides, then the core remains spherical in shape concentric with the specimen. It is assumed that no overall shrinkage occurs, consequently the radius of the outside particle surface remains at R_0 throughout the reaction while the radius and surface area of the shrinking core are, respectively, R_c and A_c. Neglecting minor sensible heat requirements of the shell, all the heat flowing into the specimen is consumed by the decomposition reaction which occurs at the very thin interface A_c between the core and shell. The core is therefore isothermal at the tempera-

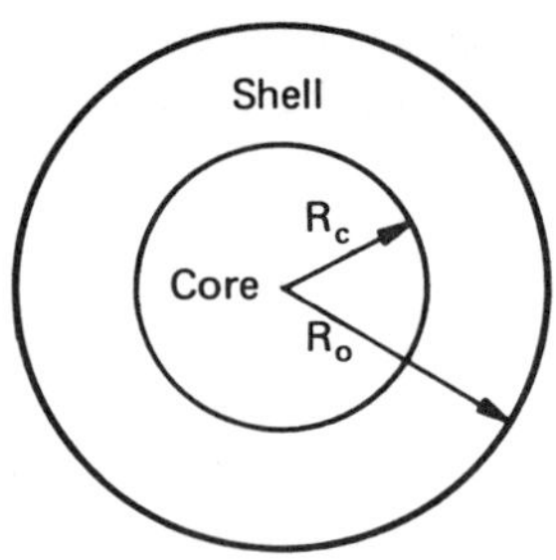

Fig. 8-1. Core and shell model of a reacting sphere. During the reaction, R_c remains unchanged.

ture of the decomposition reaction, and the temperature gradient within the specimen is confined to the porous shell. Reactions of this type in solids, in which a shell of product is separated from the unused core by a more or less sharply defined reaction surface, are called topochemical.

The amount of product formed at the core surface, namely r moles of shell solid per unit core area and per unit time, is by definition $dn/A_c d\theta$. It is obvious that A_c equals dV_c/dR_c, and since dn_c equals $\rho_c dV_c/M_c$, where ρ_c/M_c is moles of reactive material per unit core volume,

$$r = \frac{dn}{A_c d\theta} = -\frac{\rho_c dR_c}{M_c d\theta_c} \tag{8-1}$$

Three cases of interest are now discussed.

1. Kinetics controls. When reaction rate is slow relative to heat and mass transfer, then the sphere is isothermal. Chemical reaction rate now controls, with the core material reacting at the constant rate of α moles per unit core area and per unit time. Upon substitution of α for r in Eq. (8-1), integrating over the time during which the core shrinks from the initial radius R_0 to R_c yields the following:

$$R_0 - R_c = \frac{\alpha M_c \theta}{\rho_c} \tag{8-2}$$

Thus R_c diminishes linearly with time. The time required to decompose a particle completely, so that R_c becomes zero, is obviously proportional to the initial radius of the spherical specimen.

2. Diffusion out through shell controls. Assume that 1 mole of gaseous product forms per mole of solid reactant decomposed, and that the moles of gas forming per unit time equal $dn_G/d\theta$, or $\dot{n}_G$. The entire sphere again remains substantially isothermal at the core temperature T_c, for which the corresponding constant decomposition pressure of the gaseous product formed at the core-shell interface is p_c. Mass transfer through the film around the particle is assumed great as compared to pore diffusion rates, hence the product's partial pressure at the particle surface equals that in the atmosphere around the particle, namely p_∞.

The instantaneous flux of product gas outward through any spherical surface of radius R, located within the shell of this specimen, is

$$\dot{n}_G = -4\pi R^2 D' \frac{dp}{dR} \tag{8-3}$$

where D', the effective coefficient of diffusion per unit of total cross-sectional area, is taken here as constant, independent of time and radius. At any fixed instant of time, $\dot{n}_G$ is the same through all spherical surfaces of radius R, where R lies between R_0 and R_c, hence Eq. (8-3) can be integrated:

$$D'(p_c - p_\infty) = \frac{-\dot{n}_G}{4\pi} \int_{R_0}^{R_c} \frac{dR}{R^2} \tag{8-4}$$

or for this instant:

$$\dot{n}_G = \frac{4\pi R_c R_0 D'(p_c - p_\infty)}{(R_0 - R_c)} \tag{8-5}$$

Over differential time, a thin shell of core material of volume $4\pi R_c^2 dR_c$ decomposes, containing (ρ_c/M_c) moles of reactant per unit volume. Assuming 1 mole of gas generated per mole of solid decomposed, then moles of gas generated must be $\dot{n}_G d\theta$:

$$\dot{n}_G d\theta = -\frac{4\pi \rho_c R_c^2 dR_c}{M_c} \tag{8-6}$$

Combining Eqs. (8-5) and (8-6) and rearranging:

$$\frac{-R_c(R_0 - R_c)dR_c}{R_0} = \frac{M_c D'(p_c - p_\infty)d\theta}{\rho_c} \tag{8-7}$$

Integrating between the limits of R_0 and the core radius R_{cf} finally attained at θ time:

$$\theta = \left\{\frac{R_0^2}{6} - \frac{R_{cf}^2}{2} + \frac{R_{cf}^3}{3R_0}\right\} \frac{\rho_c}{(p_c - p_\infty)M_c D'} \tag{8-8}$$

Inspection of Eq. (8-8) shows that the time required for complete decomposition (when R_{cf} becomes zero) is proportional to the square of the initial radius of the specimen.

3. Heat transfer through shell controls. Consider a spherical specimen of material undergoing an endothermal decomposition into a solid plus a gas. The temperature of the environment and of the specimen's external surface is T_0, while the temperature of the reaction surface between core and shell is T_c. Assume that reaction rates and mass transfer processes, as well as heat transfer to the sphere surface from the surroundings to be relatively rapid, so that heat transfer through the shell is rate-limiting. At any given instant, assuming heat flowing inward through any spherical surface within the shell to be negative,

$$q = -4\pi R^2 k \frac{dT}{dR} \tag{8-9}$$

where q is the total heat flow per unit time, and k represents thermal conductivity. Since accumulation of heat within the shell is negligible, then at any fixed instant of time, q is the same through all spherical surfaces within the shell, and equals the heat flow through the external surface of the pellet. At any fixed instant of time, therefore, with constant q, the foregoing equation becomes

$$\frac{q}{4\pi}\int_{R_0}^{R_c}\frac{dR}{R^2} = -k(T_c - T_0) \tag{8-10}$$

or

$$q = \frac{-4\pi k R_0 R_c (T_0 - T_c)}{(R_0 - R_c)} \tag{8-11}$$

For inward flow of heat, T_0 is greater than T_c, and so q is negative, while q is positive for outward flow of heat occurring when T_0 is smaller than T_c.

In the time interval $d\theta$, the heat accepted by a differential layer of core at the spherical shell-core interface is $qd\theta$, and so

$$qd\theta = 4\pi R_c^2 \Delta H \frac{\rho_c}{M_c} dR_c \tag{8-12}$$

where ρ_c/M_c is again moles of reactant per unit volume of specimen, and ΔH is the enthalpy of the reaction, in heat units per mole. Combining the two foregoing equations and rearranging:

$$\frac{-R_c(R_0 - R_c)dR_c}{R_0} = \frac{k(T_0 - T_c)M_c d\theta}{\rho_c \Delta H^\circ} \tag{8-13}$$

Integrating between R_0 and the final core radius R_{cf}, and remembering that $(T_0 - T_c)$ is now constant,

$$\left\{\frac{R_0^2}{6} - \frac{R_{cf}^2}{2} + \frac{R_{cf}^3}{3R_0}\right\}\frac{\rho_c \cdot \Delta H^\circ}{(T_0 - T_c)M_c k} = \theta \tag{8-14}$$

Inspection shows that the time required for complete decomposition, when R_{cf} is zero, is again proportional to the square of the initial radius. Note the similarity between Eqs. (8-14) and (8-8).

EXAMPLE 8-1

By inspection at various stages during calcination, it is found that limestone decomposition to lime follows the core-and-shell model shown for a spherical lump in Fig. 8-1. Since the molecular volume of $CaCO_3$ at 25°C is 27.1 cc/g mole as compared with 16.6 cc/g mole for CaO at the same temperature, it is not surprising to find that the shell formed during decomposition is porous. That is, calcining causes even large $CaCO_3$ crystals to break up into a multiplicity of tiny CaO crystals with relatively large spaces between neighbors. Little overall lump shrinkage occurs during calcining when operating at up to 1650°F, at which temperature the equilibrium dissociation pressure for $CaCO_3$ becomes 1 atm. On the other hand, increasing lump shrinkage develops as calcining temperatures are further increased and heating times are prolonged. This shrinkage reflects the fact that the porous microcrystalline shell of CaO slowly collapses while hot. This collapse occurs far below the melting temperature of CaO (2576°C), and appears to be caused by surface migration of atoms rather than by melting, even though it is noticeably accelerated by the presence of impurities like silica which form low-melting compounds with lime. The reactivity of the quicklime produced depends upon the size and spacing of the CaO crystallites, maximum reativity developing with small crystallites and unshrunk lumps.

Because of the differences in density of CaO and $CaCO_3$ and the shrinkage just discussed, strains develop within lumps of limestone undergoing decomposition. Cracks often form in the CaO shell, exposing some of the underlying $CaCO_3$, making the simple core-and-shell model of decomposition only a first approximation, even for spherical particles. As reaction proceeds, cracks formed earlier often close again, followed by development of holes within the lump between the core and shell.

Knibbs (77) investigated the calcining times required for total decomposition of limestone of lumps of various sizes. He reports that with a surface temperature of 1980°F, a 1-in. lump requires 1 hr for total decomposition, a 2-in. lump requires 3 hr, and a 3-in. lump requires 8.5 hr. Are these relative times of decomposition consistent with the simple core-and-shell model for decomposition, with heat transfer controlling?

Answer: By Eq. (8-14), for fixed values of surface and core temperatures, decomposition times are proportional to the initial radius squared. Thus, if a 1-in. lump decomposes in 1 hr, a 2-in. lump should decompose in 4 hr and a 3-in. lump in 9 hr. These times do not compare unfavorably with Knibb's values of 3 hr for the 2-in. lump and 8.5 hr for the 3-in. lump. It is possible that some of the disagreement between the observed and calculated times of decomposition is due to the development of internal cracks and holes, and to the increased heat conductivity of the solid lime formed.

Solid = gas. The conversion of a solid to a gas was illustrated in Chap. 2 by the decomposition of NH_4Cl, reaction (2-9). As with $CaCO_3$, decomposition rates are limited by kinetics at low temperature, by diffusion at intermediate temperatures when an inert gas like air is present and the decomposition pressure is below the total pressure of the system, and by heat transfer at higher temperatures where the decomposition pressure exceeds the total pressure of the system. However, ammonium chloride decomposition differs from calcium carbonate decomposition in that no shell of solid product forms as the reaction proceeds because the products are gaseous, and the core of unreacted solid is consequently always exposed to the surroundings.

When kinetics control, then Eq. (8-2) applies with R_0 and R_{cf} representing the radii of the exposed core at the beginning and at the end of the time period considered. When heat transfer controls, then heat is received directly from the surroundings by radiation, conduction, and convection. If the surroundings temperature and surface temperature are, respectively, constant at T_0 and T_c, then the inward flow of heat in differential time becomes

$$qd\theta = 4\pi R_c^2 U(T_0 - T_c)d\theta \tag{8-15}$$

where U is the overall coefficient of heat transfer. This heat serves to supply the enthalpy of reaction of the solid in the thin shell of core material decomposed in differential time. The volume of this shell is $4\pi R_c^2 dR_c$, consequently, the heat requirements for decomposition are

$$qd\theta = -\frac{\rho_c \Delta H^\circ}{M_c} 4\pi R_c^2 dR_c \tag{8-16}$$

where (ρ_c/M_c) again represents the moles of solid reactant per unit volume of core. Combining Eqs. (8-15) and (8-16), and integrating for the case in which U is constant,

$$(R_0 - R_c) = \frac{M_c U(T_0 - T_c)\theta}{\rho_c \Delta H} \tag{8-17}$$

Thus, when heat transfer controls, the radius changes linearly with time.

The diffusion-controlled case resembles the one to be discussed in a later section in connection with Eq. (8-24).

Solid + solid = gas. Solid-solid reactions yielding only gaseous products were found to be monovariant in Chap. 2. One illustration of this class of reactions is the carbon reduction of zinc oxide to form zinc vapor, in which both of the following reactions occur simultaneously:

$$ZnO(c) + C(c) = Zn(g) + CO(g) \tag{8-18a}$$
$$2ZnO(c) + C(c) = 2Zn(g) + CO_2(g) \tag{8-18b}$$

The foregoing equations imply that these are solid-solid reactions, but as shown by Bodenstein (10), the actual mechanism involves an intermediate reaction sequence involving gases. That is, CO molecules at the ZnO surface react to form CO_2 and gaseous zinc, after which the CO_2 diffuses back to the carbon surface where it is reduced to CO again, and the cycle repeats itself indefinitely. This actual mechanism is best represented by the following two reactions, which can be directly derived from Eqs. (8-18a) and (8-18b):

$$ZnO(c) + CO = Zn(g) + CO_2 \tag{8-19a}$$
$$CO_2 + C = 2CO \tag{8-19b}$$

In other words, zinc vapor results from the cycling of the carbon oxides between solid carbon and solid ZnO. Each of the two foregoing sets of equations can, of course, be used to calculate equilibrium compositions, and each leads to the same Phase Rule result, namely, that this is a monovariant system.

To improve the rate of reactions in Eqs. (8-19a) and (8-19b) in industrial operations, ZnO and carbon in finely divided form are briquetted, thus bringing the two different solid surfaces into close proximity and minimizing the diffusional path length of the carbon oxides. The effectiveness of such briquetting is demonstrated by the fact that, in industrial zinc distillation, the internal temperatures of the briquettes lie within 20° of 900°C, which is the temperature at which the equilibrium decomposition pressure for the ZnO–C system reaches 1 atm. As expected, when operating at these relatively high temperatures, the rate of zinc distillation is controlled by the rate of heat supply, and the reaction proceeds topochemically within a briquette.

Gas-solid reactions. Discussion so far has centered on the decomposition or interaction of solids, with one of the reaction products being a gas. Consideration will now be given to reactions in which one of the reactants is a gas, as in the combustion of solid carbon with air. For a gas to react with a solid, molecules of this gas must collide with the solid surface. If reaction occurs immediately upon collision, then the rate will be proportional to the collision frequency, and consequently, proportional to the partial pressure of the reacting gas. If the gas molecules must first be adsorbed by the solid, after which some fraction of these adsorbed molecules react, then the rate becomes proportional to the fraction of the solid surface covered by adsorbed molecules, which in turn is roughly proportional to the partial pressure to some exponent.* In either case, only those molecules react which have the necessary activation energy, hence the rate expression contains the usual exponential term $e^{-E/RT}$. When kinetics control, therefore, the moles of gas reacting per unit area and per unit time are as follows:

$$\frac{\dot{n}_G}{A} = -\frac{\rho dR_c}{M_c d\theta} = B \cdot p_c^x \cdot e^{-E/RT} \tag{8-20}$$

Solids of the same composition but different history sometimes have very dissimilar reaction rates. Thus, wood charcoal prepared by low temperature heating is far more reactive than when made by prolonged high temperature calcining. This difference appears due not only to the greater surface area per g of the former material, but to the greater roughness and reactivity of this surface. Similarly, freshly prepared lime "burned" at relatively low

*This assumes the adsorption to follow the Freundlich-type isotherm. Modifications can easily be made for Langmuir-type adsorption.

temperatures reacts much more rapidly with water to form slaked lime, namely $Ca(OH)_2$, than does another sample of this same lime after being kept at high temperature for several hours. Part of this reduction in the reactivity of lime is evidently due to the loss of internal surface, since the porous lime made by decomposing $CaCO_3$ at low temperature rapidly becomes less porous upon exposure to high temperature. Changes of the surface itself, such as in its roughness, also appear to contribute to the loss in reactivity. Similar changes in reactivity of solid catalysts following exposure to high temperature or adsorption of traces of various gases have already been encountered. It is worth noting that many solids show this type of behavior.

Solid + gas = gas. A pioneering exploration of the rates of this class of reaction was made by Hottel (151) in studies of carbon combustion. Experimental procedure involved suspending a sphere of nonporous and relatively ash-free electrode carbon about 1 in. in diameter in hot air, and causing the sphere to rotate slowly to achieve uniform combustion. Reaction rate, determined by continuously weighing the specimen, was found to be a function of the surface temperature of the carbon sphere (as read by an optical pyrometer), and of the velocity and composition of the surrounding gas. Inspection of the typical results presented in Fig. 8-2 shows that at low temperature, the rate expressed as g of carbon lost/(cm^2)(sec) is independent of gas velocity, changes rapidly with temperature, and is directly proportional to p_∞, namely, the partial pressure of oxygen in the surrounding atmosphere. At high temperatures, on the other hand, the rate is proportional both to

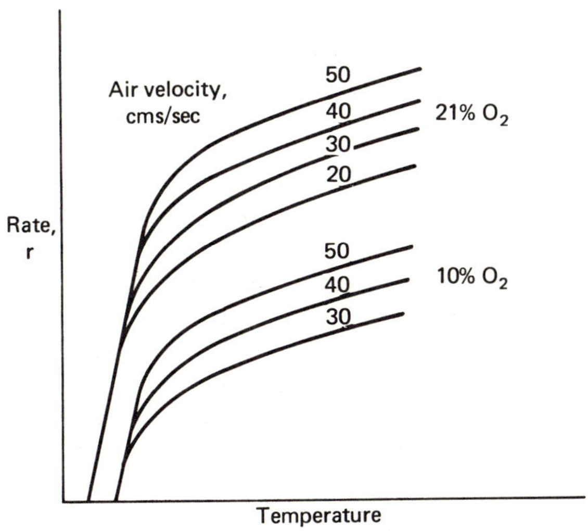

Fig. 8-2. Specific reaction rate of a carbon sphere vs. temperature (151).

the air velocity to a power of about 0.8 and to p_∞, the partial pressure of oxygen, but now varies with the square root of the absolute temperature. Obviously, chemical kinetics control at low temperature and diffusion controls at high temperatures. Inspection of Fig. 8-2 shows that, as is to be expected, a transition region exists at intermediate temperatures where diffusion and chemical kinetics are both important.

Quantitative formulation. Quantitative formulation of the rate expressions applying to the combustion of this sphere of ash-free carbon can now be undertaken. Equation (8-20) applies directly at low temperature where kinetics control. Under these conditions, diffusion is relatively rapid, so that the partial pressure of oxygen in the bulk gas, namely p_∞, is substantially the same as the carbon surface, namely p_c. Over the temperature range studied, the exponent on the oxygen partial pressure in Eq. (8-20) was found experimentally to be unity, suggesting that a fixed proportion of the collisions of oxygen molecules on the carbon surface are violent enough to result in direct combination. For this case, therefore, Eq. (8-20) becomes

$$\frac{\dot{n}_G}{A_c} = \beta p_c e^{-E/RT} \tag{8-21}$$

At high temperatures where chemical rates are very fast, diffusion of oxygen through the stagnant gas film around the particle is the rate-limiting step, with $(p_\infty - p_c)$ being the partial pressure driving force. This film thickness is an inverse function of gas velocity, and is here assumed constant for any given case at $(R_f - R_c)$. In stagnant gas, in which the diffusivity for oxgyen is D, then $\dot{n}_G$, the moles per sec of oxygen flowing inward through the gas film, at a point where the distance away from the particle center is R may be approximated by the following expression:

$$\frac{\dot{n}_G}{A} = \frac{D}{R_g T}\frac{dp}{dR} \tag{8-22}$$

Here dp/dR is the pressure gradient through the film around the particle at radius R, A or $4\pi R^2$ is the area of a spherical shell passed through this gas film at radius R, and R_g is the gas constant. The oxygen partial pressures corresponding to R_f and R_c, the outer and inner radii of the gas film are, respectively, p_∞ and p_c. At any instant, the quantity $\dot{n}_G$ is constant through all spherical shells drawn through this gas film, hence the foregoing equation can be rearranged:

$$\dot{n}_G \int_{R_c}^{R_f} \frac{dR}{R^2} = \frac{4\pi D}{R_g T}(p_\infty - p_c)$$

integrating:

$$\dot{n}_G\left\{\frac{R_f - R_c}{R_f \cdot R_c}\right\} = \frac{4\pi D}{R_g T}(p_\infty - p_c)$$

When R_c is small compared to R_f, then the preceding expression reduces to

$$\dot{n}_G = \frac{4\pi R_c D}{R_g T}(p_\infty - p_c)$$

Dividing by $4\pi R_c^2$, which equals A_c, the surface area of the carbon sphere

$$\frac{\dot{n}_G}{A_c} = \frac{D}{R_c R_g T}(p_\infty - p_c) \tag{8-23}$$

The oxygen partial pressure at the carbon surface, namely p_c in Eq. (8-23), is of course zero at high temperature, where reaction rates are great. At intermediate temperatures, both diffusion and chemical rates are of comparable magnitude, consequently the partial pressure of oxygen at the carbon surface lies somewhere between zero and p_∞. This unknown term p_c can be eliminated by combining Eqs. (8-21) and (8-23):

$$\frac{\dot{n}_G}{A_c} = \frac{p_\infty}{\dfrac{R_g R_c T}{D} + \dfrac{1}{\beta e^{-E/RT}}} \tag{8-24}$$

For larger particles and for finite gas velocities relative to the particle, the term $(D/R_c R_g T)$ in Eqs. (8-23) and (8-24) is replaced by the appropriate mass-transfer coefficient whose magnitude varies with the gas velocity raised to a power not far from unity. Inspection shows Eq. (8-24) now to be consistent with the curves of Fig. 8-2. That is, at low temperature, the term R_c/DR_gT in the denominator of Eq. (8-24) is small as compared to the second term in the denominator, hence this equation reduces to Eq. (8-21), since p_c and p_∞ are now identical. At high temperatures, on the other hand, the second term in the denominator on the right-hand side of Eq. (8-24) becomes negligible and this equation now reduces to Eq. (8-23), since the term p_c in Eq. (8-23) is zero because of the relatively high reaction rates.

Solid + gas = solid. The kinetics of a gas-solid reaction which produces a solid product depend on the porosity of the reacting particle. When the initial particle is dense and without pores, then a thin impervious surface coating of the solid product may form around the particle which prevents further reaction. Thus, a dense sphere of lime develops such a coating of $CaCO_3$ when exposed to an atmosphere in which the CO_2 partial pressure is above the decomposition pressure of limestone at the temperature of the operation. Even highly porous lime made by careful calcining at low temperatures to minimize thermal shrinkage may not react rapidly or completely, due to the formation of a $CaCO_3$ film on the pore walls throughout the particle. When such impermeable films form, then the rate expressions derived in the foregoing paragraphs cannot be used without modification.

Solid + gas = solid + gas. The rates of reactions of this type introduce no new factors. If the solid product does not seal off the unused reactant in the particle core, then an equation of the type of Eq. (8-24) applies directly.

Reaction sequence. This case was illustrated in Chap. 2 with a reaction sequence involving the gaseous reduction of Fe_2O_3 to Fe_3O_4, which was then reduced in turn to FeO and finally to metallic Fe by the reactions of Eqs. (2-15), (2-16), and (2-17). Analysis of the rates of such reactions proceeds along the lines just discussed, but is sometimes complicated by the presence of several concentric shells of solid products within each reacting solid particle. Thus, when reducing a sphere of Fe_2O_3 at high temperature with H_2, several concentric shells can be present, the outside one being metallic iron, followed by an inner shell of FeO, a shell within this one of Fe_3O_4, and a core of Fe_2O_3. Such topochemical layer formation is typical of sequential reactions involving solids.

Multiple particle systems. So far, the rate behavior of only single particles has been considered. In industrial gas-solid reactions, such as the combustion of powdered coal or the hydrogen-reduction of iron oxide ore, multiple particle systems are usually involved. Complications develop because the sizes of these particles all differ, with the larger ones taking longer to react than the smaller ones. Thus, in a piston flow system, as reaction proceeds in a moving slug of gas containing solids in suspension, there is usually a simultaneous change in the number of particles, in the particle surface area at which reaction occurs, and in the gas composition. In addition, the temperature of the slug will not remain constant unless heat exchange is carefully controlled.

Each of the factors just listed can have an important effect upon the amount of material reacting per unit time in these multiple particle systems. To illustrate the importance of just one of these factors, namely, particle size distribution, consider a system of nonporous plate-like particles, all having the same face area but differing in thickness. As shown in Fig. 8-3, the dimensions of the large faces of these particles, namely w and s, are assumed much greater than the thickness L. Obviously, if the reaction rate (mole per unit area and per unit time) is the same for each face, the percentage change in dimensions w and s will be trifling compared to that in the thickness. Thus, as these particles react to form a gaseous product and diminish in thickness,

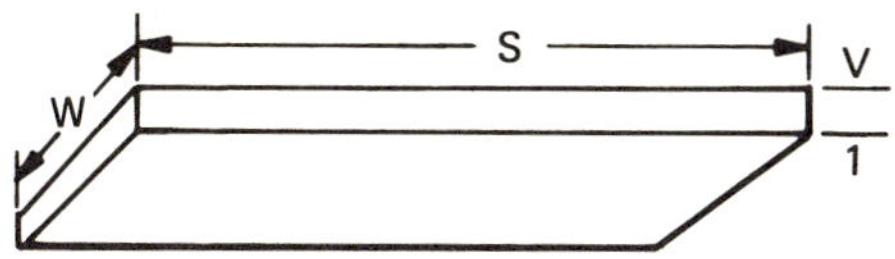

Fig. 8-3. A plate-like particle. The area of the four sides is negligible compared to the two large faces.

the area of the two large faces of each particle may be assumed constant, while the areas of the four side faces are negligible.

These platelets are now gasified in an isothermal operation while suspended in a gaseous environment of unchanging and uniform composition, as might be approximated by reacting dense ash-free carbon flakes with H_2 to form CH_4 in a shallow fluid bed. Under these conditions, regardless of whether kinetics or heat transfer control, the thickness of each particle changes at the same linear rate with time as both opposite faces simultaneously react:

$$\frac{-dL}{d\theta} = G \tag{8-25}$$

or

$$L_0 - L = G\theta \tag{8-26}$$

where G is a constant, L_0 is the initial thickness and L is the thickness at time θ. Assume that there are N_0 particles present initially, each with the same face area $2ws$. Two arbitrarily chosen examples of the many possible relations between the initial thickness and particle number are as follows:

Case I. All particles have the same initial thickness L_0, consequently the initial volume of the bed solids is $w \cdot s \cdot L_0 N_0$.

Case II. Each particle is greater than the next one by a constant increment of thickness, or

$$L = \frac{L_0}{N_0} N \tag{8-27}$$

where N is the number of particles initially present whose thickness equals or is smaller than L. The initial volume of all particles whose initial thickness is L or less is then

$$V = ws \int_0^L N \, dL \tag{8-28}$$

The total initial bed volume is obtained by combining Eqs. (8-27) and (8-28) and integrating between L_0 and zero, or V_0 equals $(wsL_0N_0/2)$.

Case I. These identical particles all diminish in thickness equally with time as shown by Eq. (8-26), hence they all disappear simultaneously at the very end of the reaction period. For this case, then, the volume of all the particles present at any time θ equals $wsN_0(L_0 - G\theta)$, hence it follows that the bed particle volume diminishes linearly with time.

Case II. At any time θ after the start of the operations, it follows from Eq. (8-26) that those N particles in the original bed which were thinner than L have disappeared, where L equals $G\theta$. The original volume of these vanished particles has already been shown to equal $wsNL/2$. The number of surviving particles is $(N_0 - N)$, but these too have all diminished in thickness by L, and so have lost a volume of $ws(N_0 - N)(L)$. Thus V_θ, the volume of the bed particles surviving at θ, is the initial volume of all the bed particles minus the volume of the particles which have disappeared and minus the volume shrinkage of the surviving particles, or

$$V_\theta = \frac{wsN_0L_0}{2} - \frac{wsNL}{2} - ws(N_0 - N)L \tag{8-29}$$

Since both particle number and thickness diminish with time, it is obvious from inspection of Eqs. (8-26), (8-27), and (8-29) that unlike case I, bed volume here does not diminish linearly with time.

As demonstrated by these two simple cases, depending on particle size distribution, the change of bed volume or mass with time can vary widely, even in the simple isothermal operation involving platelets as discussed here. Obviously, since particle porosity, shape, and composition are usually not constant, rate equations are complex, and operations are rarely isothermal save in beds of fluidized solids and great variations may be encountered in the changes of bed mass with time for solid gas reactions.

Moving beds. One convenient method for carrying out gas-solid reactions in a steady-state is in a moving bed. By this procedure, a vertical cylindrical shaft could, for example, be filled with solids, then fed continuously with fresh solids at the top while withdrawing product solids at the bottom. When operations are to be countercurrent, the gas is passed continuously up through the descending bed of solids, while cocurrent operations are attained by causing the gas to move in the downward direction.

Perhaps the commonest example of a moving bed operation is the solid fuel-fired grate furnace. A solid fuel such as coke (typically 90% carbon, 10% ash by weight) is continuously dropped onto the fuel bed from above, and ash is withdrawn from below. Air enters through the grate bars and flows up through the burning bed. In operation, an ash layer a few inches thick is maintained at the bottom of the bed, and these ashes are cooled by the entering air, the ash layer serving both to protect the grate from excessive temperatures, and to preheat the entering air. The zone of "primary" combustion is located immediately above the ash zone, and here the exothermal combustion reaction occurring is as follows:

$$C + O_2 = CO_2 \qquad \Delta H = -94 \text{ kcal} \tag{8-30}$$

A plot of gas composition versus distance up from the grate bars for a coke-fired furnace is presented in Fig. 8-4. Inspection shows that the oxygen concentration drops sharply with distance up through the bed, while the CO_2 concentration rises correspondingly. For a given size of coke, an increase in air velocity results in no shift in the two curves of Fig. 8-4 within this primary combustion zone, indicating that reaction here is mass transfer controlled and that the coefficient of mass transfer is (roughly) proportional to the gas velocity. An increase in the size of the coke particles, of course, broadens the depth of this primary combustion zone.

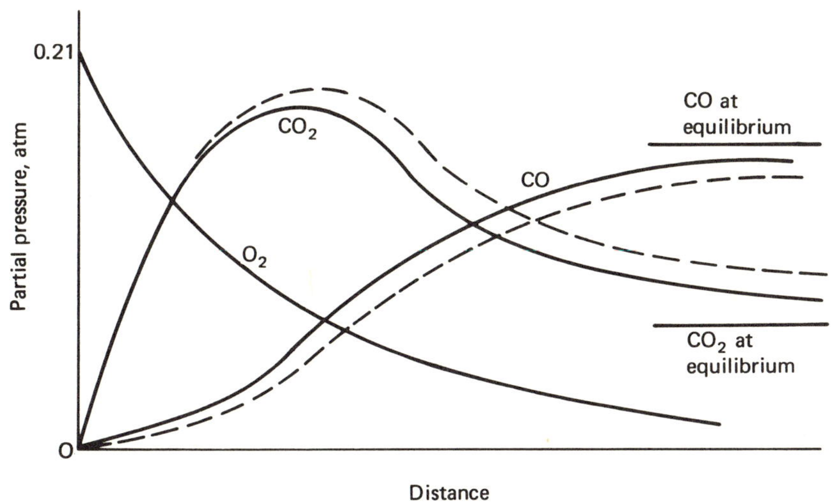

Fig. 8-4. Gas compositions in a burning coke bed (107). The horizontal axis represents the distance above the grate bars up through the solid fuel bed (an ash layer is usually maintained on the grate bars). Dotted lines represent the shifts to be expected with increased gas velocities.

Further inspection of Fig. 8-4 shows a second zone above the primary zone, in which the CO_2 concentration falls while the CO concentration rises because of the following endothermal reaction:

$$CO_2 + C = 2CO \qquad \Delta H = +40 \text{ kcal} \tag{8-31}$$

Given sufficient bed depth, the reaction of Eq. (8-31) goes to equilibrium. This second zone overlaps the primary combustion zone somewhat, in that a little CO obviously forms early in the bed. As shown by the dotted lines, a change in gas velocity changes the gas composition at any given plane within this second zone, indicating that the rate of reaction in Eq. (8-31) is controlled by chemical kinetics.

This example of the coke-fired grate furnace, while complicated by the occurrence of the sequential reactions of Eqs. (8-30) and (8-31), serves to illustrate two types of behavior of reaction zones of moving beds. That is, when chemical rates are very high as for the high temperature primary combustion reaction, then mass transfer typically controls. The reaction zone is then narrow, since mass transfer is rapid both through the ash shell around the burning lump of carbon and between the surface of this ash layer and the gas stream. Of course, if the ash in the system were to melt, imposing a serious resistance to mass transfer in the form of a molten layer around the particle, then the zone would not be so compressed. On the other hand, when kinetics are slow enough to be rate-controlling, as for the reaction of Eq. (8-31), then the reaction zone is deeper and its depth becomes a function of gas velocity.

ADIABATIC COUNTERCURRENT REACTORS

Continuous countercurrent operations in adiabatic systems are common. An example is the vertical shaft kiln, to which solids are fed at the top, as shown in Fig. 8-5. The bed of these solids which fills the kiln moves slowly downward through the shaft as solids are withdrawn below, while gas flows up through this moving mass from below. Heat is released or consumed within this "moving bed" kiln as reaction occurs between components of the gas and solid.

Operation of this system is illustrated by the hydrogen reduction of cuprous oxide, for which the equilibrium constant is large at the temperatures considered:

$$Cu_2O + H_2 = 2Cu + H_2O(g) \qquad \Delta H = -30{,}000 \text{ Btu/lb mole } Cu_2O$$

Above 400°C, rates are high and so reaction occurs within a relatively narrow horizontal reaction zone, within which the temperature is taken as uniform. In the simplified analysis presented here, heat and mass transfer coefficients are assumed to be so related that gases and solids always leave the reaction zone at the same temperature. Assuming the solid particles to be porous enough so that all the Cu_2O present is reasonably accessible to the gas, then the solids below the reaction zone contain no Cu_2O, and the gas above the reaction zone contains only H_2O.

It is obvious that the reaction zone will remain stationary only if the

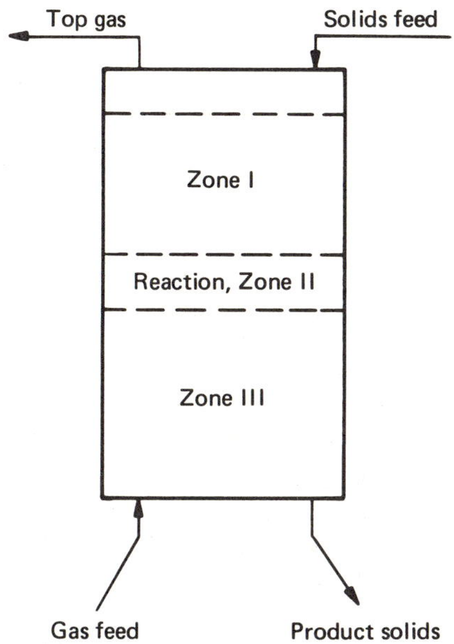

Fig. 8-5. Adiabatic kiln containing a bed of solids moving countercurrently to the gas.

feed rates of Cu_2O and H_2 are in exact stoichiometric balance, that is, only if one atom of oxygen is fed (as Cu_2O) per mole of H_2. If Cu_2O feed rates are increased relative to H_2 feed rates, then the reaction zone will travel downward through the kiln shaft, while the reverse is true if the Cu_2O feed rate is diminished. It is generally possible to control feed rates so that the reaction zone remains stationary at the center of the kiln height.

The zones above and below the reaction zone in Fig. 8-5, respectively labeled I and III, serve as heat exchangers. If the bed of solids in the shaft has a fair depth, then zones I and III may, as a first approximation, be regarded as infinite exchangers through which the gas and solid move countercurrently. As discussed in Chap. 5, the stream having the smaller heat capacity always leaves the kiln at the same temperature as that at which the other stream enters. If the two streams both enter the kiln at the same temperature, namely T_e, then the stream having the lower total heat capacity also leaves the kiln at T_e. The heat evolved within an adiabatic kiln always leaves as sensible heat in the stream having the greater total heat capacity. Finally, when both gas and solid streams enter at the same temperature T_e, and when the reaction occurring is exothermal, then because of heat exchange in zones I and III, the temperature in the reaction zone is higher than that of any of the streams leaving and entering the tower. These points are now illustrated for the system shown in Fig. 8-5.

EXAMPLE 8-2

The adiabatic tower of Fig. 8-5 is fed at the top with chunks of partially oxidized copper, in which the ratio of atoms of oxygen to atoms of copper is 1: 5. Hydrogen fed to the bottom of the tower completely reduces the solid phase in the reaction zone at the center of the tower, and then leaves the tower as water vapor. Both the solid and the gas enter the tower at 220°F, in the proportions of one atom of combined oxygen in the solid per mole of hydrogen gas.

(a) To keep the reaction zone stationary at the center of the tower in steady-state counterflow operations, how much H_2 must be fed per atom of oxygen fed in the solid-phase?
(b) What are the temperatures of the gas stream and of the solid stream leaving the kiln?
(c) What is the temperature of the reaction zone?

Notes: The heat capacities of Cu, Cu_2O, H_2, and gaseous H_2O are 7, 17.5, 7, and 10.5 Btu/(lb mole)(°F), respectively. The total heat capacity of the solid feed to the tower is then 38.5 Btu/°F for every atom of combined oxygen. The reduction reaction is assumed to be rapid and to go to completion.

Answer: (a) Obviously, the reaction zone will be stationary when 1 mole of H_2 is charged per atom of combined oxygen (or per five atoms of copper) in the solid feed. (b) The total heat capacities of the solids moving through both zones I and III of the kiln are much greater than those of the gases, with the heat capacity ratios being 38.5/10.5 in zone I and 35/7 in zone III. Thus, since there is temperature equality at that end of an infinite exchanger where the stream with the greater heat capacity enters, then the water vapor leaving zone I at the top of the kiln must have the same temperature as the entering solid, namely 220°F. Obviously, the solid stream leaving the bottom of the kiln will remove all of the heat generated in the kiln and, if its temperature is T_s, then by an overall energy balance on the kiln,

$$35(T_s - 220) = 30{,}000$$

or

$$T_s = 1080°\text{F}$$

(c) Taking a balance around zone III, and recognizing that the gas and solid must have an equal temperature at the top of this zone, namely T_t, then,

$$35(T_t - 1080) = 7(T_t - 220)$$

or

$$T_t = 1300°\text{F}$$

The temperature in the narrow reaction zone must equal T_t, namely, 1300°F.

Consider now that the reaction zone is moved to the bottom of the bed by temporarily increasing the flow of solids relative to the gas. When the zone reaches the bottom of the bed, it is maintained in this position by again feeding the gas and solids in stoichiometric proportions. As a result, zone III in Fig. 8-5 has been eliminated.

EXAMPLE 8-3

The reaction zone of the system in example 8-2 is moved to the bottom of the bed as just described, thus eliminating zone III in Fig. 8-5, after which the original feed conditions are restored.

(a) What is the kiln off-gas temperature? (b) What is the reaction zone temperature?

Answer: (a) It will again be assumed that zone I in Fig. 8-5 functions as an infinite exchanger. Since the total heat capacity of the solids remains higher than that of the gas, the top gas temperature must again equal the entering solids temperature, namely, 220°F. (b) since zone III in Fig. 8-5 does not exist, the solid leaves the kiln at the temperature of the reaction zone, namely, T_R. By using an overall energy balance on the kiln,

$$35(T_R - 220) = 30{,}000$$

Solving the reaction zone temperature is found to be 1080°F.

A somewhat different condition exists when the reaction zone is maintained at the top of the bed, as illustrated in the next example.

EXAMPLE 8-4

The reaction zone of example 8-2 is moved to the top of the bed as described in the preceding paragraphs, thus eliminating zone I of Fig. 8-5, after which the original feed conditions are restored. (a) What is the kiln off-gas temperature? (b) What is T_s, the discharge temperature for the solids?

Answer: (a) With the reaction zone at the top of the bed as in Fig. 8-5, it is evident that the kiln off-gases leave at the temperature of the reaction zone, T_R. In addition, the solids must leave the reaction zone at T_R. Since zone III acts as an infinite exchanger, then in view of the relative heat capacities of the gas and solid streams, the gas entering the reaction zone is obviously also at T_R. By using an energy balance around the reaction zone,

$$30{,}000 = (38.5)(T_R - 220)$$

or

$$T_R = 100°\text{F}$$

(b) By using an energy balance around the whole kiln,

$$30{,}000 = 10.5(1000 - 220) + 35(T_s - 220)$$

or

$$T_s = 844°\text{F}$$

It is assumed in this discussion that the top and bottom boundary surfaces of the reaction zone are plane and horizontal. These boundary surfaces can become severely distorted by channeling, which is caused by nonuniform flow of either the gas or the solid through the kiln. Channeling is best prevented by careful sizing of the solid, and by feeding gas and solids uniformly across the kiln cross section.

It is evident that extreme internal temperatures in the kiln will develop when the total heat capacities of the solid and gas streams approach equality. Indeed, it is possible to create situations in which the total heat capacity of the solid exceeds that of the gas in zone I of this kiln, with the reverse situation in zone III. Such conditions in theory permit no heat to escape from an adiabatic kiln, and lead to breakdown.

Chemical Reactors 9

Materials undergoing reaction in a correctly designed reactor are subject to predetermined conditions of temperature, pressure, concentration, and contact with other phases and catalysts, if any. Departure from such conditions is more or less disadvantageous, depending upon the sensitivity of the reaction under study to changes in operating conditions. It is often impossible to design so that all parts of the reactant stream undergo the same history of concentration change and temperature during their travels through a reactor. Thus, because of unavoidable velocity gradients, some parts of a reagent mixture moving through a flow reactor have different residence times than other parts. Again, temperature histories will vary depending on the path followed through the reactor, since there is always a temperature gradient through the gas or liquid film on the reactor's cooling surfaces, regardless of the vigor of agitation or the velocity of flow. Contact between two phases, such as between a reactant gas and a solid reagent or solid catalyst, will vary depending upon the development of channeling. Final reactor design is, therefore, a compromise, in which adequate allowance must be made for departures from the desired conditions of operation.

Reactors differ widely in construction and function, coming in many different sizes and shapes, some for batch, others for continuous operation. Reactors for homogeneous gas phase and for liquid phase reactions were briefly discussed in Chap. 6. Examples of four other general types of reactors will now be considered, namely, gas-solid reactors, gas-liquid reactors, liquid-solid reactors and gas-liquid-solid reactors. Discussion is far from exhaustive for the few examples cited, and there are many other reactors in each category which are not described. The purpose here is primarily to establish some perspective on the factors of importance in reactor selection, scale-up, and operation.

Gas-liquid reactors. Aside from mass transfer considerations, important questions governing the selection of a gas-liquid reactor type are: (a) Is countercurrency required?, (b) What gas and liquid residence times are needed and how uniform must these be?, (c) How much heat energy must be added or removed? Problems of materials handling are usually not difficult, since gases and most liquids are readily pumped into and out of reaction vessels.

Bubble reactor. In a bubble reactor, gas is bubbled up continuously through a pool of liquid usually containing reagents or catalyst in solution or suspension. The gas is introduced through holes in the vessel bottom or through a "sparger," namely, a network of perforated pipe often lowered into the vessel from above and designed for easy removal and maintenance. The various factors of importance in bubble reactor design, including mass transfer coefficients, have been extensively discussed in the literature, and recently reviewed (31).

Bubble reactors are common in the organic field, being used for example in the production of acetaldehyde both by the oxidation of ethylene in the Wacker process, and by the hydration of acetylene in the older German process. In the latter operation, acetylene at slightly over atmospheric pressure is bubbled up through a deep pool of aqueous mercuric sulfate solution at about 100°C. The mercuric ion functions as catalyst, the acetylene conversion per pass is perhaps 50% and the acetaldehyde produced leaves in the off-gases to be recovered by condensation. Temperature control in a steady-state operation is readily attained even with this vigorously exothermal reaction, since the reaction heat is removed by the vaporizations of water and acetaldehyde. The water in the off-gases is returned to the reactor after condensation and separation from the acetaldehyde.

Gas flow up through the liquid phase in a bubble reactor is conveniently characterized by its superficial velocity, namely, the gas volume per sec per unit cross-sectional area. Depending upon bubble appearance and behavior, two flow regimes can be distinguished, namely quiescent flow at low gas velocities and turbulent flow at higher velocities, with a transition at about 0.2 ft/sec superficial velocity in aqueous systems. Gas bubbles remain separate and well-defined in the quiescent regime, rising smoothly with little interference. Bubble sizes in this regime are a complex function of orifice diameter and gas flow rate through the orifice, as well as of liquid density and surface tension. In typical water-air systems, bubble diameters range from 0.2–1.2

cm as hole diameters increase from 0.02–0.4 cm, while rate of rise of bubbles of this size is roughly constant at 0.8 ft/sec. Residence times can be estimated from this finding since the expansion of the liquid in depth, like the fraction voids, increases more or less linearly with superficial gas velocity and reaches about 0.3 at superficial gas velocities of 0.2 ft/sec. In the turbulent regime, on the other hand, the bubbles are no longer clearly defined, with dimensions apparently little affected by orifice size. Liquid depth remains constant or even decreases a trifle with increases in gas velocity, consequently average gas residence times are now inversely proportional to gas velocity. It is assumed in this discussion that the liquid under consideration does not have serious foaming problems.

Even at superficial gas velocities as low as 0.05 ft/sec in aqueous systems, liquid mixing in the axial direction is fairly good in tubes as narrow as 1 in. or so in diameter, and improves rapidly with increasing diameter. In vessels of large diameter, therefore, it is generally reasonable to assume complete liquid mixing and temperature uniformity even with relatively rapid flow of liquid through the reactor, consequently countercurrency is not attainable in a bubble reactor. Gas mixing, on the other hand, is not complete in either the quiescent or the turbulent regime, with a widespread range in gas residence times indicated. This is evident from the residence time curves of Fig. 9-1, showing the off-gas analysis after CO is suddenly substituted for nitrogen,

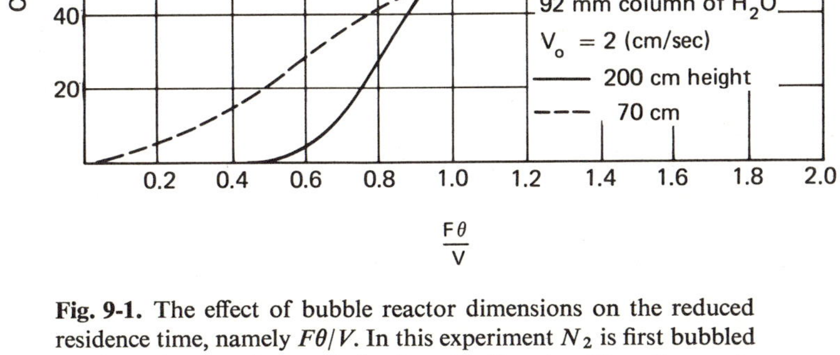

Fig. 9-1. The effect of bubble reactor dimensions on the reduced residence time, namely $F\theta/V$. In this experiment N_2 is first bubbled up through water in a cylinder 2 cm in diameter. The nitrogen is replaced by CO, and the off-gas is analyzed with results as shown. The superficial gas velocity is 2 cm/sec. Two liquid heights are reported: 70 cm and 200 cm (79).

blown up through the porous bottom of a cylinder 9.2-cm in diameter and filled to various heights with water. Operations are obviously far from piston flow, even at a relatively large height/diameter ratio. Other studies show these curves to be independent of the diameter at a given height/diameter ratio.

Heat transfer surfaces can be installed within these reactors. Heat transfer coefficients are good, increasing with the superficial gas velocity to about the 0.2 power. Often, of course, no heat transfer surfaces are used, cooling being by evaporation and reflux of a volatile component of the liquid in the reactor.

Bubble reactors have found use with explosive gas mixtures, since the bubbles of gas can be made small enough to inhibit ignition. The Bessemer converter and the newer oxygen-blown BOF converters in the steel industry are again bubble reactors, with the advantage here that metal need not be pumped, as would be the case if an attempt were made to use tower reactors described in the next section. Bubble reactors are operable with solid particles suspended in the liquid phase, but at lower gas velocities, settling problems develop.

Tower reactors. When countercurrency is required in gas-liquid systems, then bubble cap towers or packed towers like those in absorption and fractionation are used as reactors (129,145). Packed tower reactors, also called trickle bed reactors, suffer from the disadvantage common to all systems of broken solids through which liquid and gas must flow, namely that channeling and bypassing may develop. Aside from such complications, both gas and liquid are approximately in piston flow through properly operating packed bed reactors. Similarly, piston flow is approximated by both phases in plate towers, recognizing that the liquid follows a zigzag course in some plate columns as it flows across one tray, descends to the next, across which it moves in the opposite direction, etc. Limitations on liquid and gas flow rates are similar to those in fractionating and absorption towers.

A typical application of the packed tower as a reactor is in the absorption of SO_3 from converter off-gases in sulfuric acid plants. Ceramic packing is preferred to metal-plate tower construction because corrosion problems are thereby eliminated. The liquid absorbent used is 98.5–99.0% sulfuric acid, since only within this narrow concentration range are the vapor pressures over sulfuric acid of both SO_3 and H_2O negligible. The desirability of low SO_3 pressures is obvious, while low water vapor pressures are important to avoid formation of sulfuric acid mists. The absorption operation is vigorously exothermal as the dissolved SO_3 reacts with the water in the absorbing acid to form additional H_2SO_4. The heat evolved is removed from the tower as the sensible heat of the absorbing acid. This acid is in part recycled after cooling to keep tower temperatures low. While a bubble reactor could be used for this service, a tower reactor is preferred because the pressure drop for the gas is smaller, and because of its lower costs for a given capacity.

A typical example of a plate tower as reactor is in the removal of CO_2 from a CO_2-H_2 mixture by reaction with ethanolamine. Another example is in ethyl acetate manufacture, made by esterification of ethanol with acetic acid in a fractionating column, the acetic acid being introduced several plates above the ethanol feed point, the ester leaving in the overhead vapor and water as bottoms. With strongly exothermal reactions, internal cooling is sometimes attempted in these towers by installing cooling coils submerged in the liquid on the trays. Such construction is costly and involves maintenance problems, consequently, cooling by heat removal as sensible heat is preferred where possible. In packed towers, special trays are occasionally installed for cooling.

Reactors for gas-solids systems. Of the many types of reactors used for gas-solid reactions, five common varieties are briefly discussed here. Consideration in each case is given particularly to questions of solids handling, heat transfer, and gas-solids contacting.

***Fixed bed reactor* (41).** A fixed bed reactor is typically a vertical cylinder, filled with irregular lumps of solid such as would be obtained in crushing or grinding, or else shaped as spheres or cylinders as from a catalyst-pelletizing operation. Particle sizes range from perhaps $\frac{1}{8}$ in. to about 2 in., smaller dimensions usually being avoided because pressure drop increases inversely with particle size for a given bed depth. Particles are hopefully packed in such a way that serious channeling and bypassing are avoided, making gas flow reasonably uniform and allowing piston flow to be approximated. Bed cross section is determined by the allowable pressure drops and gas velocities, while bed depth depends upon the gas residence time required. Gas flow is often down through the bed, thus avoiding the possible lifting of smaller particles in the top bed layers, causing these to rub together, and thereby generating fines and dust.

No provision is made for solids movement or for frequent bed replacement in a fixed bed reactor. These reactors are typically used with solid catalysts having a relatively long life, but not for solids which are rapidly consumed in chemical reaction. Occasionally, a fixed bed reactor contains a solid reagent which removes trace components from a gas, but in these cases solid consumption rates are small enough so that bed replacement is infrequent. Fixed beds are also used when in situ regeneration is feasible.

Beds of solid particles are excellent heat insulators, since contact areas between touching particles are small, thus affording little area for conduction. For example, a bed of randomly packed copper spheres, all of the same diameter, has a volume fraction for solids of about 60%, but the thermal conductivity at room temperature is perhaps one thousandth that of bulk copper. To transfer heat into or out of any fixed bed of solid particles without the development of regions of excessively high or low temperature, all particles must be located close to heat exchange surfaces. To satisfy this requirement, these reactors are filled with a multiplicity of closely spaced tubes containing pellets of solid and surrounded with heating or cooling fluid. Alternatively, the heat transfer liquid may be inside the tubes and the solid pellets outside. In the original Weiss and Downs converter for the highly exothermal oxidation of naphthalene to phthalic anhydride, the catalyst pellets were packed into tubes $\frac{1}{4}$–$\frac{1}{2}$ in. in diameter surrounded by boiling mercury. Because of the highly exothermal nature of this oxidation and the poor heat transfer through the packed solids, the temperature of particles in the center of the larger tubes was as much as 100°C higher than at the tube walls. In any given system operating with a constant mass flow rate of reagents, the temperature difference between the center and the periphery of catalyst-packed tubes varies about as the square of the tube diameter, consequently large radial temperature variations are difficult to avoid. In cooling, to keep the tube center at a reasonable temperature in an exothermal reaction, the particles near the cooled walls may actually be below the reaction kindling temperature. Thus, the fluid moving near the walls may not react at all, while the fluid in the tube center may be exposed to very severe conditions.

The fixed bed is obviously advantageous where solid replacement is infrequent, and is especially simple in adiabatic operation where no heat transfer need be provided for. Attrition is minimal in a fixed bed, and attention requirements are small.

EXAMPLE 9-1

The space velocity (SV) of a reactor is defined as Fv/V_c, where Fv is the volume of the feed entering per unit time, and V_c is the volume of the catalyst bed. The quantity (Fv) is defined as the volume of gas flowing per unit time as measured either at reactor entrance conditions or at standard temperature and pressure. It is important to establish the definition of Fv when discussing space velocities. For a catalytic system in which chemical reaction rate controls, scale-up can obviously be carried out by keeping space velocity constant. (a) It is proposed to scale up a small adiabatic catalytic bed in a pilot reactor to a larger size, keeping the approach velocity, temperature, pressure, and composition of the feed gas unchanged. The same conversion is to be attained in the two cases. If scale-up is to be based upon a constant space velocity, what will be the dimensions of the new reactor relative to the old? (b) It is stated in a conference that in a reactor of given diameter, when increasing velocity and bed depth in the same proportion to keep (SV) constant, conversion per pass remains unchanged. Is this always true?

Answer: (a) For the same feed gas conditions and approach velocity, an unchanged space velocity obviously requires the bed depths of the two reactors to be equal, and bed cross sections to be in the ratio of total gas flows. (b) By combining the definition of space velocity, namely, Fv/V_c, with Eq. (7-11), and remembering that AL equals V_c, then

$$(SV) = \rho v \Big/ \int \frac{dx}{r} \tag{9-1}$$

where ρ is the catalyst bed density and r is some function of reagent concentrations, such as in Eq. (7-9). It will be remembered that r does not change with gas velocity when chemical kinetics or pellet diffusion control, but does change with gas velocity when mass transfer to the exterior of the particles controls. Inspection of Eq. (9-1) shows that, as long as r is independent of gas velocity, the statement made in conference is correct.

Moving bed reactors. Typically, a moving bed reactor is a vertical shaft filled with broken solids which move slowly and more or less continuously downward through the unit under the action of gravity. Gas is generally blown up through these beds counter to the descending solids, as in the vertical shaft lime kiln and the blast furnace. Solids are charged at the top of the shaft, and withdrawn below by a suitable mechanism, such as a screw conveyor, a vibrator, or a star valve. Solids handling is obviously rather awkward, especially at the solids discharge end, and solids movement can cause severe erosion of the kiln lining. The weight of the column of solids in the reactor tends to cause crushing at the base, making some off-size material and fines which, in part, leave as dust in the top gas. There is some variation in residence times of the solid particles travelling through the reactor, depending on the location of the solid discharge ports, but this variation is usually not very great.

Shaft diameters are preferably large relative to particle diameter to minimize jamming and bridging by solids. If the product becomes pasty at elevated temperature, then the tendency to bridge becomes more serious. Channels are even more apt to develop than in fixed beds, and these channels can grow to be very large when the solids are gasified in the reaction being carried out in the bed. Particularly when channeling is severe, the temperature-time history of the gas stream across the shaft is far from uniform, and large deviations from piston flow occur. Because of the bed movement and associated erosion, installation of cooling tubes within the shaft is impractical. Since it is not feasible to accomplish much cooling through the walls of the shaft, a moving bed operation is typically adiabatic, with adjustment of the operating temperatures being made by control of the quantities, temperatures, and compositions of the feed streams.

Despite its limitations, the shaft kiln lends itself well to countercurrent gas-solids operations. A common example, of course, is a coal-fired grate furnace in which fresh coal is supplied from above, ash is removed from below, and air for combustion is blown up through the bed.

Rotary kilns. A rotary kiln is a long narrow cylinder, resting on its side on rollers, which is caused to rotate slowly about a slightly inclined axis. In modern cement plants, kilns as large as 500 ft long and 12 ft in diameter are operated, while smaller kilns are common in other industries. Because of kiln inclination ($\frac{1}{4}$–$\frac{3}{4}$ in. per foot) and rotation (usually 0.4–2 rpm), solids fed to the higher end tumble slowly down through the kiln and discharge at the lower end. The solids fed in metallurgical processing are typically $\frac{1}{4}$–3 in. in size, though much smaller-sized materials are fed in other industries, such as in cement or in polymer processing. The feeding and especially the discharging of solids is simpler to accomplish in rotary kilns than in shaft kilns. The solids usually occupy about 10% of a rotary kiln's cross section (see Fig. 9-2), the time the solids travel through a kiln is reasonably constant from particle to particle, and little axial mixing of solids occurs. The larger particles tend to segregate around the outside of the tumbling mass, displacing the smaller particles into the center, where they are said to form a "kidney" (see Fig. 9-3). In calcining operations, this kidney formation may cause "over-burning" of the large pieces, leaving the smaller pieces in the kidney less extensively

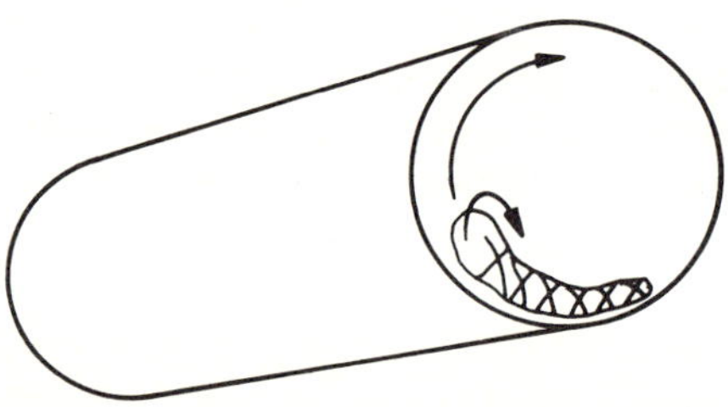

Fig. 9-2. Rotary kiln showing a tumbling bed.

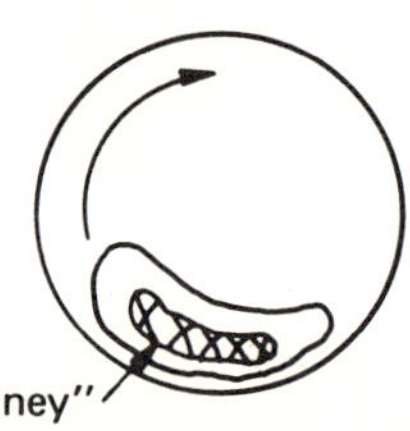

Fig. 9-3. Rotary kiln end showing a "kidney" of smaller particles in charge.

reacted because they do not "see" the radiation from the flame in the gas space. Attrition in the tumbling mass is severe, and there can be extensive dusting.

The mass slowly rolls over itself as it travels down the kiln, but only the surface particles make significant contact with the gas. To improve gas-solids contact and to reduce segregation of solids by sizes, these rotaries are often equipped with "flights" (see Fig. 9-4), which lift the charge and then let it fall back through the gas space. Obviously, kiln maintenance problems are increased by use of such flights.

Rotary kilns are usually operated countercurrently. In cement manufac-

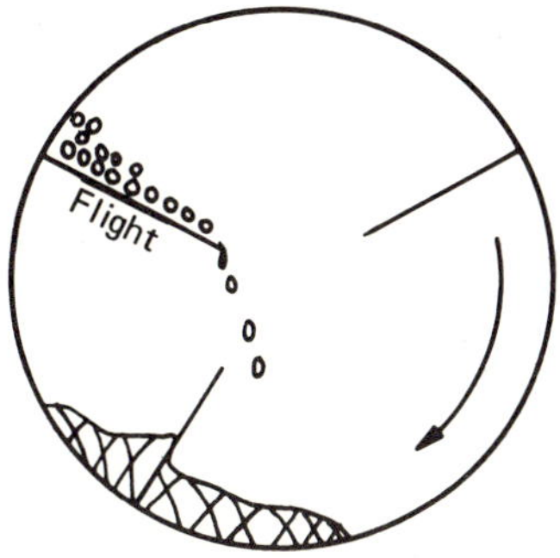

Fig. 9-4. Rotary kiln with flights, showing solids dropping through gas space.

ture and lime burning, the fuel used may be coal, oil, or natural gas. The fuel is introduced at the discharge end for solids in the kiln, and the products of combustion travel counter to the descending solids. To protect heat-sensitive materials, as in drying, cocurrent firing is sometimes practiced. Thermal efficiencies of rotary kilns are low, since they cannot be well insulated due to the limitations of the materials of construction. That is, the steel shell cannot be insulated externally, since such insulation would increase the shell temperature, and since steel loses its strength when hot, the kiln would buckle. More insulation on the inside of the shell is as yet impractical, since no materials have been developed which will withstand the battering of the tumbling charge well enough to provide a proper base for high temperature brick in these kilns.

A frequent difficulty encountered in high temperature operations is the formation of a growing ring of sintered solids within the kiln, perhaps 20–40 ft from the flame end. Such ringing is caused by the softening or melting of some component in the charge, and if unchecked will usually grow and force kiln shutdown. Prevention of ringing is difficult and no completely satisfactory cure exists despite the variety of corrective measures used, which range from short heavy chainlengths suspended within the critical section in the kiln, to shotgun blasts.

The rotary kiln has the advantages of large capacity, countercurrency, ease of feeding and discharge, and low operating labor requirements. It is used where large tonnages of low unit value solids must be processed, and where fuel is cheap.

Tray furnace. A tray furnace, frequently called a Herreshoff or Wedge furnace in this country, is sketched in Fig. 9-5. It consists of a vertical cylinder lined with fire brick, containing 8–15 hearths or trays. Each tray

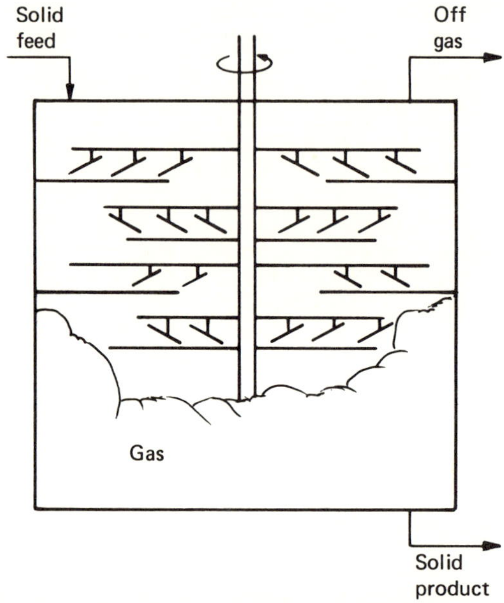

Fig. 9-5. Schematic diagram of a tray or "Herreshoff" type furnace.

is perforated with spill hole openings through which the solid being processed falls to the next lower tray. The top tray has spill holes near the center, the next lower one has spill holes at the periphery, and the one below it has holes at the center again, etc. Solid is moved across the trays by hoes attached to arms projecting from a central shaft. As this shaft slowly rotates, solid is raked inward on the top tray, falls through the central spill holes to the next tray where it is slowly raked outward, then falls through the peripheral spill holes onto the next tray where the rakes again move the solid inward, etc. As the solid proceeds outward and inward down through the reactor, air moves over these same hearths in the opposite direction, consequently countercurrent operation is attained. The hoes, arms, and central shaft of the rake mechanism are hollow, and cooling air is blown through these parts to protect them from overheating.

A relatively long residence time is afforded for both gas and solid in furnaces of this type, but gas-solids contact is rather poor, being restricted to the surfaces of the solid layers on the trays. No heat transfer surfaces for cooling or heating the charge within the furnaces are installed because of the obvious problems involved. The solids processed are finely divided, but because of the relatively gently agitation of the hoes, dust losses are not great.

Fluidized solids. A so-called "fluidized solids" system develops when a gas is blown up through a bed of dry, powdered solid. This bed may be contained in any suitable chamber, such as a vertical cylinder provided with a perforated base plate which serves to distribute the gas under the bed. As the gas velocity upwards through a bed of solids is increased from zero, frictional drag increases on each particle, and the whole bed expands slightly. A velocity is soon reached at which the pressure drop across the bed equals the bed weight per unit area. The particles, which are now supported by the gas, are observed to be in violent chaotic motion, mixing rapidly. The bed is said to be fluidized, and has a volume which is perhaps 10–15% greater than that of the original bed before gas passage. Further study shows that part of the gas moved up through the bed in the form of rapidly rising bubbles, down through which solid particles rain. The remainder of the gas moves up more slowly through the so-called "emulsion" phase, in which the density of solids is much greater than in the bubbles. The emulsion phase itself undergoes violent turbulent mixing because of the passage of the gas bubbles. As the gas velocity is increased, a greater proportion of the gas travels up through the bed in the bubble phase, and at the same time, the interface between the fluidized bed and the gas space above it becomes less well defined as these bubbles break and toss a spray of particles upward. The entire bed resembles a boiling liquid.

A limiting velocity exists, above which the bed is blown out of the vessel, and dusting is increasingly severe as this velocity is approached. Fluidization, then, develops between the velocity of incipient fluidization and the velocity at which the bed "takes flight." This velocity range is a complex function of the properties of the gas and solid particles used, but characteristically might lie between $\frac{1}{2}$–10 ft per sec for a material like limestone having a particles size distribution between 12- and 100-mesh.

Most powders can be fluidized, although difficulties develop with particles shaped like needles or plates. Particles having a ratio of maximum to minimum dimension not exceeding 3 or so usually fluidize readily.

Since gas moves through a fluid bed more rapidly as bubbles than it does through an emulsion phase, piston flow of the gas is not achieved. Gilliland (42) and associates found that when gas B was suddenly substituted for A, curves of off-gas compositions versus time reminiscent of those in Fig. 9-1 for bubble reactors were obtained showing wide variations in residence times of the gas traveling up through the bed. Some gas backmixing was also found, in that when a tracer gas was injected into the upper part of a bed, small concentrations of this gas were found below the injection point.

Solid particles in a fluid bed mix very rapidly and thoroughly, consequently a fluid bed may be viewed as a stage contacting device as far as solids are concerned. Fluid bed systems are always substantially isothermal, again because of the rapid motion and mixing of solid particles. The very high rate of heat flow through these beds, and the large heat transfer coefficients between the bed particles and submerged cooling or heating surfaces are a consequence of this motion of the particles.

Solids can readily be fed and discharged continuously from a fluid bed through ports in the reactor walls. Bed depth becomes self-regulating, if the unit is provided with an overflow port through which solids can spill as displaced by fresh feed. A cyclone, with fines return to the bed, is often used to remove dust from the tail gas.

EXAMPLE 9-2

Mixing of solids in a well-fluidized bed is excellent. When feeding and withdrawing solids at a constant rate in a steady-state operation, each of the particles within the bed system has, at each instant, an equal chance of leaving through the overflow pipe, regardless of its residence time.

Catalyst particle life in a certain fluid bed reactor is known from batch experiments to be 10 min. A continuous operation of this catalytic reactor is now planned, with catalyst feed and withdrawal to be at such a rate that only 30% of the particles in the bed will have been present for 10 min or more. To attain such an operation, what would be the value of W/F in minutes, where F is the solid feed rate in pounds per minute and W is total weight of catalyst within the reactor?

Answer: Assume, as a model, that the bed operates continuously with a constant feed of red-colored particles entering and leaving of F lb/hr. The bed initially contains only red particles. The solid feed is now suddenly switched to blue particles, which but for color are identical in all respects to the red particles. Let "x" be the weight fraction at any moment of red particles in the solid stream leaving the reactor. By material balance on the red particles over any differential time, interval $d\theta$ after changeover:

$$\text{Input} - \text{Output} = \text{Accumulation}$$

or

$$0 - Fx\,d\theta = d(Wx)$$

The term $d(Wx)$ is identical with $W\,dx$, since W is constant, so by rearranging and integrating:

$$\theta = \frac{W}{F}\ln\left(\frac{x_0}{x}\right) \tag{9-2}$$

where x_0 and x are the concentrations of red particles in the bed and in the solid overflow at the moment of changeover to blue particles and at time θ, respectively. It is obvious that in this example x_0 and x are both unity at changeover, when θ is zero.

The red particles present at θ min are the only particles which have been in the bed for θ min or more, because feeding of red particles was stopped θ min ago. If the bed (and overflow) is to contain only 30% of red particles after 10 min, then x is 0.3 when θ is 10, and upon substituting into Eq. (9-2), W/F is 8.3 min. Therefore, if bed dimensions are so chosen that W/F is 8.3 min, then only 30% of the particles present at any instant have been in the bed for 10 min or more.

EXAMPLE 9-3

At elevated temperature, pure solid A in granular form is known to undergo a gradual phase change to B, and the rate of this phase change is studied by fluidizing a batch of A in an inert hot gas and analyzing from time to time. In any batch operation, of course, all bed particles have the same residence times and temperature-composition history, regardless of particle size, etc.

For isothermal operations, the reaction rate expression was found to be

$$-\frac{dx}{d\theta} = kx$$

where x is the mole fraction (or weight fraction) of A in the solid particles at time θ.

This bed is next operated continuously and in steady state at a constant temperature. Solid A is fed, and solid product is withdrawn, at F lb-min, while the bed weight remains constant at W lb. The solid product and the bed itself obviously have the same average composition x_w, which is determined by analysis of a sufficiently large sample of the bed or overflow material. Develop a general expression relating x_w to F and W.

Answer: In this steady state operation, from Eq. (9-2), the fraction of particles present for θ min or more is

$$f = e^{-F\theta/W}$$

The weight fraction of A in any particle present in the bed for θ min is

$$\frac{x}{x_0} = e^{-k\theta}$$

Combining the foregoing two equations, and recognizing that x_0 is unity in this case, the fraction particles present whose composition is x or less is

$$f = x^{F/kW}$$

The size of the fraction of these particles whose composition lies between x and $(x + dx)$, and so has an average composition of x, is

$$df = \frac{F}{kW}\{x^{(F/kW-1)}\}\,dx$$

By definition, x_w may be written

$$x_w = \int_0^{1.0} x\, df$$

Combining the foregoing two equations and integrating:

$$x_w = \frac{F}{kW} \int x(x^{F/kW-1})\, dx$$

or

$$x_w = \frac{F}{kW} \left(\frac{x^{F/kW+1}}{F/kW + 1} \right) \Bigg]_{x=0}^{x=1}$$

Substituting limits as indicated,

$$x_w = \frac{1}{1 + kW/F} \qquad (9\text{-}3)$$

The difference between feed composition and x_w in the bed and in the product stream becomes smaller as (kW/F) diminishes in size.

Liquid-solid-gas reactors. Reactions are sometimes encountered in which gaseous, liquid, and solid phases are simultaneously involved. The reaction vessels used are generally modifications of those already discussed, as illustrated in the following examples.

(a) Bubble reactor. Ammonium sulfate is made in the by-product coke industry by bubbling coke over gas containing about 1% ammonia up through an aqueous solution of sulfuric acid saturated with ammonium sulfate. This liquid, which contains freshly precipitated ammonium sulfate crystals in suspension, is violently agitated by gas blowing up through the vessel. Solid ammonium sulfate precipitates continuously, and is recovered by cycling the slurry through a filter. Gas enters the reactor, called a "saturator," through an open pipe whose end is submerged several inches below the surface of a pool of acid. Despite the short contact time of the gas as it bubbles up through the liquid phase, ammonia removal is substantially complete because of high reaction and mass transfer rates. It is noteworthy that no special effort is made to produce small bubbles, other than that the gas entry pipe may have a sawtooth end. Heat of reaction is effectively removed by volatilization of water from the liquid phase in the saturator into the tail gas stream.

(b) Tower reactors. Bubble plate or packed towers are used only occasionally for gas-liquid-solid reactions because of plugging problems caused

by solid settling. Thus, tower reactors are used in the Solvay industry for precipitating sodium bicarbonate from an ammoniacal solution of sodium chloride in water by contact with gases containing CO_2. The "equilibrium limited" reaction is carried out countercurrently in a plate column. As in many precipitations, the freshly formed solid, sodium bicarbonate in this case, tends to stick to the internal tower surfaces, and to plug all small holes. Thus, each plate is provided with a single large bubble cap instead of a multiplicity of smaller caps, to minimize the number of small size channels. Each precipitation tower is taken offstream on a fixed schedule for removal of deposited sodium bicarbonate by washing with hot water.

Fluid bed. A fluid bed can in some cases be used for a gas-liquid-solid reaction. Thus, a strong solution of aluminum nitrate may be sprayed continuously into a bed of Al_2O_3 particles fluidized by hot combustion gases. The solution droplets coat onto the bed particles, but the amount of solution present at any moment is so small that drying occurs without sticking. The aluminum nitrate so deposited decomposes to nitrogen oxides and Al_2O_3, the latter remaining as a coating on bed particles already present. Alumina particles are drawn off to maintain a bed of constant mass, while nitrogen oxides leave in the off-gases for recovery and reuse (21).

Spray chamber. Spray chamber operations involve "atomization" of a liquid into a hot gas stream. Thus, the aluminum nitrate decomposition just discussed under fluid bed operation is also carried out in spray chambers. The fine solid particles produced are carried along in the moving gas stream, and recovered in a cyclone or settling chamber of some type. The solid product particles are often less dense than those from fluid bed operation. Ferric chloride solutions are similarly converted to iron oxide and HCl in spray chamber operations, using hot combustion gases directly as a source of heat. The droplets dry and then undergo oxidation and hydrolysis while suspended in the gas phase. The iron oxide particles are recovered by cyclones or gas filters and HCl is removed from the off-gases by water scrubbing.

Moving belt reactors. In some reactors, moving endless belts are used to carry a layer of solids continuously through a reactor space. Countercurrent operations and controlled uniform residence times for the solids are readily attained, and sticky solids can be handled. Belt maintenance and corrosion problems are often severe, however, and heat transfer other than by radiation is poor. Gas-solids contacting is good only when the gas is blown up or down through the layer of solids on the belt.

The so-called Dwight-Lloyd machine, a moving belt reactor, is used in the metallurgical industry when it is desirable to sinter finely divided ores into larger porous chunks for easy use in subsequent processing. Thus, in copper smelters, a layer of sulfide ores 2–3 in. thick is continuously charged onto the moving belt of a Dwight-Lloyd machine, ignited by passing under a gas flame, and then partially oxidized to SO_2 and metal oxide by air sucked down through the charge and the belt. Sufficient quantities of heat develop so that the reacting ore fuses slightly, forming porous masses of solid "sinter." Belt speeds and ore depths are so adjusted that reaction is completed as the ore reaches the discharge end of the belt.

Consideration of the foregoing discussion makes it evident that the reactor is perhaps the least standardized of all types of industrial equipment. The reactors described in the preceding paragraphs represent only a few of the many different designs encountered in industry. This diversity of design is needed to solve industry's wide variety of materials handling problems and rate problems.

Reactions Having an Unfavorable ΔF° 10

As discussed earlier, two issues are important in the initial study of an unfamiliar chemical reaction. First, the temperature range is established over which equilibrium becomes favorable. Second, this temperature range is explored to determine where the kinetics of the reaction become favorable. When rates are found to be poor at all conditions of favorable equilibria, then a catalyst search is intiated, with an increasing chance of success as theoretical understanding of catalysis advances. Reactions are known, however, for which no temperatures exist at which rates and equilibria are simultaneously favorable, and others for which equilibria remain unfavorable at all industrially attainable conditions of temperature and pressure. Such reactions are not rare in the organic field, but are perhaps more often inorganic, and represent a special challenge to the industrial chemist. Some techniques for dealing with such situations are now discussed.

Free energy versus work. As already discussed, a reaction's equilibrium constant is "favorable" when its magnitude is greater than unity, but rapidly becomes unfavorable as this constant falls below unity. By Eq. (1-9), a value for K of less than unity corresponds to a positive value of ΔF°, the reaction's standard free energy change at the temperature in question. The quantity ΔF° represents the isothermal work involved when a reaction goes to completion under reversible conditions, with both the reactants originating and the products obtained pure at 1 atm pressure. A positive value of ΔF°

shows the reversible isothermal work which must be supplied to make the reaction proceed under these idealized conditions, while a negative ΔF° denotes the isothermal, reversible work which could be generated by the reaction.

It was Van't Hoff who first showed, by use of his clever equilibrium box, that any chemical reaction could be made to go to completion, regardless of the magnitude of its equilibrium constant. To accomplish this, it was, of course, necessary to supply or remove work from the reacting system, with the magnitude of this work requirement being related to K by Eq. (1-10). Unfortunately, despite its enormous usefulness in theoretical analysis, the Van't Hoff box is not a practical device for carrying out reactions, largely because the semipermeable membranes required for box operation are not available for most materials. It has been necessary, therefore, to find other methods of supplying free energy to a reacting system of unfavorable ΔF°. Thus, free energy is supplied electrically by electrolysis, mechanically by high or low pressure operations, and chemically by "reaction coupling" and by the "Solvay technique," all of which are now discussed.

Electrolysis. By operating in an electrolytic cell of proper design, a chemical reaction can be made to proceed to completion despite so large a positive value for ΔF° that equilibrium conversion at the temperature and pressure in question is negligible. Thus, the room temperature reaction of sodium chloride with water to form caustic soda, chlorine, and hydrogen has a very unfavorable value of ΔF°:

$$2NaCl + 2H_2O = 2NaOH + H_2 + Cl_2 \qquad \Delta F^\circ_{25^\circ C} = 102 \text{ kcal} \quad (10\text{-}1)$$

This reaction can, nevertheless, be made to proceed electrolytically at 25°C with excellent yields, the necessary deficit in free energy being supplied directly, if somewhat inefficiently, as electrical energy. Similarly, the electrolysis of water (to which caustic soda has been added to improve conductivity) can be successfully carried out at 25°C, as follows:

$$H_2O(l) = H_2(g) + \tfrac{1}{2}O_2(g) \qquad \Delta F^\circ_{25^\circ C} = 56.7 \text{ kcal} \qquad (10\text{-}2)$$

Again, molten magnesium chloride can be electrolyzed to magnesium metal and chlorine gas at elevated temperatures:

$$MgCl_2(l) = Mg(l) + Cl_2(g) \qquad \Delta F^\circ_{1000^\circ C} = 110 \text{ kcal} \qquad (10\text{-}3)$$

While these examples show electrolysis to be a powerful tool for causing reactions to go to completion despite an unfavorable value for ΔF°, yet the technical requirements for successful operations in a cell unfortunately cannot be fulfilled by most reactions. As will be discussed in more detail in

Chap. 11, electrolysis can occur only when reactants and products can be made to participate in two complementary and more-or-less reversible half-cell reactions, whose sum equals the desired overall reaction. This difficult requirement eliminates many inorganic reactions and most organic reactions, leaving only a relatively small number suitable for electrolysis.

Economics places a second severe limitation upon industrial electrolysis. That is, relatively low-cost, nonelectrolytic processes which can replace electric cell operations are, in most cases, known and operated, and only a handful of electrolytic operations has survived this competition, e.g., in caustic soda-chlorine, aluminum, magnesium, and zinc production. The electrorefining of copper and the limited electrolytic production of materials like permanganates and hydrogen peroxide also deserve mention. It is possible that the large present research effort devoted to the fuel cell may strengthen the competitive position of electrolytic processes, uncover new fields, and bring about further growth in industrial electrolysis. Further stimulus may also come if, as seems likely, the cost of electrical energy continues to fall relative to the price level.

Elevated pressure. Operation at pressures above atmospheric tend to be costly, consequently cannot often be justified. Elevated pressure is usually used either because of beneficial effects upon equilibrium or reaction rate, or because it makes possible the maintenance of a liquid phase. Discussion here focuses primarily upon the effect of pressure on equilibrium conversion.

Pressure versus equilibrium. Both the standard free energy change and the equilibrium constant for any reaction are, of course, functions of temperature alone, independent of total pressure. The equilibrium conversion attained at a given temperature in a homogeneous gas reaction, on the other hand, varies with total pressure, providing there is a change in the number of moles as reaction proceeds. Thus, when pressures are low enough so that the perfect gas laws apply, the equilibrium constant expression of Eq. (1-6) for gas reaction (1-1) may be restated as follows:

$$K_p = \frac{y_M^m \cdot y_N^n}{y_A^a \cdot y_B^b \cdot y_C^c} \pi^\delta \tag{10-4}$$

where y is the mole fraction, π is the total pressure, and δ is defined by Eq. (1-7) as the sum of the algebraic coefficients of reaction (1-1):

$$\delta = (m + n - a - b - c) \tag{10-5}$$

In this chapter, δ is always stated per mole of desired product M, in which case m is taken as unity. By inspection of Eq. (10-4), it is evident that an isothermal increase in π will increase the mole fractions of M and N in the equilibrium gas mixture when δ is negative, decrease M and N when δ is positive, and have no effect when δ is zero.

Reactions with real gases. It was pointed out in Chap. 1 that in the real gas region, the equilibrium constant for reaction (1-1) should be expressed in terms of fugacities:

$$K = \frac{f_M^m \cdot f_N^n}{f_A^a \cdot f_B^b \cdot f_C^c} \tag{10-6}$$

The Lewis and Randall rule states that f_A, the fugacity of component A in the mixture, is related to $f_{\pi A}$, the fugacity of this component at the same temperature and total pressure of the mixture, as follows:

$$f_A = f_{\pi A} y_A \tag{10-7}$$

This rule is subject to error, especially near the critical temperature and pressure of any of the components, but is used here to simplify discussion. Combining the foregoing two relations, and recognizing that the activity coefficient, γ_A, is defined as $f_{\pi A}/\pi$, then

$$K = K_\gamma \frac{y_M^m \cdot y_N^n}{y_A^a \cdot y_B^b \cdot y_C^c} \pi^\delta \tag{10-8}$$

where, by definition,

$$K_\gamma = \frac{\gamma_M^m \cdot \gamma_N^n}{\gamma_A^a \cdot \gamma_B^b \cdot \gamma_C^c} \tag{10-9}$$

Comparison of Eqs. (10-4) and (10-8) shows that conversions in real gas systems differ from those attained in ideal gas systems by the presence of the term K_γ, defined in Eq. (10-9). When the total presssure is small so that the perfect gas laws are obeyed, then the activity coefficients and consequently K_γ approach unity, and Eq. (10-8) reduces to the perfect gas Eq. (10-4). Even at elevated pressure, the magnitude of K_γ remains in the vicinity of unity, usually lying between about 0.5 and 2.0. It again follows from Eq. (10-8) that, as with ideal gases, isothermal increases in pressure increase equilibrium conversion in reactions between real gases when δ is negative, reduce conversion when δ is positive, and have little effect when δ is zero.

Ammonia synthesis from the elements is the classical example of the use of high pressure to increase conversion. The standard free energy of formation of ammonia at 25°C is —4.0 kcal/g mole, and operations at this temperature would show a quite favorable equilibrium conversion even at atmospheric pressure. Unfortunately, commercial catalysts kindle at the relatively high temperatures of about 325°C, while operations are run typically at 450°C and higher. The standard free energy of formation of ammonia is +7.2 kcal/g mole at 450°C, corresponding to an equilibrium constant for the synthesis reaction of 6.5×10^{-3}. Operation at elevated pressure is undertaken to offset the low equilibrium conversion resulting from atmospheric pressure operation at 450°C.

Free energy gain by pressure. The standard free energy change of a reaction involving ideal gases is related to its equilibrium constant K by Eq. (1-10). Combining with Eq. (10-4),

$$\Delta F^\circ + \delta RT \ln \pi = -RT \ln \frac{y_M^m \cdot y_N^n}{y_A^a \cdot y_B^b \cdot y_C^c} \tag{10-10}$$

A moment's reflection will show that the term on the right-hand side of the foregoing equation is the isothermal free energy change when both the reactants originate and the products finish at the same pressure as in the reactor itself, namely π atm. The term ($\delta RT \ln \pi$) represents the algebraic sum of the isothermal free energy changes when separately compressing each of reagents A, B, and C from 1 to π atm, and then separately expanding each of the products M and N from π to 1 atm. In the perfect gas region, therefore, the contribution made by total pressure to the free energy change of a reaction per mole of product M produced is represented by the term ($\delta RT \ln \pi$). For atmospheric pressure operation, of course, this term ($\delta RT \ln \pi$) has a value of zero.

It is easily shown that, in the real gas region, combination of Eqs. (1-2) and (10-8) yields

$$\Delta F^\circ + (\delta RT \ln \pi + RT \ln K_\gamma) = -RT \ln \frac{y_M^m \cdot y_N^n}{y_A^a \cdot y_B^b \cdot y_C^c} \tag{10-11}$$

where the terms in parentheses are now the reversible, isothermal work of compressing the reactants and expanding the products between 1 and π atm. When term $RT \ln K_\gamma$ is small enough to be neglected relative to $\delta RT \ln \pi$, then Eqs. (10-10) and (10-11) are identical.

The contribution made by term $\delta RT \ln \pi$ to a chemical reaction's free energy needs can be illustrated by the ammonia synthesis. It will be remembered that at 450°C, where ΔF° of formation is +7.2 kcal, equilibrium conversion to ammonia in a nitrogen-hydrogen mixture is vanishingly small at atmospheric pressure. At 500 atm, however, term $\delta RT \ln \pi$ equals $(-1) \cdot (1.99)(450 + 273)(10^{-3})(\ln 500)$, or -9 kcal, consequently, the left side of Eq. (10-11) becomes $(+7.2 - 9)$ or -1.8 kcal. Equilibrium conversion to ammonia is easily shown to be quite favorable for this net negative value of the free energy change.

An estimate can be made of the maximum value of term $\delta RT \ln \pi$, as attainable in present commercial practices, when δ is minus unity. The extremes of temperature and pressure which could be used industrially are, of course, dictated by the materials of construction available, but might be fixed at 1000 atm and 1200°K. Such conditions, when incorporated into term $\delta RT \ln \pi$, yield a value of $(-1)(1.99)(1200)(10^{-3})(\ln 1000)$ or -16.5 kcal. This, then, is the maximum contribution which pressure can be expected to make to the free energy needs of a reaction, consequently, when δ is minus unity, pressure cannot benefit equilibrium conversion of reactions for which ΔF° values are much larger than 16.5 kcal.

Heterogeneous reactions. As with homogeneous reactions, elevated pressure will benefit heterogeneous reactions when there is a reduction in the number of moles in the gas phase. This generalization applies, regardless of the number of moles of solids or liquids formed or disappearing. Thus, the equilibrium constant Eq. (2-4) for the heterogeneous reaction (2-2) can be converted directly to the following form by use of the perfect gas law:

$$K_{het} = \frac{y_N^n}{y_B^b \cdot y_C^c} \pi^{\delta'} \qquad (10\text{-}12)$$

where δ' now refers to the change in the number of moles in the gas phase in reaction (2-2), and equals $(n - b - c)$. Inspection shows that when δ' is negative, conversion must increase with pressure.

Since pressure has a favorable effect upon the equilibrium of any reaction in which there is a reduction in volume, homogeneous liquid phase reactions showing a volume shrinkage also benefit from pressure operation. At most, however, volume changes in liquid systems due to reaction are very small, so that even enormous changes in pressure have only a small effect on equilibrium composition. The same generalizations apply to homogeneous solid-phase reactions, consequently, with few exceptions pressure is applied industrially to effect equilibrium conversion only when a gas phase is present.

It is interesting to note that high pressure research today is being pushed to several million psi and to temperatures over 1000°C. This frontier is of potential commercial importance for effecting phase transformations in solids leading to the production of new materials with novel properties. The recently developed diamond synthesis is a dramatic example of activity in this area.

Pressure to maintain a liquid phase. It was stated earlier that pressure operations are resorted to for reasons other than to improve equilibrium conversion, such as to maintain a liquid phase needed either as a reaction medium, or to transport products as a slurry. Thus, when extracting nickel oxide from Cuban laterites with aqueous ammonia, operations are carried out at 150°C to attain reasonable rates of reaction. At this temperature, the vapor pressure is considerably in excess of 1 atm, consequently elevated pressure operation is necessary to maintain the liquid phase required for dissolving the ammonia-nickel complexes away from the large bulk of inert ore constituents.

Another reason for maintaining a liquid phase is to keep a dissolved catalyst in solution, or to hold the solution at the proper concentration. Thus, in the Shell process for ethanol production, ethylene reacts with water at 570°F using 85% of phosphoric acid as catalyst. The vapor pressure of water over 85% phosphoric acid at 570°C is about 35 atm and, if the partial pressure of water is allowed to exceed 35 atm at this temperature, the acid becomes too dilute to support reasonable rates of reaction, while at the lower partial pressures corresponding to higher acid strengths, undesired side reactions become important. Controlled operation at elevated pressure is obviously essential in this process.

Pressure versus reaction rate. Rates of homogeneous gas reactions are often markedly influenced by pressure. An interesting example is the oxidation of nitrous oxide, which proceeds homogeneously as follows:

$$2NO + O_2 = 2NO_2 \tag{10-13}$$

The rate of NO disappearance, expressed as usual in mole/(unit volume) (unit time), is written

$$r_{NO} = -kp_{NO}^2 \cdot p_{O_2} \tag{10-14}$$

It is easy to show that, as a consequence of this expression, the time for the attainment of a given degree of conversion varies inversely as the square of the total pressure. Because of this fact, absorption towers for NO_2 in nitric acid plants are often operated at pressures as high as 8 atm.

The kinetics of heterogeneous reactions between gases and solids are likewise often benefited by pressure, since the collision frequency of gas molecules on a solid surface is directly proportional to pressure. The problems of charging and discharging solids in pressurized systems have been obstacles to the application of pressure in gas-solid reactions, but headway is being made, notably in current work on coal gasification with hydrogen.

The kinetics of gas-liquid reactions may also be benefited by pressure, since an increase in pressure increases the concentration of gases in the liquid phase where rates are sometimes greater than in the gas phase.

Low pressure operation. Just as high pressure increases the equilibrium conversion of a reaction involving a net decrease of moles in the gas phase, subatmospheric pressure increases conversion when there is a net increase in a reaction's total moles in the gas phase. Vacuum, like elevated pressure, is costly and is resorted to only if atmospheric pressure operation is for some reason disadvantageous or impractical.

Homogeneous gas reactions. Reduced pressure processes are operated on a large scale for reactions like the dehydrogenation of paraffins to olefins and diolefins. Thus, 1,3-butadiene is made by the catalytic dehydrogeneration of butanes and butenes at about 0.2 atm pressure and 1100°F, as follows:

$$nC_4H_{10} = nC_4H_8 + H_2 \qquad K_p = 0.176 \tag{10-15}$$

$$nC_4H_8 = C_4H_6 + H_2 \qquad K_p = 0.167 \tag{10-16}$$

These reactions are endothermal, hence, their equilibrium constants improve with increasing temperature. For atmospheric pressure operation, equilibrium conversion to butadiene is small up to 1100°F, and higher temperature operations are disadvantageous because of extensive cracking. Since δ is positive for these reactions, equilibrium conversions are obviously improved by operating at subatmospheric pressure, with the further benefit in this case that the ratio of butadiene to butene can be controlled by change of pressure (75).

Low partial pressures are obtained industrially in one of two ways. The first, involving the use of vacuum pumps to reduce the total system pressure, requires extensive investment in equipment and precautions to avoid air leaks, especially when working with hydrocarbons at high temperature. The second way to reduce partial pressures is by diluting the reagents with an inert gas, steam being widely used in the petroleum industry, since it is easily separated from hydrocarbons by condensation and decantation. By use of an inert gas, the total system pressure can be maintained at atmospheric, without losing the benefits of high equilibrium conversions attributable to low partial pressure. The contribution made by the enthalpy of the inert gas is often a further advantage when operations are adiabatic.

Heterogeneous systems. Vacuum is used to obtain larger equilibrium yields not only for gas reactions having a positive value for δ, but also for heterogeneous reactions producing an increased number of moles in the gas phase. Thus, in the Pidgeon (115) process for magnesium metal production, dolomite ($CaCO_3 \cdot MgCO_3$) is calcined to the oxides, ground, and mixed with finely divided ferrosilicon (over 75% of silicon), and then heated batchwise in evacuated ceramic retorts. Operations are at 1150°C, where the dissociation pressure of magnesium is roughly 10 mm for the following reaction (80):

$$2(CaO \cdot MgO)(c) + Si(c) = 2Mg(g) + Ca_2SiO_4(c) \tag{10-17}$$

Ceramic retorts collapse under vacuum at above 1150°C, while the dissociation pressure of magnesium over this system falls off rapidly to impractically small values as temperature is reduced.

The contribution to the free energy change of the generalized heterogeneous reaction (2-2) resulting from an increase or decrease in pressure is derived by combining Eqs. (10-12) and (1-10):

$$\Delta F^{\circ}_{\text{het}} + \delta' RT \ln \pi = -RT \ln \frac{y_N^n}{y_B^b \cdot y_C^c} \tag{10-18}$$

where δ' is the increase in moles within the gas phase. The similarity of Eqs. (10-10) and (10-18) should be noted.

The magnitude of the free energy contribution which can be made by vacuum operation is well illustrated by the Pidgeon reaction in Eq. (10-17). For this reaction, δ' has the value of plus unity, consequently, at 1150°C and 10 mm of mercury, the term ($\delta' RT \ln \pi$) becomes $(+1)(1.99)(1150 + 273)(10^{-3}) \ln (\frac{10}{760})$ or -12.3 kcal. In industrial operations, it seems unlikely that temperatures much in excess of 1150°C can be used at present, but total pressures might be lowered to perhaps 1 mm, whereupon the term ($\delta' RT \ln \pi$) becomes (-18.5 kcal), which represents the maximum contribution to free energy which can be expected from vacuum operation when δ' is minus unity.

Reaction coupling. The equilibrium yields of some chemical reactions can be improved by a technique which is called here reaction coupling, whereby a by-product of the first reaction is consumed in a second reaction. In order for this second reaction to occur, a new reagent must be introduced into the system, and for yields to improve, this second reaction must involve a relatively favorable free energy change. By way of illustration, consider the production of M by the following reaction having an unfavorable ΔF°:

$$aA + bB = M + nN \tag{10-19}$$

A second reaction having a favorable ΔF°, involving by-product N and an auxiliary reagent C, is as follows:

$$cC + nN = P \tag{10-20}$$

Upon mixing A, B, and C, reaction (10-20) will couple with (10-19) in that by-product N produced in the first of these reactions will be consumed in the second. The overall reaction is the sum of Eqs. (10-19) and (10-20), or

$$aA + bB + cC = M + P \tag{10-21}$$

If ΔF° for reaction (10-20) is a large negative number, then the equilibrium yield of M will be greater by reaction (10-21) than when reaction (10-19) occurs alone.

The use of reaction coupling in industry is illustrated by the reduction of metal oxides with carbon. Thus, the dissociation of FeO to the elements shows an unfavorable standard free energy change at 1000°K:

$$FeO = Fe + \tfrac{1}{2}O_2 \qquad \Delta F^\circ_{1000^\circ K} = 47.55 \text{ kcal} \qquad (10\text{-}22)$$

Reaction of carbon with oxygen to form CO, however, has a favorable standard free energy change:

$$C + \tfrac{1}{2}O_2 = CO \qquad \Delta F^\circ_{1000^\circ K} = -47.94 \text{ kcal} \qquad (10\text{-}23)$$

Upon mixing FeO with carbon and heating, reactions (10-22) and (10-23) couple, with the resultant overall reaction becoming

$$FeO + C = Fe + CO \qquad \Delta F^\circ_{1000^\circ K} = -0.39 \text{ kcal} \qquad (10\text{-}24)$$

The values of ΔF° for the foregoing three reactions are shown graphically in Fig. 10-1. Inspection shows that ΔF° for reaction (10-24) becomes zero at

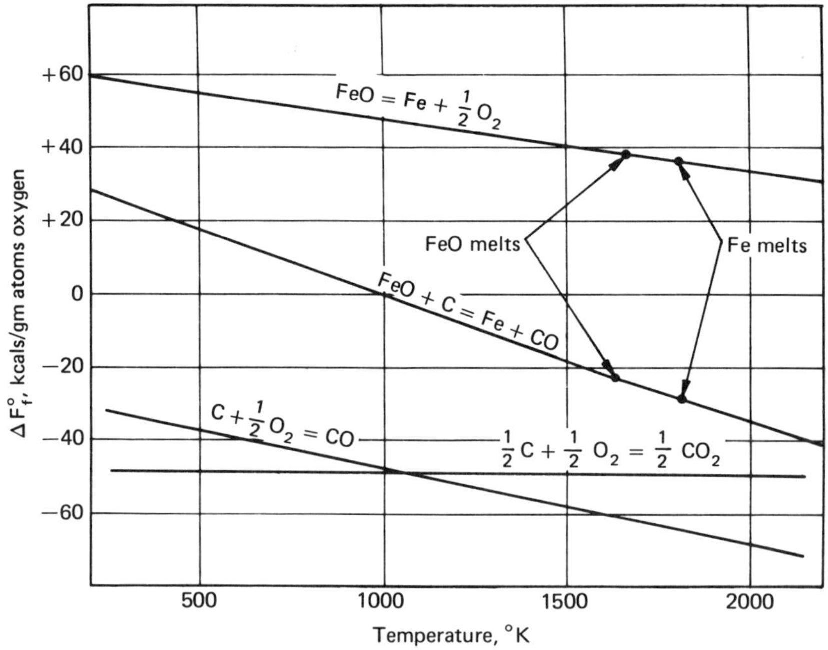

Fig. 10-1. Standard free energy changes vs. temperature for iron reduction reactions.

about 980°K, at which temperature the dissociation pressure of CO over the system Fe-FeO-C must, therefore, reach 1 atm. This dissociation pressure for CO obviously becomes larger rapidly as the temperature is further increased.

EXAMPLE 10-1

When reducing FeO with carbon in practice, the following reaction occurs simultaneously with reaction (10-24):

$$2FeO + C = 2Fe + CO_2 \qquad \Delta F°_{1000°K} = +0.49 \text{ kcal} \qquad (10\text{-}25)$$

Obviously, reaction (10-25) is a sum of two sequential reactions, namely, reactions (10-22) and (10-26):

$$C + O_2 = CO_2 \qquad \Delta F°_{1000°K} = -94.61 \text{ kcal} \qquad (10\text{-}26)$$

Reaction (10-25) again has a much more favorable free energy than reaction (10-22) because of the large (negative) magnitude of $\Delta F°$ for reaction (10-26). When heating FeO and carbon at 1000°K, what is the total pressure and the gas composition when equilibrium is attained in the presence of these solid phases?

Answer: For reaction (10-24) $\ln p_{CO} = -\Delta F°/RT = 390/(1.987 \times 1000)$, consequently, $p_{CO} = 1.2$ atm.

For reaction (10-21) $\ln p_{CO_2} = -\Delta F° = RT = -490/(1.987 \times 1000)$ consequently, $p_{CO_2} = 0.78$ atm. The total pressure is, therefore, $(1.217 + 0.78)$ or 2 atm.

Charts of free energies of formation versus temperature for the metal oxides, chlorides, nitrides, carbides, etc., such as Fig. 10-1, are now widely published.

Characteristics of coupled reactions. The use of an auxiliary material like carbon to cause a coupled reaction to occur involves added expense for raw material. For this reason, the auxiliary reagent is preferably air or a low-cost fuel, although other auxiliary reagents are also utilized. Obviously, kinetics of both the main reaction and the secondary reaction must be favorable for operations to be successful. The coupled reaction forms secondary products, such as the CO and CO_2 of reactions (10-20) and (10-21), which are often of low value or may even represent a disposal problem. Three further examples of coupled reaction systems are presented below:

1. Deacon reaction. Large quantities of by-product HCl are obtained in organic chlorination reactions. One possible means of disposal of this HCl would be by reconversion to elemental chlorine, but the simple decomposition of HCl to its elements has an unfavorable equilibrium constant over the normal temperature range. Thus, at 900°K,

$$HCl = \tfrac{1}{2}H_2 + \tfrac{1}{2}Cl_2 \qquad \Delta F°_{900°K} = 23.9 \text{ kcal}$$

The free energy change for hydrogen oxidation to water, on the other hand, is favorable at these temperatures:

$$\tfrac{1}{2}H_2 + \tfrac{1}{4}O_2 = \tfrac{1}{2}H_2O \qquad \Delta F°_{900°K} = -23.4 \text{ kcal}$$

Thus, by mixing air or oxygen with HCl, these reactions couple, yielding an overall reaction called the Deacon reaction, which is the sum of the two preceding ones:

$$HCl + \tfrac{1}{4}O_2 = \tfrac{1}{2}Cl_2 + \tfrac{1}{2}H_2O \qquad \Delta F°_{900°K} = 0.5 \text{ kcal} \tag{10-27}$$

The equilibrium constant of reaction (10-27) becomes more favorable as temperature falls, having a value of unity at about 870°K. Unfortunately, reaction rates also diminish rapidly with decreasing temperature, and available catalysts, such as copper chloride, are ineffective much below 350°C.

Air is usually used as the oxygen source in the Deacon reaction, but pure oxygen appears preferable despite its high costs because of the increased ease of chlorine recovery from nitrogen-free off-gases. Disposal of the secondary product, namely water, presents no difficulty in this case.

2. Isoprene. Dehydrogenation of isopentane to isopentene and isoprene proceeds under vacuum, under conditions and limitations similar to those of reactions (10-15) and (10-16) for dehydrogenating butanes to butenes and butadiene. Equilibrium conversions are small at atmospheric pressure up to 600°C, and higher temperature operation results in severe cracking. Oxidative dehydrogenation, such as by the following reactions, results in improved yields at atmospheric pressure:

$$\text{iso } C_5H_{12} = nC_5H_{10} + H_2 \tag{10-28}$$

$$H_2 + \tfrac{1}{2}O_2 = H_2O \tag{10-29}$$

$$\text{iso } C_5H_{12} + \tfrac{1}{2}O_2 = nC_5H_{10} + H_2O \tag{10-30}$$

Selective catalysts found to date for these reactions are not as yet entirely satisfactory (74).

3. Phosphoric acid. The production of wet-process phosphoric acid is accomplished by reacting sulfuric acid with the ore appatite, which is a complex, highly insoluble calcium fluorosilicate. During this operation, fluorine is volatilized as hydrogen fluoride, calcium sulfate precipitates, and the various impurities go into solution for subsequent removal from the phosphoric acid formed. Sulfuric acid is used here as supplementary reagent because appatite will obviously not hydrolyze directly to $Ca(OH)_2$ and phosphoric acid. The secondary product formed in this case is calcium sulfate, which has little value.

Solvay technique. Ernest Solvay in 1860 developed a procedure for making soda ash from salt and limestone, whereby the following overall reaction (showing an unfavorable free energy change at all reasonably attainable operating conditions) is nevertheless made to occur:

$$2NaCl + CaCO_3 = CaCl_2 + Na_2CO_3 \qquad \Delta F^\circ_{25^\circ C} = +9.6 \text{ kcals} \qquad (10\text{-}31)$$

Solvay's process involves separation of the main reaction (for which no conditions exist at which rates and equilibria are simultaneously favorable) into a series of sequential reactions, whose algebraic sum equals the main reaction. These reactions are so chosen that it is possible to find for each an industrially attainable temperature and pressure at which both equilibrium and rate are simultaneously favorable. Since the operating conditions for each reaction are different than for the next one, each reaction is usually carried out in a different piece of apparatus.

This general approach, called here the Solvay technique, has been found applicable not only to soda ash production, but also to many other reactions of unfavorable free energy. A relatively simple example is the production of carbon tetrachloride from coke and chlorine, which upon first consideration appears feasible since the equilibrium constant for this reaction is greater than unity below about 400°C. Unfortunately, the rates of this direct com-

TABLE 10-1. CARBON TETRACHLORIDE VIA CARBON DISULFIDE

	ΔF°, kcal	
	25°C	1023°C
$C + S_2(g) = CS_2(g)$	+15.5	−25.0
$2Cl_2 + \frac{2}{3}CS_2(g) = \frac{2}{3}CCl_4(g) + \frac{2}{3}S_2Cl_2(g)$	−30, approx.	large
$\frac{2}{3}S_2Cl_2(g) + \frac{1}{3}CS_2(g) = \frac{1}{3}CCl_4(g) + 2S(c)$	−0.8, approx.	large
$C + 2Cl_2 = CCl_4(g)$	−15.3	+16.6

bination are negligible even at temperatures much higher than 400°C and, since no good catalyst is known, synthesis from the elements has not as yet proven practical. Production from the elements is therefore by a three-stage process employing the Solvay technique, with reactions as in Table 10-1. Carbon is first converted to CS_2 by reaction with sulfur vapor at 1000°C in externally-heated iron vessels. The CS_2 is next allowed to react with chlorine at normal temperatures to form CCl_4 and S_2Cl_2. After separation, the S_2Cl_2

is made to react with more CS_2 to form additional CCl_4 and elemental sulfur. Inspection shows that the first three reactions of Table 10-1 sum up to the last reaction, which is the direct synthesis of CCl_4. This reaction scheme fulfills the requirements of the Solvay technique, namely the separation of the main reaction into a series of coupled reactions, each of which proceeds with a good rate and with a favorable equilibrium constant at different but industrially attainable temperatures and pressure.

It might be noted in passing that the fact that carbon cannot be directly chlorinated is disadvantageous in the manufacture of carbon tetrachloride but fortunate in other respects, as when the inertness of carbon makes it the preferred material of construction for apparatus in which corrosion is severe. Thus, when a halogen is evolved at the anode of an electrolytic cell operation, no really satisfactory low-cost electrode material except graphite is known.

Recycle. An important feature of the Solvay technique is the need to recycle at least one of the reagents involved, as with sulfur in the CCl_4 synthesis just discussed. In the United States, of course, CS_2 was sometimes made in one manufacturing establishment and then shipped great distances before conversion to CCl_4 in a second plant. In these cases, the sulfur produced in the third reaction of Table 10-1 was not itself returned to the first plant, but instead was used locally for other purposes. Fresh sulfur was then purchased in the plant of origin to make more CS_2. The introduction of sulfur in this fashion into the recycle loop was a matter of economics, and does not alter the fact that the entire operation may be viewed as an example of the Solvay technique. This carbon disulfide route is no longer very important, since large quantities of carbon tetrachloride are obtained as a coproduct in other chlorination reactions.

Butadiene production. The low-pressure dehydrogenation of *n*-butane to butene and butadiene by reactions (10-15) and (10-16) has already been discussed. At best, conversions per pass obtained in this process are low, and yields are not great because of complicating side reactions which produce undesired by-products whose separation from butadiene is difficult. Improved yields and performance appear in prospect for a Solvay-type scheme, based on the finding that iodine is an effective dehydrogenating agent for paraffins and monoolefins (74). The reaction sequence, as presented in Table 10-2, depends upon an iodine promoted dehydrogenation carried out in the presence of NiO, followed by regeneration of I_2 from the NiI_2 formed by air oxidation.

Reaction (A) of Table 10-2 has a favorable equilibrium constant, but reaction (B) has an unfavorable equilibrium at the low operating temperatures used (about 550°C) to avoid side reactions. Reaction (B) is, nevertheless, forced to go to completion by injection of granular nickel oxide into the system, causing reaction (C) to proceed. Iodine is regenerated by removing the solid NiI_2 product for oxidation with air at over 600°C by reaction (D). Both the I_2 and NiO formed by this reaction are recycled. Obviously, the NiO serves not only to force reaction (C) to completion, but also acts to recover iodine from HI for recycle. Because of high iodine costs, losses of this reagent must be kept to a minimum.

TABLE 10–2. IODINE PROMOTED OXIDATIVE DEHYDROGENATION OF BUTANES

$$C_4H_{10} + I_2 = C_4H_8 + 2HI \quad \text{(A)}$$
$$C_4H_8 + I_2 = C_4H_6 + 2HI \quad \text{(B)}$$
$$2HI + NiO = NiI_2 + H_2O \quad \text{(C)}$$
$$NiI_2 + \tfrac{1}{2}O_2 = NiO + I_2 \quad \text{(D)}$$

Solvay soda process. Reaction (10-31) is the algebraic sum of the six constituent reactions of the Solvay process for making soda ash from limestone and salt, as presented in Table 10-3. The calcining of limestone by reaction (A) is the first step in this operation, followed by slaking to form lime hydrate in reaction (B), which is then added to an aqueous solution chloride and heated to release ammonia as in reaction (C). This ammonia is absorbed in a strong NaCl solution, CO_2 is added, and reactions (D) and (E) occur to precipitate sodium bicarbonate. After recovery by filtration, the bicarbonate is calcinated in reaction (F) to yield the final product, namely, soda ash.

TABLE 10–3. SOLVAY PROCESS REACTIONS

Reaction	Temperature, °C	
$CaCO_3 = CaO + CO_2$	1000	(A)
$CaO + H_2O = Ca(OH)_2$	100	(B)
$Ca(OH)_2 + 2NH_4Cl = CaCl_2 + 2NH_3 + 2H_2O$	120	(C)
$2NH_3 + 2H_2O + 2CO_2 = 2NH_4HCO_3$	60	(D)
$2NH_4HCO_3 + 2NaCl = 2NaHCO_3 + 2NH_4Cl$	60	(E)
$2NaHCO_3 = Na_2CO_3 + H_2O + CO_2$	200	(F)

The flow sheet for the Solvay process, also often called the ammonia-soda process, is shown in Fig. 10-2, and the various pieces of apparatus involved are easily identified with the reactions of Table 10-3. A shaft lime kiln with a mixed feed of coke and limestone is used for reaction (A), while the slaker for reaction (B) may be a rotating drum in which lime is tumbled while being doused with an appropriate amount of water. The ammonia still is a plate column operated with open steam. Since NH_4HCO_3 in aqueous solution decomposes by simply boiling it at atmospheric pressure, reaction (D) is reversed in the top part of this still. Simple boiling will not decompose NH_4Cl

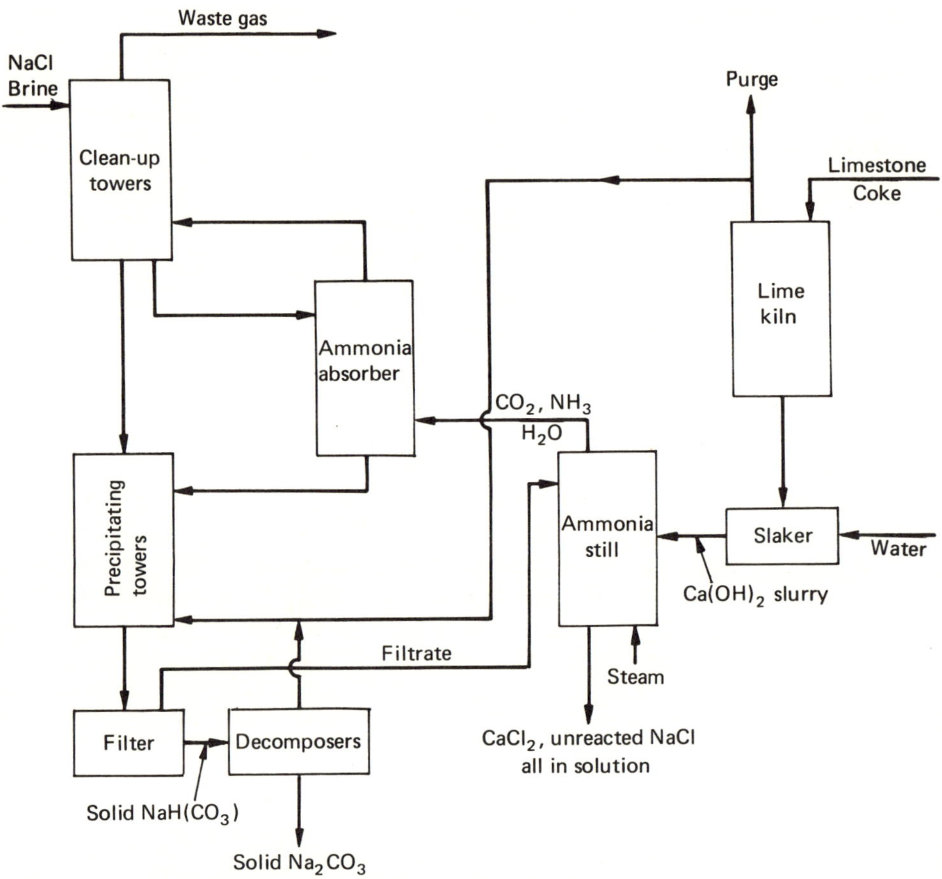

Fig. 10-2. Solvay process flow sheet.

in aqueous solution, consequently, ammonia is recovered from this material by treatment with lime hydrate in the lower half of the ammonia still, causing reaction (C) to occur. The so-called precipitators for reaction (E) are plate columns, with each plate carrying a single large bubble cap. The plates of the precipitators, as already discussed, are designed to reduce the rate at which the column plugs with sodium bicarbonate, which when freshly formed tends to stick to solid surfaces. The ammonia absorbers and cleanup towers serve to minimize ammonia losses from the system. The precipitated sodium carbonate is removed from the slurry leaving the bottom of the precipitating towers on a battery of rotating drum filters, is washed, dried, and calcinated to soda ash in externally heated rotary kilns.

Energy sources in the Solvay technique. Each reaction of a Solvay sequence is carried out at conditions different from the next one and so chosen that $\Delta F°$ for each reaction is favorable. The energy requirements for the overall reaction are supplied by fuel consumed in providing the sensible heats and heats of reaction involved in each step. In Solvay's soda process, the necessary energy comes primarily from the fuel burned in the lime kiln, but also from the steam supplied to the ammonia still, the electrical energy for operating the blowers, etc.

Solvay's technique for dealing with reactions of unfavorable $\Delta F°$ gives the industrial chemist wide scope for ingenuity and invention. A rather complex plant is usually required, however, involving a series of reactors and recycle loops. As always, the final industrial operation and survival of a Solvay-type process depends upon competitive economics.

11 Industrial Electrochemical Operations

Electrolysis, which denotes all those processes occurring when electricity passes through an electrolyte, is carried out in a reaction vessel called a cell. An electrolytic cell in its simplest form, as shown in Fig. 11-1, consists of two electrodes separated by an electrolyte. The current is carried by ions through the electrolyte, and by electrons through the electrodes. When the electrolyte is a liquid, as is usually the case, then a suitable insulating container for the cell elements must be provided. Most cell operations are isothermal and at atmospheric pressure, and may be either batch or flow (45, 119, 156).

A chemical reaction is always made to occur when current passes through a cell, and at least part of the reaction's free energy becomes directly interconvertible with electrical energy. By operating isothermally in a cell of suitable design, reactions with a positive free energy change accept this free energy directly as electrical energy, while reactions having a negative free energy will release this energy directly as electrical energy. In either case, as pointed out in the preceding chapter, the reaction occurring in the cell can be made to proceed to completion regardless of the magnitude of its equilibrium constant. Of course, the amount of electrical energy actually consumed or produced differs from the reaction's free energy because of the cell's operating inefficiencies. Thus, a battery or fuel cell delivers less than the theoretical

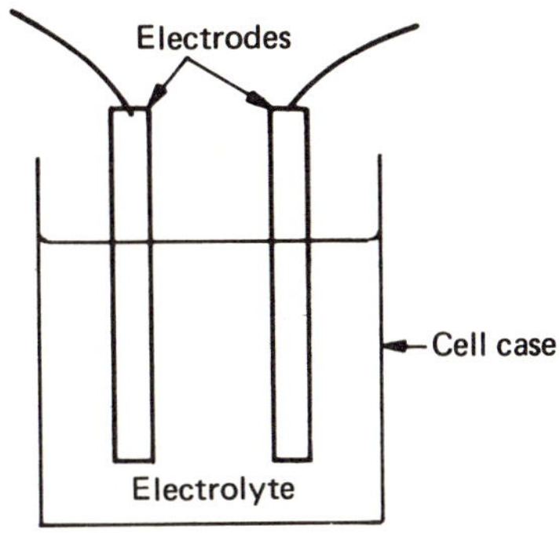

Fig. 11-1. A simple cell.

free energy because of various losses within the cell itself, while the energy requirements of a reaction having a positive free energy change are in practice always greater than the theoretical requirements.

Electrochemical reaction characteristics. The direct interconvertibility of a reaction's free energy with electrical energy can be illustrated by the cell in Fig. 11-2 both by the decomposition and synthesis of HCl from its elements, for which the overall reactions are

$$\mathrm{HCl}\ (1\ N) = \tfrac{1}{2}\mathrm{H_2} + \tfrac{1}{2}\mathrm{Cl_2} \qquad \Delta F_{25^\circ\mathrm{C}} = +31.6\ \mathrm{kcal} \qquad (11\text{-}1)$$

$$\tfrac{1}{2}\mathrm{H_2} + \tfrac{1}{2}\mathrm{Cl_2} = \mathrm{HCl}\ (1N) \qquad \Delta F_{25^\circ\mathrm{C}} = -31.6\ \mathrm{kcal} \qquad (11\text{-}2)$$

Assume one electrode of this cell to be a platinum plate, the other a platinum plate coated with platinum black. These electrodes are connected electrically though a resistance, an ammeter, and a voltmeter as shown. Chlorine gas is bubbled up through the electrolyte (1 N HCl) past the uncoated electrode, and hydrogen gas is bubbled up past the coated electrode, these two gases being kept from mixing by a porous membrane. When the electrical resistance between the electrodes is infinite (corresponding to open circuit) the voltmeter registers 1.36 V. As this resistance is lowered, the voltage falls progressively, while a current is observed to flow whose magnitude is more or less proportional to the reciprocal of this resistance.* The hydrogen electrode is found to be the negative pole in this operation. Reaction (11-2) is observed to occur whenever current flows with part of the H_2 and Cl_2 gases bubbling up past the electrodes being consumed as the HCl content of the electrolyte increases. Under these conditions, the cell obviously functions as a fuel cell, in that electrical energy is generated.

*If no H_2 and Cl_2 gases are supplied when the potential is reduced below 1.36 V, then no current flows.

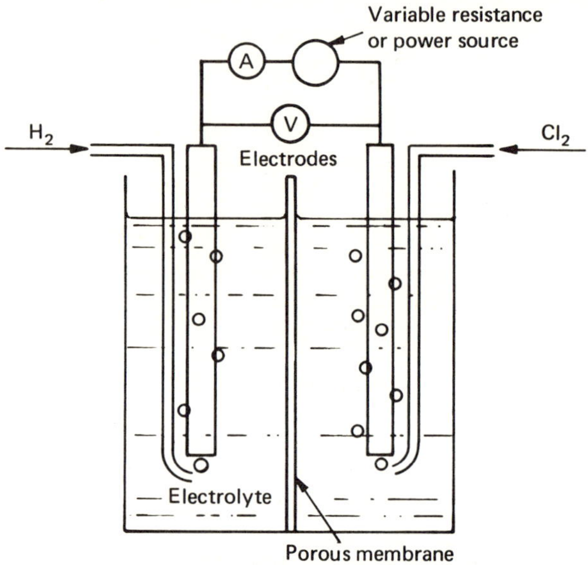

Fig. 11-2. A cell illustrating either electrolysis of HCl to the elements, or a fuel cell operation in which HCl is synthesized from the elements.

Without interrupting gas flow, a potential greater than 1.36 V is applied across the electrodes, with the negative pole of the exterior power source connected to the hydrogen electrode. The current is now obviously forced to reverse its direction through the cell. The hydrogen chloride in the electrolyte decomposes by reaction (11-1) and evolution rather than consumption of hydrogen is found to occur at the electrode past which the hydrogen is being bubbled, with simultaneous chlorine evolution at the other electrode.* The rate of this gas evolution is found to be entirely independent of the rates at which these gases are bubbled up past the electrodes from below. Needless to state, the electrodes remain unaltered during all these operations.

Careful measurements show that the rate of HCl production or decomposition in this cell is directly proportional to the current flowing. Further study shows that operations are reversible at 1.36 V, in that a greater potential causes reaction (11-1) to occur, a smaller potential causes reaction (11-2) to take place. In the limit, the electrical energy consumed in HCl decomposition approaches that produced in synthesis, and this electrical energy, when converted to heat units, equals 31.6 kcal/mole of HCl. This quantity of energy equals numerically the free energy of reaction (11-1).

*In theory, oxygen not chlorine, should be evolved at the chlorine electrode in this experiment. No oxygen is evolved here, however, because of the relatively high overvoltages of oxygen against smooth platinum. Overvoltages are discussed later in this chapter.

The foregoing experiment demonstrates that the free energy of a reaction occurring within a cell can be directly supplied or released isothermally as electrical energy and that the cell reaction approaches reversibility at some voltage (1.36 V in this case). This reversible voltage is called the cell's theoretical voltage.

Limitations of electrolysis. It might be argued that it should be possible to make any chemical reaction the basis of an electrolytic cell operation, but unfortunately, few reactions will meet the necessary conditions. To be amenable to electrolytic operation in a cell, the overall reaction must be separable into two complementary half-cell reactions, each involving ions and electrons, whose algebraic sum equals the overall reaction in question. The rates of these reactions must be high. It is customary to write these half-cell reactions separately, as is illustrated below for a cell operating on reaction (11-1):

Cathode: $$H^+ + e = \tfrac{1}{2}H_2 \tag{11-3a}$$

Anode: $$Cl^- = \tfrac{1}{2}Cl_2 + e \tag{11-3b}$$

Overall: $$H^+ + Cl^- = \tfrac{1}{2}H_2 + \tfrac{1}{2}Cl_2 \tag{11-1}$$

The reverse of reactions (11-3a) and (11-3b) takes place at the electrodes when the current direction is reversed.

Most chemical reactions cannot be separated into two such complementary half-cell reactions. For example, organic materials generally will not ionize at all under normal conditions, because their bonding is largely covalent. Other materials ionize but enter into electrode reactions reluctantly, and irreversible side reactions tend to occur. These technical factors place important limitations upon the scope of industrial electrolysis.

Half-cell reactions. The two electrode or half-cell reactions of any cell operation, as illustrated by Eqs. (11-3a) and (11-3b) always occur at the interfaces between the electrodes and the electrolyte, and always involve a transfer of an electron between the electrode and an ion or molecule in solution. A half-cell reaction is said to involve oxidation when the positive valence of one of the reaction partners increases due to transfer of an electron to the electrode, as in the typical half-cell reactions proceeding from left to right listed in Table 11-1.

TABLE 11–1. TYPICAL HALF-CELL REACTIONS

$\tfrac{1}{2}H_2 = H^+ + e$

$Fe^{++} = Fe^{+++} + e$

$MnO_4^{--} = MnO_4^- + e$

$Cl^- = \tfrac{1}{2}Cl_2 + e$

$OH^- = \tfrac{1}{2}H_2O + \tfrac{1}{4}O_2 + e$

Reduction is said to occur when there is a decrease in the positive valence of a reaction partner, as when the reactions of Table 11-1 proceed from right to left. In these reactions, electrons always travel from the electrode to an ion or molecule in the electrolyte solution. By definition, the anode is that electrode at which oxidation occurs, while the cathode is that electrode at which reduction occurs. Thus, the half-cell reaction (11-3a) is a reduction occurring at the cathode, while reaction (11-3b) is an oxidation occurring at the anode. When the current is allowed to reverse as in fuel cell operations, then these half-cell reactions also reverse. Hydrogen is now consumed at the anode and chlorine at the cathode.

In electrolysis, electrons flow through the circuit external to the cell from the anode to the cathode, that is, from the negative to the positive pole. It should be remembered that, by long-standing convention, electricity moving through metallic conductors is treated in the electrical engineering literature as a positive fluid moving from the positive to the negative pole. This rather misleading convention continues to be observed, and should be kept in mind when working in the field.

Cell reactions. It is necessary to distinguish between the primary half-cell reactions occurring at the electrodes, as just illustrated, and any subsequent reactions of the primary products formed with the electrodes, the electrolyte, or other materials present in the electrode environment. Thus, when the anode in the foregoing example is made of a corrodible metal like iron, then part of the evolved chlorine reacts to form iron chlorides. Some of the chlorine evolved will also react with the water in the electrolyte to form hypochlorous acid. Similarly, any small amounts of oxygen dissolved in the electrolyte tends to react with part of the newly formed cathodic hydrogen to yield water. The possible occurrence of such secondary reactions must always be taken into account, recognizing that they are disadvantageous in some cases, and desirable in others.

The electrolytes used industrially are usually either aqueous solutions of materials which ionize in water, or else ionized melts. Low temperature operation using aqueous electrolytes is generally preferable to melt electrolysis, since problems of corrosion and materials of construction are less severe. Aqueous electrolysis may nevertheless be impracticable because of limited ionization, low reagent solubility, the occurrence of undesirable side reactions, etc. Thus, when electrolyzing water soluble salts of magnesium or aluminum, hydrogen rather than the metal forms at the cathode because the free energy of decomposition of water is smaller than that for the decomposition of the salts of these metals. It is generally true that, when several ionic species are present, barring unusual obstacles to electrolysis (see the following discussion of overvoltages), the predominant electrolytic reaction will be the one having the smaller free energy.

Current requirements. Faraday discovered that in any electrolysis the number of chemical equivalents of primary anode and cathode products are always equal. Moreover, the quantity of current involved/equivalent of primary reaction product at either electrode is always the same in all cells, namely, 96,454 C (amp sec). This quantity of current is called the Faraday ($\mathscr{F}$), and in industrial operations is more conveniently expressed as 26.79 amp hr/g equivalent, or 12,163 amp hr/lb equivalent. Faraday's laws are always correct, applying regardless of temperatures, current densities, voltages, etc., but they may be obscured by secondary reactions of the primary products with the solvent, the electrode material, etc.

It follows from the foregoing that when the time is known over which current of a given magnitude is supplied to a cell, the amount of primary products produced at each electrode can always be calculated. The kinetics of the primary half-cell reactions are, therefore, determined and controlled by the flowing current. Of course, if the current can in part travel electronically as well as ionically through the electrolyte (as with certain molten salts), or if an internal short develops, then current utilized in the electrolysis is correspondingly reduced. Moreover, as already pointed out, the net yield of primary products may be reduced by the occurrence of secondary reactions. In view of these various ways in which current can be lost, the concept of current efficiency has developed, as follows:

$$\text{Efficiency} = \frac{\text{Equivalent of recoverable product}}{\text{Faradays of current consumed}} \times 100$$

This efficiency is not necessarily the same for the anode as for the cathode, as when the anode product reacts with the anode itself to form an undesired by-product while the cathode product does not undergo any such reaction. In industrial operations especially, where energy costs must be minimized for competitive reasons, cells are very carefully designed to maximize current efficiency.

Voltage. The power requirements of any cell cqual the product of current and voltage. The voltage is made up of four components: the theoretical voltage of the overall operation, the *IR* drop through the electrolyte, the so-called "concentration polarization," and overvoltages at the two electrodes. Each of these is now discussed.

Theoretical voltage. The free energy change associated with the chemical reaction carried out isothermally in a cell is, of course, identical with (minus) the electrical work theoretically obtainable from the cell. When the activities of all the reaction partners charged to and leaving the cell are unity, then the free energy change in the cell becomes ΔF°, namely, the standard free energy change of the reaction occurring. The electrical work is thus related to ΔF° as follows:

$$\Delta F^\circ = -n\mathscr{F}E^\circ \tag{11-4}$$

The standard voltage of this cell, namely E°, is defined by Eq. (11-4), in which n is the number of equivalents involved and $\mathscr{F}$ is the Faraday number. When ΔF° is expressed in W-sec then $\mathscr{F}$ equals 96,454 C/g equivalent as pointed out above. When ΔF° is expressed in calories, $\mathscr{F}$ has the units of cal/(V) (equivalent) and numerically equals 23,010. Thus, for reaction (11-1), ΔF° is -31.6 kcal and n is unity, hence by Eq. (11-3), the theoretical voltage for this reaction equals 31,600/23,010 or 1.36 V.

When the reaction partners in the cell are not all at unit activity, then the cell's free energy change and voltage are no longer standard. However, the relation between the cell's theoretical voltage and the free energy change for the existing terminal conditions is as before,

$$\Delta F = -n\mathscr{F}E \tag{11-4a}$$

The difference between Eqs. (11-4) and (11-4a) is obviously the reversible isothermal work in expanding or compressing the reaction partners from the nonstandard terminal conditions of Eq. (11-4a) to unit activities, or

$$\Delta F - \Delta F^\circ = RT \ln \frac{a_C^c \cdot a_D^d}{a_A^a \cdot a_B^b} \tag{11-5}$$

Substituting Eqs. (11-4) and (11-4a) into Eq. (11-5),

$$E^\circ - E = \frac{RT}{n\mathscr{F}} \ln \frac{a_C^c \cdot a_D^d}{a_A^a \cdot a_B^b} \tag{11-6}$$

Theoretical voltages of cells when operating with reagents at other than unit activities can, therefore, be calculated from Eq. (11-6). For this equation E° and E have the units of V, R is 1.987 cal/(g mole) (°K) and $\mathscr{F}$ is 23,010 cal/(V) (g equivalent).

EXAMPLE 11-1

The cell of Fig. 11-2 is now to operate at 25°C and 1 atm pressure. The gas bubbling over the anode is 50% H_2, 50% N_2, while air (21% O_2, 79% N_2) bubbles over the cathode. The depletion of hydrogen and oxygen from these feed gases is to be negligible, hence gas compositions are substantially unchanged in passing through the cell. The electrolyte is 0.1 N KOH. What is the theoretical voltage of this cell?

$$H_2(g) + \tfrac{1}{2}O_2(g) = H_2O(1) \qquad \Delta F^\circ{}_{25^\circ C} = -56.69 \text{ kcal}$$

Answer: Obviously, the electrolyte's concentration has no influence on the theoretical voltage of this operation. From Eq. (11-4a), since two equivalents are involved,

$$E^\circ = \frac{56{,}690}{2 \times 23{,}010} = 1.23 \text{ V}$$

Substituting into Eq. (11-6) and assuming that water in the electrolyte is at unit activity,

$$E = 1.23 - \frac{1.987 \times 298}{2 \times 23{,}010} \times \ln \frac{1}{0.5 \times (0.21)^{1/2}}$$

or

$$E = 1.23 - 0.019 = 1.21 \text{ V}$$

Actual cell voltages. It has already been pointed out that, when current flows through a cell at a finite rate, voltages greater or smaller than the theoretical voltage develop across the cell, depending on whether the overall reaction occurring has a positive or a negative free energy change. In other words, more than the theoretical voltage is required when energy must be supplied to cause the cell electrolysis reaction to proceed, while a fuel cell always delivers electric energy at less than the theoretical voltage. Three factors contribute to this difference between the theoretical and operating voltage, namely the electrolyte's resistance, the concentration polarization, and the phenomenon of "overvoltage." These will now be discussed with particular reference to reactions having a positive free energy change.

Electrolytic resistance. As in all conductors, both the electrolyte and the electrodes offer resistance to the flow of current. Since resistance of the electode is usually small as compared to the electrolytic resistance, electrode resistance is not further discussed here. Electrolytic resistance is conveniently expressed as specific resistance, namely, the resistance of a cube of electrolyte measuring 1 cm on a side, with units of ohm cm. Another common measure of the electrolytic resistance is its specific conductance or conductivity, which is the reciprocal of the specific resistance, and has the units of reciprocal ohm cm or mhos/cm. Conductivities of aqueous electrolytes, of which typical examples are presented in Table 11-2, vary roughly with concentration. This is to be expected, since ions in an electrolyte are the current carriers and the number of ions present is more of less proportional to the electrolyte's concentration.

The conductivities of even saturated aqueous salt solutions are at least 10^5-fold smaller than for the metals of Table 11-3, and about 100-fold smaller than for the molten salts of Table 11-4. Voltage drops in aqueous electrolytes

TABLE 11–2. SPECIFIC ELECTRICAL CONDUCTANCES AT 25°C OF AQUEOUS SOLUTIONS*

Compounds	Concentrations, equivalents/l		
	0.001	0.01	0.1
HCl	4.21×10^{-4}	4.12×10^{-3}	3.91×10^{-2}
LiCl	1.12×10^{-4}	1.07×10^{-3}	0.96×10^{-2}
NaCl	1.24×10^{-4}	1.18×10^{-3}	1.07×10^{-2}
$MgCl_2$	1.24×10^{-4}	1.15×10^{-3}	0.97×10^{-2}
$AlCl_3$			

*Expressed as mho/cm.

TABLE 11-3. SPECIFIC CONDUCTIVITIES AT 25°C OF METALS

Metals	mho/cm
Aluminum	3.75×10^5
Copper	5.80×10^5
Magnesium	2.17×10^5
Silver	6.33×10^5
Sodium	2.41×10^5

TABLE 11–4. SPECIFIC ELECTRICAL CONDUCTIVITIES OF MOLTEN SALTS

Salts	Melting T, °C	mhos/cm	
		700°C	800°C
$MgCl_2$	709	1.00	1.17
NaCl	808	3.33	3.55
$CaCl_2$	782	1.66	1.99
$BaCl_2$	962	1.11	1.40
KCl	776	1.98	2.09
LiCl	610	7.51	7.61
$AlCl_3$	194	0.56×10^{-6}*	

*Aluminum chloride develops a sublimation pressure of 1 atm at 160°C, and of 2.5 atms at its melting temperature, 194°C.

resulting from their low conductivities can be significant when operating with current densities of commercial interest. Thus, with flat parallel electrodes spaced one inch apart and separated by an aqueous electrolyte having a typical conductivity of 0.1 mho/cm, a cell would have an *IR* loss of about 0.4 V at a current density of 144 amp/sq ft. Substitution of a molten salt with its higher specific conductance for an aqueous solution of the same salt would obviously reduce this voltage drop significantly at any given current density.

Electrolytic resistances in general decrease with increasing temperature. Electrolytes not containing H^+ and OH^- ions show an increase in conductivity of about 3% per °C at 40°C, and an even greater rate of increase if H^+ and OH^- are present. The conductivities of molten salts again increase with temperature, as is evident from inspection of Table 11-4.

Concentration polarization. In any electrolysis, the positively charged ions (cations) travel toward the cathode, the negatively charged ions (anions) travel toward the anode. The linear rate of ion travel is slow, usually around 1 cm/hr in stagnant electrolyte for the voltage drops and concentrations encountered in industrial electrolysis. Simultaneously with this ion travel, as a consequence of the half-cell reactions occurring, change occurs in the number or in the charge of the ions around each electrode. Since electrical neutrality must always be maintained microscopically as well as macroscopically in the electrolyte, concentration gradients develop which tend to be offset by convective and diffusional mixing. In industrial operations, mixing is often deliberately undertaken to minimize these gradients which, however, always persist to some degree during electrolysis.

Any isothermal solution in which concentration varies from one point to another is not at equilibrium, and work can in theory always be obtained as the solute distributes itself uniformly through such a system. Thus, consider a simple cell as in Fig. 11-1, containing two copper electrodes and filled with copper sulfate solution. When the concentration is made higher around one electrode than the other, a potential results which will cause a current to flow when the electrodes are connected by a conductor. Like diffusion and convection, this electrochemical process will work to correct concentration differences, in that copper ions will form at the electrode in the weaker solution and discharge at the electrode in the stronger solution.

When operating any electrolysis under either a steady-state or a batchwise condition, depending on ion migration velocities relative to the electrode reaction requirements, the concentration of the electrolyte around the anode and the cathode tends to differ from that in the bulk of the solution. Because of these concentration gradients, a voltage develops which always opposes the net operating voltage between the cell electrodes. Such a "back voltage" is known as concentration polarization. Concentration polarization can be reduced by vigorous stirring of the electrolyte, as when gases like H_2, O_2, and Cl_2 are evolved at the electrodes.

The existence of a back *EMF* is readily demonstrated. Consider a cell containing two copper electrodes separated by a copper sulfate electrolyte of uniform composition. Upon passing a current through this cell, copper ions form at the anode and discharge at the cathode, with the net result being the transfer of metal from anode to cathode. In theory, energy requirements for such a transfer are zero. In practice, a potential must be applied to overcome both the *IR* drop through the electrode and the concentration polarization. This polarization develops because the concentration of copper sulfate becomes higher at the anode relative to the cathode. The existence of this polarization can be shown by suddenly disconnecting these two electrodes from the power source, whereupon a voltmeter will show a back voltage between them, which decays rapidly as concentration gradients disappear.

Indeed, one general method of monitoring cell performance is to measure the back *EMF* during a brief interruption of current flow.

Overvoltage. The voltage of most operating cells is greater than the sum of the theoretical voltage, the *IR* drop, and the concentration polarization. This excess voltage, called the overvoltage, develops because of the reluctance of ions and electrons to enter into half-cell reactions. Measurement of the overvoltage of an anode is illustrated in Fig. 11-3 for a cell in which water is being decomposed between two platinum rod electrodes coated with platinum black.

A reference electrode (such as a calomel electrode) is included in a secondary circuit as shown, with a capillary tube bridge connecting the electrolyte in the reference cell to the electrolyte in the main cell. No current flows through this secondary circuit, hence when operating the cell under load, the anode's voltage relative to the constant reference electrode voltage can be determined directly.

Overvoltages measured for oxygen against different anode materials in 1.0 *N* KOH over a range of current densities on the main electrode are shown in Fig. 11-4. Inspection shows that the overvoltages observed vary greatly

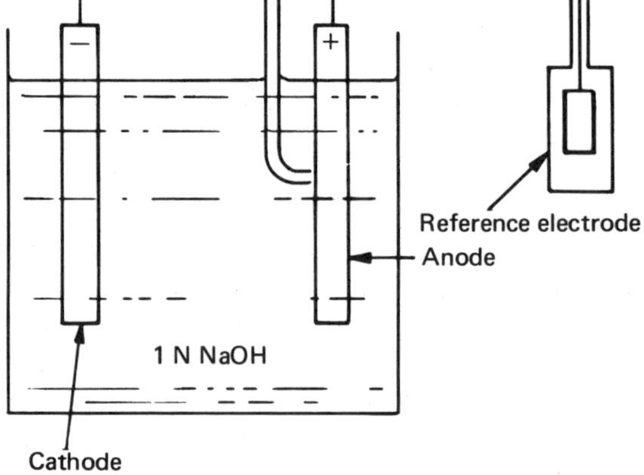

Fig. 11-3. A cell illustrating the measurement of overvoltage.

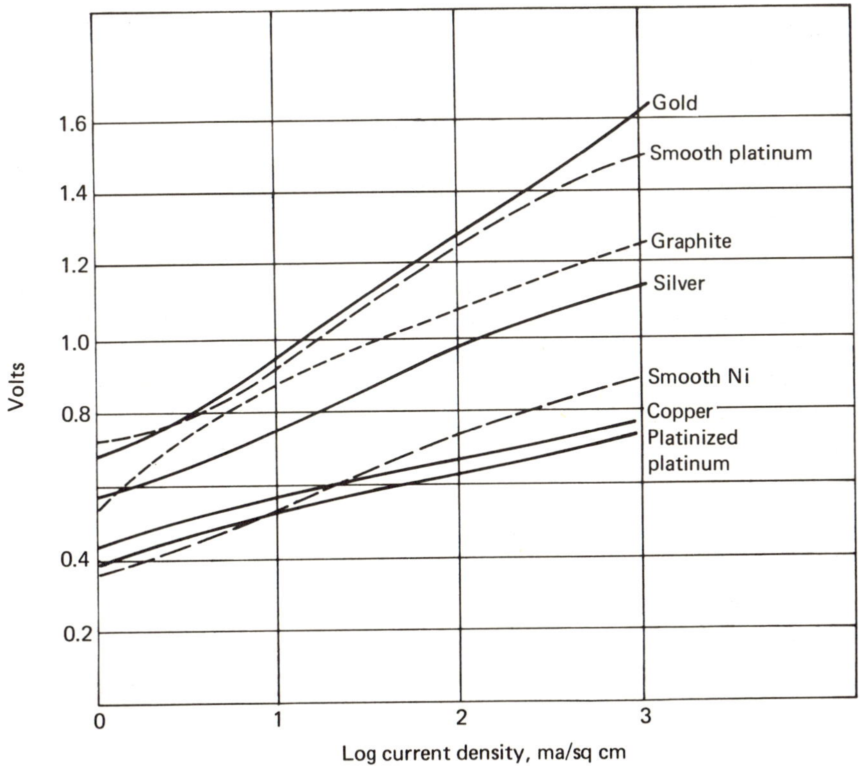

Fig. 11-4. Oxygen overvoltages at 25° C (reference 61, vol. VI, page 340).

from one material to the next, and for any one material always increase with increasing current density. Overvoltages of oxygen are higher than for most other materials involved in electrode reactions, with hydrogen a close second. Overvoltages are usually quite small for the halogens and for most metal ions.

Electrodes. Electrical conductivity of an electrode should be high and, unless deliberately designed to be consumed in the cell reactions, the electrode should be inert to the reagents with which it makes contact. Thus, for inertness, graphite is used for anodes at which chlorine is evolved, lead is used in sulfate solutions when nitrate-free, and iron is used for cathodes in aqueous caustic soda electrolysis for hydrogen evolution. Platinum must be used in strong acid for both electrodes when complete inertness is essential, as in hydrogen peroxide production, but its high cost and limited supply critically limit its use. Newer electrode materials, like silicon carbide and titanium, are also used where acid resistance is important, but these materials do not match platinum in catalytic activity. For maximum effectiveness, metal catalysts may be incorporated onto the electrode in a finely divided form, such as platinum black. Electrodes are rarely completely inert, and must, therefore, be replaced from time to time.

In metal refining, as indicated in the discussion of concentration polarization, the metal being refined is itself used for both electrodes. The anode is made of impure metal, which dissolves during electrolysis, as further described in a later section, while pure metal deposits on the cathode. Molten metal electrodes are sometimes preferred in metal winning and refining cells because of the ease with which liquids can be handled, but such operations are complicated by problems of corrosion, etc., introduced by the elevated temperatures required, and by the molten salts used as electrolytes.

Recent activity in the fuel cell field has focused particularly on half-cell reactions involving consumption of hydrogen at the anode and of oxygen at the cathode. Overall reactions in acid and basic electrolytes are as follows:

$$\begin{array}{rcl} \tfrac{1}{2}\mathrm{H_2} &=& H^+ + e \\ e + \tfrac{1}{4}\mathrm{O_2} + \mathrm{H^+} &=& \tfrac{1}{2}\mathrm{H_2O} \\ \hline \tfrac{1}{2}\mathrm{H_2} + \tfrac{1}{4}\mathrm{O_2} &=& \tfrac{1}{2}\mathrm{H_2O} \end{array} \qquad \begin{array}{rcl} \tfrac{1}{2}\mathrm{H_2} + \mathrm{OH^-} &=& \mathrm{H_2O} + e \\ e + \tfrac{1}{4}\mathrm{O_2} + \tfrac{1}{2}\mathrm{H_2O} &=& \mathrm{OH^-} \\ \hline \tfrac{1}{2}\mathrm{H_2} + \tfrac{1}{4}\mathrm{O_2} &=& \tfrac{1}{2}\mathrm{H_2O} \end{array}$$

In all these cases, the gaseous reagents must be supplied to the electrode and caused to enter into the indicated half-cell reactions. The problem is difficult with an aqueous electrolyte at relatively low temperature since these half-cell reactions, when uncatalyzed, are very slow below several hundred degrees centigrade. The best catalyst known for the hydrogen half-cell reactions is platinum black, while both platinum black and silver are good catalysts for the oxygen half-cell reaction in basic electrolytes.

Electrodes for these half-cell reactions are usually in the form of a sheet of porous conducting materials containing the catalyst particles. The reagent gas is supplied to one side of this electrode sheet from a suitable gas supply chamber, while the other side of the electrode is in direct contact with the electrolyte. Under proper operation, the electrode pores are only partly filled with electrolyte, with the meniscus kept stationary somewhere within the pores by incorporating into the electrode a material, such as Teflon particles, which do not "wet," that is, against which the electrolyte has a high contact angle. The half-cell reactions then take place within the electrolyte at the pore walls, near the gas-electrolyte meniscus (88).

Depolarization. Reagents called depolarizers may be added to either or both anode and cathode compartments of a cell to react with the primary products of the electrolysis. When rates of depolarization reactions are great, then the concentrations (or back pressures) of the primary products are reduced accordingly. In the limit, the free energy change of the cell now equals the algebraic sum of the free energies of the primary and polarization reactions, with overall voltages of corresponding magnitudes. Thus, in some operations, an electrode is deliberately made of a depolarizing material chosen to react with the primary product of a half-cell reaction, as in aluminum metal production where the oxygen evolved at the anode combines with the carbon electrode to form an off-gas containing CO and CO_2. The cell voltage is reduced below what it would be with an inert anode, as is evident from the following comparison of the free energy requirements at 1300°K for the overall cell reactions.

With anodes made of carbon to act as depolarizer:

$$Al_2O_3 + 3C = 2Al + 3CO \qquad \Delta F^\circ_{1300^\circ K} = +139 \text{ kcal}, \; E^\circ_{1300^\circ K} = 1.0$$
$$Al_2O_3 + \tfrac{3}{2}C = 2Al + \tfrac{3}{2}CO_2 \qquad \Delta F^\circ_{1300^\circ K} = +159.6 \text{ kcal}, \; E^\circ_{1300^\circ K} = 1.15$$

With inert anodes:

$$Al_2O_3 = 2Al + \tfrac{3}{2}O_2 \qquad \Delta F^\circ_{1300^\circ K} = +301.7 \text{ kcal}, \; E^\circ_{1300^\circ K} = 2.17$$

The interruption in production resulting from electrode replacement is disadvantageous, hence, in larger installations, electrodes are often fed continuously to the cells. The "Soderberg" carbon anode used in some aluminum cells is made continuously in the cell by pouring a suitable pitch-carbon mixture through a form suspended above the cell. This mixture decomposes and bakes as it comes near the hot electrolyte, continuously forming a carbon rod which is fed into the electrolyte as fast as the electrode is consumed. Offsetting the advantage of such continuous production is the relatively high resistance of electrodes made in this way, due to porosity and incomplete graphitization. Gassing of these electrodes when submerged, due to incomplete baking, is also a disadvantage. An alternative to the Soderberg technique, of course, is to use the more highly conductive prebaked electrodes with threaded ends. These are fitted onto those already in use, and the combination is fed continuously, as required.

Cathodes, as well as anodes, can be depolarized by a secondary reaction. Thus, the half-cell reaction (11-3a), shown again below, could be depolarized with oxygen, using a porous gas diffusion type electrode:

$$H^+ + e = \tfrac{1}{2}H_2 \qquad (11\text{-}3a)$$
$$\tfrac{1}{2}H_2 + \tfrac{1}{4}O_2 = \tfrac{1}{2}H_2O \qquad (11\text{-}8)$$
$$\overline{H^+ + \tfrac{1}{4}O_2 + e = \tfrac{1}{2}H_2O} \qquad (11\text{-}9)$$

Combining the half-cell cathode reaction (11-9) with reaction (11-3b), the overall reaction becomes

$$H^+ + Cl^- + \tfrac{1}{4}O_2 = \tfrac{1}{2}H_2O + \tfrac{1}{2}Cl_2 \qquad (11\text{-}10)$$

Since ΔF° for reaction (11-2) is 31.6 kcal at 25°C and for Eq. (11-8) it is -28.9 kcal, then ΔF° for reaction (11-10) at this temperature is $+2.7$ kcal, equivalent to a theoretical voltage of only 0.12 V. The advantages of reduced voltage and power gained by successful depolarizing of reactions at an electrode are obvious. Reaction (11-10) is the Deacon reaction, here being carried out electrolytically instead of thermally.

Cell membranes. Any two half-cell reactions, such as those listed in Table 11-1, can in theory be combined regardless of the direction in which each is to occur. In those half-cell combinations in which the overall ΔF is positive, electrical energy must be supplied to the cell for the indicated reaction to proceed, while when ΔF is negative electrical energy can be generated. In practice, it is simpler and, therefore, preferable to operate with two half-cells using the same electrolyte, as in the HCl electrolysis involving reactions (11-3a) and (11-3b) described earlier. When combinations of half-cell reactions are chosen having different and incompatible anolytes and catholytes, such as solutions of copper sulfate and of sodium hydroxide, then irreversible and undesirable reactions occur where these electrolytes come into contact. The physical chemist avoids this difficulty by inserting a layer of KCl solution between incompatible anolytes and catholytes. The industrial chemist cannot use such intermediate liquid layers because of problems of control and product recovery. Instead, diffusional and convective mixing is minimized by inserting a membrane between the two electrode compartments. Asbestos paper sheets are extensively used for this purpose, with ion travel occurring through the liquid-filled pores of the paper. More recently, nonporous sheets of ion-exchange material have found use as membranes. Thus a cationic resin membrane will allow passage only of cations, and its use entirely eliminates bulk mixing between electrode chambers. Such a cation-permeable membrane could be made from the hydrogen polystyrenesulfonate described in Chap. 3.

Cell construction. Design of a cell depends on many factors, such as whether the electrodes are liquid or solid, on whether gaseous, liquid, or solid products are obtained, on the problems of keeping the products of one electrode separate from those of the other electrode, and on the temperature of operation. The widely different conditions encountered in the various electrolytic operations of industry have resulted in the development of many different cell types, in which great ingenuity is displayed. A representative group of such cells is now described.

1. Copper refining. In copper refining, current is made to pass between an impure "blister" copper anode and a pure copper cathode using an acidified aqueous copper sulfate solution as electrolyte. The following half-cell reactions occur at each electrode, proceeding to the right at the anode and to the left at the cathode:

$$Cu = Cu^{++} + 2e$$
$$Cu = Cu^{+} + e$$

The free energy needed for transfer of metal between electrodes in this cell is obviously zero, but in practice an operating voltage must be applied to overcome the *IR* drop through the electrolyte and also any concentration polarization. Each cell is typically an asphalt-lined cement tank 10 ft long, 4 ft wide, and 4 ft deep.

Alternate sheets of anode and cathode material are suspended vertically in this tank from above, dipping into the electrolyte. The blister copper anodes, initially about 2 in. thick, slowly go into solution to produce copper ions, leaving impurities, such as gold, silver, and a few other relatively inert materials undissolved. These impurities fall to the bottom of the cell, forming "anode mud" which is removed at appropriate intervals to be worked up for precious metal recovery. When operating at about 60°C with the relatively low current density of 20 amp/sq ft obtained with 0.25 V between electrodes, a coherent copper deposit about $\frac{1}{2}$-in. thick builds up on the cathode in perhaps an interval of 25 days. The product cathodes and spent anodes are replaced as needed with fresh electrodes and the cycle is repeated. A large plant will contain hundreds of tanks with perhaps 22 electrode pairs per tank. Warm electrolyte is slowly circulated through all of these cells, the electrolyte being reconditioned as required.

2. Aluminum (114). Unlike copper, aluminum cannot be recovered by electrolysis of aqueous solutions because water will decompose to form hydrogen at the cathode at lower voltages than those at which aluminum salts decompose. This difficulty is avoided by operating with a fused salt electrolyte. Low-melting aluminum chloride, an easily-produced salt, is not a suitable electrolyte when molten because of its low conductivity (see Table 11-1), relatively high volatility, sensitivity to moisture, and tendency to creep. Another readily available aluminum compound is alumina, but this cannot be used directly as an electrolyte because of its very high melting point of 2050°C. Fortunately, alumina is somewhat soluble in molten cryolite (Na_3AlF_6) from which aluminum metal can be recovered by electrolysis, as was discovered independently by Hall in America and Herioult in France. Aluminum plants operate today with a 3–5% solution of Al_2O_3 in cryolite, whose melting point is reduced from 1006°C to under 900°C by addition of small amounts of CaF_2.

A typical industrial aluminum cell is a steel box lined with a thick layer of carbon, usually anthracite coal loaded with pitch. Exterior dimensions for a 100,000-amp cell are about 35 ft by 16 ft by 10 ft tall. A "pad" or "heel" of molten aluminum covers the bottom of the cell and serves as the cathode, making direct electrical contact with the conductive carbon lining in which conductors, called "bus-bars," are buried. A 12-in. deep layer of molten electrolyte (2.1 g/cc), only slightly lighter than molten aluminum (2.3 g/cc) at the cell-operating temperature of 950°C, floats on top of the aluminum pad. Carbon anodes suspended from above the cell dip into the electrolyte to within about 2 in. of the molten aluminum cathode. These anodes, which are either prebaked or of the Soderberg type, operate at current densities up to 1000 amp/sq ft.

Cell operation is batchwise in that alumina is added to the electrolyte at intervals and metal accumulates in the cell to be withdrawn perhaps every other day by siphoning. The primary cathode reaction is the discharge of Al^{+++} ions, but part of the metal produced forms tiny droplets dispersed in the electrolyte which find their way to the anode where they are reoxidized, reducing current efficiency to perhaps 85%. Oxygen formed at the anode combines with the carbon present to yield CO_2, which is then in part reduced to CO by the aluminum droplets in the melt. The off-gases consequently analyze perhaps 80% of CO_2, 20% of CO, plus traces of fluorine and fluorides from cryolite decomposition.

3. Zinc. Zinc metal is made by electrolysis of aqueous zinc sulfate solution. The ore, namely zinc sulfide, is roasted to the oxide and dissolved in sulfuric acid. The resulting solution is carefully purified and finally electrolyzed. The cells are typically rectangular lead- or rubber-lined tanks equipped with vertical electrodes, the anodes being $\frac{1}{2}$-in. aluminum sheets while the cathodes are sheets of lead containing from 0.5–1.0% silver to help resist corrosion. Electrode reactions are as follows:

$$\begin{array}{ll} \text{Cathode:} & Zn^{++} + 2e = Zn^0 \\ \text{Anode:} & H_2O = 2H^+ + \tfrac{1}{2}O_2 + 2e \\ \hline & Zn^{++} + H_2O = Zn^0 + 2H^+ + \tfrac{1}{2}O_2 \end{array}$$

Industrial cells operate at about 3.40 V rather than the theoretical voltage of 2.35 V because of resistance and concentration polarization in the electrolytes used. Hydrogen is not evolved at the cathode until reaching about 3.45 V, despite the low theoretical decomposition voltage of water of 1.23 V, because of hydrogen's high overvoltages against zinc. Thus, the margin between zinc deposition and hydrogen evolution in only about 0.05 V, requiring careful control of cell conditions to keep hydrogen overvoltages high. Electrolyte composition must be continuously adjusted, since increased H^+ concentrations and smaller Zn^{++} concentrations will obviously favor hydrogen evolution over zinc deposition. Because increased temperatures always reduce overvoltages, cells are cooled to below 40°C to avoid hydrogen evolution. Increased roughness of electrode surfaces also decreases overvoltages and, since deposited zinc becomes rougher as temperatures decrease, these cells are operated above 30°C. Impurities in the electrolyte, such as minute traces of tellurium, selenium, arsenic, nickel, and especially germanium, likewise reduce hydrogen overvoltages. When present, tiny specks of these materials deposit on the cathode immediately becoming centers of hydrogen evolution and causing neighboring zinc to dissolve by forming a local cell, thus lowering current efficiency and causing a poor deposit to form which is full of holes.

Other impurities, like sodium, potassium, magnesium, and aluminum, will not discharge at the operating voltage of the cell and so do not harm the electrolysis. If these metal ions are allowed to build up in the recycling cell liquors, however, zinc sulfate solubility is seriously reduced, while electrolyte viscosity and resistance are increased. Because of their lower decomposition voltages, copper, cadmium, and lead will deposit simultaneously with zinc. Since their hydrogen overvoltages are greater than that of zinc, current efficiency is not seriously reduced by their presence, but the purity of the zinc deposited is reduced. Anions other than sulfate also are a source of trouble. Fluoride ions cause the zinc deposit to stick so firmly to the anodes that removal becomes impossible. Chloride ions cause severe corrosion of the lead cathodes, and must be avoided. Silver sulfate can be added to the electrolyte to remove the chloride ion, with the silver being recovered from the sludge.

In plant operations, the solutions to be electrolyzed are carefully purified by treatment with zinc dust, thus precipitating dissolved copper, iron, lead, cadmium, tin, and other materials lower than zinc in the electromotive force series. Since, as just described, zinc electrolysis is disadvantageously affected by the presence of impurities in amounts often too small for analysis, control is difficult. The usual procedure involves testing the feed solutions continuously under conditions such that large scale operating conditions are duplicated.

4. Caustic-chlorine production (137). When an aqueous solution of sodium choloride is electrolyzed between inert electrodes, chlorine is evolved at the anode while caustic soda plus hydrogen form at the cathode. This anode gas must not be allowed to come into contact with caustic soda to avoid yield losses resulting from undesired secondary reactions. Formation of chlorine-hydrogen gas mixtures must also be avoided, both because of yield losses and explosion hazards. Cells must, therefore, be designed to meet the following apparently contradictory requirements: (a) the spacing between anode and cathode must be small to minimize energy losses resulting from resistance heating of the electrolyte, and (b) gaseous and liquid anode and cathode products must be kept separate to avoid undesired secondary reactions. The latter requirement is made more difficult because the OH^- ions formed at the cathode are attracted toward the anode where they will immediately react with the chlorine.

After many years of development two types of caustic soda-chlorine cells have evolved, namely, the diaphragm cell and the mercury cell. Both are operated today and the selection of one or the other depends upon local circumstances.

Diaphragm cells. The distinguishing feature of any diaphragm cell, as shown in the sketch of Fig. 11-5, is its division into two compartments by a diaphragm, which in caustic soda-chlorine cells is generally an asbestos fiber mat. Saturated brine at 70°C, containing 26.5% NaCl by weight, is fed

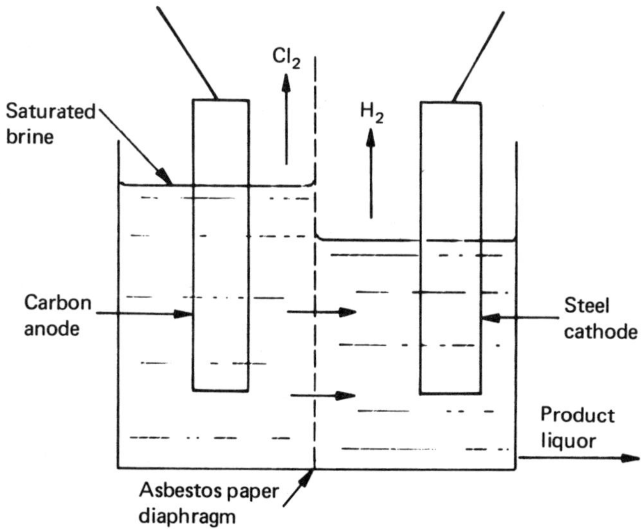

Fig. 11-5. Schematic diagram of a diaphragm cell for caustic-chlorine manufacture.

continuously to the anode compartment and flows by gravity from here through the pores of the diaphragm into the cathode compartment. High salt concentrations and cell temperatures around 100°C are used to maximize the conductivity of the electrolyte, thereby reducing the *IR* drop in the cell. The product liquor leaves the cathode compartment at about 11.5% NaOH and 16% NaCl. The anode is a graphite rod at which chlorine gas is evolved while hydrogen and caustic soda are produced at the iron cathode:

Anode reaction:	$Cl^- = \frac{1}{2}Cl_2 + e$
Cathode reaction:	$e + H_2O = OH^- + \frac{1}{2}H_2$
Overall:	$Cl^- + H_2O = \frac{1}{2}H_2 + OH^- + \frac{1}{2}Cl_2$

Unless prevented, the following undesirable side reaction will occur. In the electrolyte, chemical reactions:

$$Cl_2 + 2OH^- = Cl^- + OCl^- + H_2O$$
$$3OCl^- = ClO_3^- + 2Cl^-$$
$$CO_2 + 2OH^- = CO_3^{--} + H_2O$$

At the anode, electrochemical reactions:

$$OH^- = \tfrac{1}{2}H_2O + \tfrac{1}{4}O_2 + e$$
$$OCl^- = \tfrac{1}{2}O_2 + \tfrac{1}{2}Cl_2 + e$$
$$ClO_2^- = \tfrac{1}{2}Cl_2 + \tfrac{3}{2}O_2 + e$$

At the anode, chemical reactions:

$$C + O_2 = CO_2$$

These reactions are prevented in several ways. The OH^- ions are kept out of the anode compartment by maintaining the flow velocity of the anolyte through the membrane pores greater than the back migration velocity of the OH^- ions to the anode. Bubbles of gas are effectively kept from crossing over from the anode to the cathode compartment, and vice versa, by this same asbestos diaphragm. To minimize the solubility of chlorine in the anolyte liquor, the feed brine is acidified very slightly with HCl. As a result of all of these precautions, chlorine current efficiency is 97% or more in a properly operating cell.

Depending on the purity of the brine and the severity of chemical attack on the electrode, the membrane must be replaced periodically due to plugging with insoluble calcium and magnesium salts and with flakes of carbon from the anodes. As the diaphragm slowly plugs, the head of brine in the anode compartment must be increased to maintain constant liquor flow. The diaphragms are usually replaced every 250 days or so, a relatively disagreeable task which interrupts production.

Diaphragm cells operate at from 3.8–4 V as compared to a theoretical voltage of 2.3, and are built to take about 35,000 amp/cell.

Mercury cells. In modern mercury cells, a saturated NaCl brine is electrolyzed between a graphite anode and a mercury cathode. Chlorine forms at the anode as in the diaphragm cell, while sodium metal produced at the cathode enters into the mercury to form an amalgam:

$$\begin{array}{ll} \text{Anode reaction:} & Cl^- = \tfrac{1}{2}Cl_2 + e \\ \text{Cathode reaction:} & e + Na^+ = Na\ (\text{amalgam}) \\ \hline \text{Overall reaction:} & Na^+ + Cl^- = Na\ (\text{amalgam}) + \tfrac{1}{2}Cl_2 \end{array}$$

Fortunately, hydrogen has a high overvoltage against mercury, consequently, no hydrogen is evolved at the cathode, despite the fact that the cell operates at about 4.3 V. The solidification temperature of mercury rises rapidly as sodium dissolves in it, hence the mercury must be circulated fast enough to keep the sodium content below about $\frac{1}{4}\%$ by weight. The amalgam passes from the cells to a decomposer which consists of a tower packed with graphite, down through which the amalgam flows counter to a rising water stream. This graphite packing in contact with amalgam and caustic soda solution forms a short-circuited cell, in which the sodium reacts with water to form additional caustic soda:

$$\begin{array}{ll} \text{Anode reaction:} & Na\ (\text{amalgam}) = Na^+ + e \\ \text{Cathode reaction:} & e + H_2O = OH^- + \tfrac{1}{2}H_2 \\ \hline \text{Overall reaction:} & Na\ (\text{amalgam}) + H_2O = Na^+ + OH^- + \tfrac{1}{2}H_2 \end{array}$$

In theory, it should be possible to obtain energy from this decomposer by operating it as a fuel cell, since the free energy change of the foregoing overall reaction is large and negative. However, because of hydrogen's high overvoltage against mercury, the net external voltage available from this reaction is small at best; consequently, it has proven more economical to operate the decomposer irreversibly as described. Caustic soda concentrations up to 73% can be obtained, if temperatures are kept above 100°C to avoid solidification of the caustic soda solution and to maintain high reaction rates. More often the system is operated to produce 50% by weight of caustic soda, and a tower temperature of about 80°C is maintained by the heat generated in the reaction. The stripped mercury recycled to the cell from this tower contains about 0.1% by weight of sodium.

Flow sheets. The brine used in diaphragm cells is obtained either directly by solution mining, namely, by pumping water through the underground salt deposit, or by dissolving solid mined salt in water. Calcium and magnesium sulfates and chlorides present as impurities in the brine are precipitated with a little soda ash and caustic soda, and are removed by settling and filtration. This purified brine is fed to a battery of diaphragm cells operated in series-parallel. The hydrogen gas from the cathode is scrubbed with water to remove entrained electrolyte, and then compressed or otherwise used as required. The chlorine is water-cooled in a packed tower, dried with sulfuric acid, compressed to about 50 psi, and liquefied by refrigeration. The cell liquors (11.5% of NaOH, 16% NaCl) are concentrated in double or triple effect salting-out type evaporators. The 50% caustic soda produced contains about 1% of NaCl in solution, and may be further concentrated by evaporation to 73%. The solid salt from the evaporators is washed and recycled.

The mercury cell process starts with recycled weak brine (about 270 g of NaCl/l), which has been first dechlorinated by acidifying to a pH of 2 with HCl, followed by flashing off most of the dissolved gas into a 400-mm Hg vacuum. Any remaining chlorine is removed by passing the brine down through packed towers countercurrently to a rising air stream. Solid salt is then dissolved in this dechlorinated brine to a concentration of perhaps 310 g/1, after which the solution is purified as in the diaphragm cell process, and pumped to the cells. The chlorine gas from these cells, and the hydrogen gas from the decomposers, are handled in much the same way as in the diaphragm cell process.

Mercury cells versus diaphragm cells. The caustic soda solution from the decomposers following the mercury cells is of high concentration, pure, and chloride-free, thus meeting the rigorous specifications of important customers like the rayon industry. On the other hand, the diaphragm cell produces a relatively dilute caustic soda solution, the cost of evaporation and purification is significant, and the caustic soda product is not chloride-free. While both cells have high current efficiencies, the operating voltage of mercury cells is about 0.5 V more than that of diaphragm cells, but the resultant higher power costs of the mercury cell are at least partly offset by smaller investment requirements. The mercury cell must be supplied with solid salt for resaturation of the recycled liquor while the diaphragm cell can operate directly on brines obtained in "solution" mining. These cost factors are so nearly in balance that process selection depends upon local conditions.

Organic electrochemistry. Until recently, electrochemistry has found little application in organic industrial operations, although this may change with the advent of Monsanto's new adiponitrile process (7). Organic substances, with few exceptions, do not participate in primary half-cell reactions because (a) they usually do not ionize, (b) they are usually incapable of exchanging electrons with the electrode, and (c) they are not water soluble. Ordinarily, organic materials can be made to participate in electrochemical reactions only as depolarizers, that is, by reacting with one of the primary products of a half-cell reaction. Thus, in a fuel cell with an aqueous electrolyte, the hydrazine, methanol, or hydrocarbons fed at the anode as fuel may be viewed as depolarizers combining with oxygen formed in the primary half-cell reaction.

In early studies in this field, McKee (93) advocated the use of electrolytes containing salts of polyxylenesulfonic acids and related sulfonic acids, since these have the unusual property of increasing the solubility of many organic substances in water. These so-called McKee salts have the added advantage of ionizing extensively, hence their aqueous solutions are good ionic conductors. These salts are also relatively stable and resist oxidation and reduction in low voltage operations. It was originally hoped that electrolysis using McKee salts would provide a tool in organic operations whereby a delicate control of oxidizing or reducing conditions could be established merely by controlling cell voltage. Despite much effort, however, large scale application has appeared only recently.

Adiponitrile. Adiponitrile (ADN) is reportedly being made competitively by Monsanto from acrylonitrile (ACN) in a new electrolytic process (7). The advantage of this synthesis lies in the relatively low cost of ACN, but due to side reactions and polymerization, difficulties in obtaining high yields were encountered in the original development work.

The half-cell reactions involved might be written:

$$\begin{array}{lr} \text{Cathode:} & 2CH_2{:}\,CHCN + 2H^+ + 2e = CN(CH_2)_4CN \\ \text{Anode:} & H_2O = 2H^+ + \tfrac{1}{2}O_2 + 2e \\ \hline \text{Overall:} & 2CH_2{:}\,CHCN + H_2O = CN(CH_2)_4CN + \tfrac{1}{2}O_2 \end{array}$$

The cell, made of polypropylene, is believed to operate below 60°C, and contains lead cathodes and acid resistant alloy anodes which are separated by a membrane made of cation exchange material. The catholyte is a McKee-type salt, possibly tetraethylammonium xylenesulfonate, which is recirculated very rapidly past the cathode to minimize buildup of a product-rich film, in which ADN would undergo further undesirable reactions. The anolyte is a mineral acid like H_2SO_4 which is also circulated rapidly. It is important to maintain the acidity of the catholyte just short of neutrality, and to exclude K^+ and Na^+ ions from the system. Higher current densities, possibly in excess of 100 amp/sq ft, and concentrations of ACN over 40% by weight, are reported to improve current efficiencies through reduced by-product formation.

Notes to Problems

The following problems are chosen to illustrate typical industrial situations. A reasonable background in physical chemistry and classical thermodynamics, as well as some familiarity with chemical processing equipment needed for their solution, is assumed. The problem statements usually contain the data needed for solution, but the student is encouraged to look further afield, as required. The literature references in the problems indicate the source of the data and of some of the information presented. These references need not be read in order "to work" the problems, but do provide a source of further background for the interested reader. Further information can be conveniently obtained from the larger encyclopedias (46, 75, 152).

Many of these problems have been simplified to avoid algebraic complexity, which tends to distract from the main point being made. Thus raw material streams are often assumed freer from contaminants than they would be in practice. Perfect gas behavior is frequently assumed (an assumption which is usually justifiable). Averaged heat capacity data are presented, which are generally assumed independent of temperature and pressure. These heat capacities are often so adjusted that enthalpies of reaction remain unchanged with temperature. Because of such approximations, the data presented in these problems should not be used in the final design of actual installations until after checking for more exact values in the literature.

Offsetting the foregoing simplifications, other complications of actual literature and industrial data have been retained. Thus material compositions are sometimes reported in weight per cent, other times in weight or volume concentrations, in mole per cent, etc. Gas analyses units are often not stated, again as in the literature, following the general practice that gas analyses are usually in mole per cent, which is of course the same as volume per cent. These analyses are often on a "dry basis," i.e., on a water-free basis. Thus dry air contains 21 mole per cent oxygen, the remainder primarily nitrogen together with a little argon and traces of other rare gases.

No attempt is made to hide the fact that many different systems of units are used in the technical literature. Some references involve data in the c.g.s. system, others in the English engineering system, etc. Ample opportunity is afforded to develop facility in converting from one set to the others. A minimum number of conversion factors should be available for this purpose, and the more important of these are listed below:

Temperature: Temperature in °Kelvin multiplied by 1.8 equals temperature in °Rankine. Zero degree centigrade corresponds to 273° Kelvin, 492° Rankine, and 32° Fahrenheit. (Temperature in °C) $\times$ 1.8 + 32 = °F.

Length: There are 30.5 cm in 1 ft.

Mass: There are 454 g in 1 lb, therefore there are 454 g molecular weights in 1 lb molecular weight.

Energy: There are 778 ft-lb in 1 Btu, 252 cal in 1 Btu, and 3416 Btu in 1 kw-hr. A chu (centigrade heat unit) is the amount of heat needed to raise 1 lb of water 1°C while 1 Btu is needed to heat 1 lb of water 1°F. It follows from this that there are 1.8 Btu in 1 chu.

Gas constant: The gas constant R_g is 1.987 cal/(g mole)(°K). Using the conversion factors just presented, it is easy to show that R_g has the value of 1.987 Btu/(lb mole)(°R), or 1546 ft lb/(lb mole)(°R).

Enthalpy: An enthalpy change in cal-g mole can be converted to Btu/(lb mole) by multiplying by 1.8. This factor can of course be derived from the foregoing conversion factors.

Heat Capacity: A heat capacity in Btu/(lb mole)(°F) is numerically the same as in cal/(g mole) (°K), as can again be derived from the foregoing conversion factors.

The student's familiarity with references for molecular and atomic weights, heat capacities, densities, etc., is assumed. The concepts of a pound atom, a gram atom, a pound mole, etc., are also assumed to be understood. The usual mathematical conventions are retained. Thus log A and ln A represent logarithms of A to the base 10 and base e, respectively.

Problems

PROBLEM 1-1

The following reaction, often called the *water-gas shift reaction*, is frequently encountered in combustion and is important in hydrogen manufacture:

$$CO + H_2O = CO_2 + H_2$$

When occurring homogeneously in the gas phase, rates become appreciable above about 900°C, increase with temperature, and are very rapid at 1500°C and above. The reaction responds especially to iron and nickel as catalysts, which can be effective down to 600°F. The equilibrium constant, which changes slowly with temperature since $\Delta H°$ is rather small, is approximated as follows:

$$\ln K = \frac{8240}{T} - 4.33$$

where T is in °R, while K is dimensionless (103).

(a) Using the above equation, calculate $\Delta H°$ for the water-gas shift reaction in Btu/lb mole of CO_2 produced; in kcal/g mole and in chu/lb mole. Compare with the actual values for $\Delta H°$ of $-16{,}800$ and $-15{,}300$ Btu/lb mole at 600°F and 1200°F, respectively.

(b) The off-gases from the partial combustion of a gaseous hydrocarbon fuel in air are quenched and analyzed by gas chromatography. Since at least some water condensed out of the gas during sampling, results were reported on a dry basis as follows: 30.6% H_2, 16.8% CO, 1.9% CO_2, and the rest N_2. The gas temperature before quenching was 1800°F. Was this hot gas at equilibrium?

(c) What was the ratio of hydrogen to carbon in the fuel gas used? What was the probable analysis of this fuel? Could several different gas mixtures be proposed that would be consistent with this off-gas analysis?

PROBLEM 1-2

A fuel oil analyzes 86% C, 11% H, and 3% inerts by weight. This fuel is now burned with 50% of the dry theoretical air needed for complete combustion. After combustion, the gas mixture is brought to equilibrium at 1800°F and 1 atm. The molecular species to be considered in this equilibrium gas are CO, CO_2, N_2, H_2, H_2O, and CH_4. What is the composition of this gas mixture? It is known that no solid carbon forms under these conditions.

$$CO_2 + H_2 = CO + H_2O \qquad K_{p\ 1800°F} = 1.66$$

$$CO + 3H_2 = CH_4 + H_2O \qquad K_{p\ 1800°F} = 1.48 \times 10^{-4}$$

PROBLEM 1-3

Sulfur, a basic raw material for the chemical industry, melts at 230°F and boils at 444°C. Four molecular species are believed to be present in equilibrium with each other in gaseous sulfur, namely S, S_2, S_6, and S_8. Typical equilibrium vapor compositions at 1 atm total pressure are as follows (120), with monomeric sulfur becoming important only above 1100°C:

Temperature, °C	Partial Pressures, mm Hg		
	S_8	S_6	S_2
450	327	405	28
500	237	435	87
550	149	402	209
600	70	292	398
650	22	149	589
700	6	68	686
750	1	30	729
800	—	12	748
850	—	8	752

The heat capacity of S_2 gas has been determined spectroscopically and the heat capacities of S_6 and S_8 have been estimated (69) as

$$S_2(g): \quad C_p = 7.75 + 0.000888T \text{ (cal/g formula wt)}$$
$$S_6(g): \quad C_p = 19.25 + 0.002664T$$
$$S_8(g): \quad C_p = 25.00 + 0.003552T$$

The reactions in gaseous sulfur systems and the associated enthalpies (heats of dissociation) are

$$S_6(g) = 3S_2(g) \qquad \Delta H = 64{,}090 + 4T \text{ cal/mole } S_6$$
$$S_8(g) = 4S_2(g) \qquad \Delta H = 95{,}200 + 6T \text{ cal/mole } S_8$$

where T is in °K (147). These reactions are relatively rapid, and proceed to equilibrium.

(a) Is the value of ΔH for the formation of S_2 from S_8 as calculated at 600°C consistent with the data on equilibrium vapor compositions presented above?

(b) Sulfur vapor from a sulfur boiler-superheater is mixed with nitrogen in the ratio of 32 lb sulfur for every 2 lb moles of nitrogen. What are the partial pressures of S_2, S_6, and S_8 in this mixed gas at 600°C and 1 atm?

(c) Estimate the amount of heat needed in calories to bring a pure equilibrium sulfur vapor from 500°C to 800°C at 1 atm total pressure, starting with 1 g mole of mixed gas.

PROBLEM 1-4

Regeneration of molecular chlorine from by-product HCl represents a disposal problem in many localities. The possibility of using the so-called Deacon reaction for this purpose has been recognized for years.

$$2HCl + \tfrac{1}{2}O_2 = H_2O + Cl_2 \qquad \Delta H_{800°K} = -14 \text{ kcal}$$
$$\log K = \frac{3050}{T} - 3.55$$

The equilibrium constant has units of $(\text{atm})^{-1/2}$, while T is in °K. This exothermal reaction is relatively slow, but responds readily to catalysts like copper and iron chlorides (105) down to about 400°C.

Despite much study, the Deacon reaction has not achieved extensive commercial success, both because of corrosion and of difficulties in chlorine recovery. In one possible operating scheme, the off-gases from the reactor are scrubbed with water to remove unreacted HCl, then with strong H_2SO_4 to remove H_2O vapor, and finally compressed and cooled to condense chlorine. Condensation is made difficult by the presence of nitrogen and unconsumed oxygen. The vapor pressures of chlorine are 1 atm and 115 mm Hg at −34.6 and −70°C, respectively.

(a) The feed gas to a Deacon reactor is reported to be a mixture of 10 volumes of dry HCl gas per volume of "oxygen" (95% O_2, 5% N_2). Assuming equilibrium to be attained at 600°C and 1.2 atm pressure, what is the analysis of the product gas, both on a "wet" and "dry" basis?
(b) The product gas is water-scrubbed to remove HCl, dried with H_2SO_4, and then compressed to 50 psia. If not more than 1% of the chlorine is to be lost in the condenser off-gases, to what temperature must they be cooled?
(c) What is the temperature at which the first droplet of liquid chlorine forms in this compressed gas during the initial cooling? What recommendations would you make regarding the design of this condenser system?
(d) Refrigeration is costly. What other methods might be used for chlorine recovery? Why was so high a ratio of HCl to O_2 used in the feed?

PROBLEM 1-5

Hydrogen cyanide, a highly toxic material boiling at 25.7°C, is made from methane and anhydrous ammonia by the so-called Andrussov process. The key reaction involved is as follows:

$$CH_4 + NH_3 = HCN + 3H_2 \qquad K_p = 380 \text{ at } 1300°K$$

This reaction is highly endothermal, but the necessary heat can be provided by combustion of part of the hydrogen and methane present with air.

In an experimental operation, dry air, methane, and ammonia are metered, mixed, and fed to an adiabatic reactor containing several layers of platinum-rhodium screen as catalyst. On a dry basis, the off-gases are found to analyze 8.9% HCN, 4.5% NH_3, 6.0% CO, 9.4% H_2, and 71.2% N_2 (1).

(a) How could this gas have been analyzed? Calculate the composition of the mixed feed gas.
(b) The mole ratio of methane to ammonia in the feed is known to be unity, while moisture in the feed is negligible. Do you have confidence in the answer obtained in part (a) above?
(c) In small experimental apparatus, operations are not adiabatic, hence

it is necessary to estimate heat requirements. How much heat must be supplied or removed from this system (in Btu/lb mole of HCN product) if the feed enters at 25°C and 1 atm total pressure, and if the off-gases are at 1300°K?

(d) Do you think this operation could be "scaled up" to large plant size? What recommendations would you make for control of temperature, if an adiabatic reactor is to be used?

	Enthalpies, kcal/g mole*							
	NH_3	HCN(g)	H_2O(g)	CO	H_2	N_2	O_2	CH_4
$(\Delta H_f)_{298°K}$	−11.04	31.2	−57.8	−26.42	0	0	0	−17.89
$H_{1300°K} - H_{298°K}$	11.9	11.3	9.3	7.62	7.15	7.53	7.97	14.70

*References (78, 131).

PROBLEM 1-6

A 50% by weight solution of nitric acid in water is fed to an evaporator, wherein this acid stream is completely vaporized. These nitric acid vapors are next heated to a temperature of 400°K, whereupon the following reactions occur (20):

$$NO + \tfrac{1}{2}O_2 = NO_2 \qquad K_{p400°K} = 3.95 \times 10^3$$

$$2NO_2 = N_2O_4(g) \qquad K_{p400°K} = 0.196$$

$$3NO_2(g) + H_2O(g) = 2HNO_3(g) + NO(g) \qquad K_{p400°K} = 1.87 \times 10^{-4}$$

$$\tfrac{1}{2}N_2 + \tfrac{1}{2}O_2 = NO; \qquad K_{p400°K} = 2.5 \times 10^{-8}$$

It is of course to be recognized that NO itself is not thermodynamically stable at 400°K, and should in theory be entirely decomposed into its elements. In fact, however, rates of decomposition at this temperature are negligible (146).

(a) What is the equilibrium gas composition of the system at 400°K and atmospheric pressure?

(b) What would be the equilibrium gas composition assuming that decomposition of NO is effectively catalyzed?

PROBLEM 1-7

Commercial xylene, whether derived from the by-product coke industry or from petroleum, contains all four aromatic C_8 isomers in a mixture which is not far from thermodynamic equilibrium at 1000°K:

Compound	Commercial Xylene, Weight %	Freezing Point, °C	Boiling Point, °C	$\Delta F^{\circ *}_{298°K}$ kcal	$\Delta F^{\circ *}_{500°K}$ kcal
Ortho-xylene	23	−25.2	144.4	29.18	46.84
Meta-xylene	43	−47.8	139.1	28.41	45.89
Para-xylene	19	+13.2	138.4	28.99	46.86
Ethylbenzene	15	−94.0	136.2	31.208	48.554

*Free energy of formation from the elements (130).

Three of these four isomers are used in industrial synthesis, the ortho- being oxidized to phthalic anhydride, the para- converted to terephthalic acid, and ethylbenzene is dehydrogenated to styrene. Because differences in boiling temperatures and other physical properties are small, recovery of pure isomers is difficult, but essential because of rigid specifications. Thus the ortho-isomer and ethylbenzene are separated from the mixed xylenes in powerful fractionating towers, while the para-isomer is recovered by crystallization.

In the plant under consideration, sales outlets exist only for the ortho- and para-isomers, consequently the remaining meta-isomer and ethylbenzene are isomerized at 500°C and 1 atm pressure, followed by separation of the mixed product as before. What would be the composition of the stream leaving the reactor, if equilibrium were attained with perfect yields at 500°K and 1 atm total pressure? What would be the composition of the product stream after separation of the meta-xylene and ethylbenzene? What would be the composition of both streams, if the reactor were operated at 5 atm total pressure?

Note: In many simpler systems of simultaneous reactions such as this one, it is possible to determine equilibrium composition without resorting to the "series reactor" technique or its equivalent. That is, a little algebraic ingenuity will yield the answer directly.

PROBLEM 1-8

Methanol is produced by high pressure reaction from a "synthesis" gas mixture containing primarily H_2 and CO. Minimized by-product production in the methanol converter requires a H_2/CO ratio of 2.1 to 2.2 in the synthesis gas feed.

(a) Hot synthesis gas having a H_2/CO ratio of 2.15 is to be made by passing methane, steam, and CO_2 through a single catalytic converter under conditions such that no solid carbon can form. It is maintained that a synthesis gas of this desired composition cannot be produced from steam and methane alone, or from CO_2 and methane alone, but that a mixture of H_2O, CO_2, and CH_4 must be charged to the synthesis gas reactor to produce this gas. Do you agree? Where would this CO_2 come from? Propose an operating scheme for making this gas.

(b) It is proposed to feed a synthesis gas reactor with a mixture analyzing 50% CH_4, 10% CO_2, and the rest H_2O. What would be the composition of the final equilibrium gas when operating at 3 atm pressure and 1400°K? What would be the heat requirements in an isothermal operation at 1400°K, in Btu/lb hydrogen produced?

	Btu/lb mole, 1400°F	$K_{p\ 1400°F}$
$CH_4 + H_2O = CO + 3H_2$	$\Delta H° = 97{,}200$	0.6343×10^2
$CO_2 + H_2 = CO + H_2O$	$\Delta H° = 14{,}800$	0.83

PROBLEM 1-9

The classical routes to acrylonitrile, an important monomer for plastics are (1) the addition of HCN to acetylene, (2) the dehydration of ethylene cyanohydrin made from ethylene oxide and HCN, and (3) the reaction of HCN and acetaldehyde. A fourth process, involving the reaction of ammonia, propylene, and air oxygen, has recently become very attractive because of the development of a new catalyst, namely bismuth phosphomolybdate supported on silica (134, 155):

$$C_3H_6 + NH_3 + \tfrac{3}{2}O_2 = CH_2{:}CHCN + 3H_2O$$

This is one of the few reactions in which a catalytic surface plus three rather than two reactants are involved in making a single major product with good selectivity (140). The propylene used may contain large percentages of propane, ethane, and ethylene, since these act as inert substances, but should be free of butylenes to avoid undesired by-product formation.

Industrially, the reaction is carried out in fluid bed units (see Chapter 9 for description) equipped with cooling tubes through which water circulates. The catalyst particles in these reactors are fluidized by the reacting gases which leave the reactor at 840°F and 2 atm. In the case under consideration, pure propylene, ammonia, and air are fed continuously to the reactor in the molal proportions of 1:1.07:9. For 1 lb of propylene fed, 1.01 lb of acrylonitrile are produced, together with 0.09 lb of acetonitrile and 0.08 lb of HCN. The product gases are scrubbed with a sulfuric acid containing dissolved ammonium sulfate to remove any unreacted ammonia, then with water to remove the nitriles and HCN. Acrylonitrile is recovered from this scrubbing liquor by fractionation (142, 155).

THERMODYNAMIC DATA*

	Average MC_p, Btu/(lb mole)(°F)		Enthalpies and Free Energies of Formation at 77°F, kcal/g mole	
	77–300°F	77–840°F	$\Delta H°$	$\Delta F°$
$CH_2{:}CHCH_3(g)$	17.1	21.5	4.88	14.99
$CH_2{:}CHCN(g)$	26.7	34.0	44	45.37
$CH_3CN(g)$	13.7	16.2	21	25.2
$NH_3(g)$	8.9	10.0	−11.04	−3.981
$HCN(g)$	8.4	9.0	31.2	28.7
$H_2O(g)$	8.1	8.4	−57.8	−54.6
CO_2	9.4	10.6	−94.05	−94.26
O_2, N_2	7.0	7.2	0	0

*References (78, 130).

(a) In making a preliminary analysis of these data, it is desired to know what fraction of the ammonia fed reappears in the off-gases.
(b) No solid carbon forms in this operation. The off-gases contain CO_2 but no CO, molecular hydrogen, or hydrocarbons. Is there molecular oxygen in the off-gases?
(c) What is the reactor off-gas composition expressed in mole percent?
(d) Discuss this reactor from an equilibrium point of view.
(e) If the gases enter at 300°F, how much heat must be removed from the reactor per mole of acrylonitrile formed?

PROBLEM 1-10

Methylamines are made by passing a gaseous mixture of methanol and ammonia over activated alumina as catalyst, whereupon the following three reactions occur:

$$CH_3OH + NH_3 = CH_3NH_2 + H_2O \qquad \log K^* = 1.85$$
$$2CH_3OH + NH_3 = (CH_3)_2NH + 2H_2O \qquad \log K^* = 4.11$$
$$3CH_3OH + NH_3 = (CH_3)_3N + 3H_2O \qquad \log K^* = 6.965$$

It may be assumed that no cracking occurs in this operation. Calculate the equilibrium off-gas composition at 700°K for this system of reactions when the feed is 60 mole % ammonia and the rest methanol.

PROBLEM 1-11

With increasing sulfur prices and with stricter laws governing atmospheric pollution, recovery of sulfur from waste H_2S becomes more important. The

*At 700°K (78).

Claus process or some modification thereof, generally operated for sulfur recovery, is based upon partial combustion of the H_2S.

$$H_2S + \tfrac{1}{2}O_2 = H_2O + [S]$$

where [S] represents the equilibrium mixture of gaseous S_2, S_6, and S_8 at operating conditions. The equilibrium constant for the foregoing reaction is very large over the industrially attainable temperature range. The following reactions also occur:

$$2H_2O + \tfrac{3}{2}S_2 = SO_2 + 2H_2S \qquad K_p = 0.13\ \text{atm}^{-1/2} \text{ at } 1100°\text{K}$$

$$H_2S + \tfrac{3}{2}O_2 = H_2O + SO_2 \qquad K_p \text{ is very large at } 1100°\text{K}$$

$$S_6 = 3S_2 \qquad K_p = 11.3\ (\text{atm})^2 \text{ at } 1100°\text{K}$$

$$S_8 = 4S_2 \qquad K_p = 1.55 \times 10^5\ (\text{atm})^3 \text{ at } 1100°\text{K}$$

Operations involve passing a gaseous mixture of H_2S and air metered so that the mole ratio of H_2S/O_2 is 2, continuously through a catalyst bed comprising granules of bauxite. Rates are high above 250°C, providing no sulfur condenses, and equilibrium is attained (39). Calculate the composition (in mole per cent) and the total pressure of the equilibrium gas at 1100°K, when the partial pressure of S_2 is 0.1 atm.

PROBLEM 1-12

Carbonyl sulfide can be made at reasonable rates by passing carbon disulfide mixed with carbon dioxide over a magnesia catalyst at temperatures over 400°C (35). An equimolal mixture of these two reagents is to be passed through a reactor of proper design, whereupon total equilibrium is attained. The following reactions will presumably all occur:

THERMODYNAMIC DATA*

	K_p at 900°K	$\Delta H°_{900°\text{K}}$, kcal
$C + S_2(g) = CS_2$	13.5	−3.1
$CO_2 + CS_2 = 2COS$	9.4	−1.7
$2COS = 2CO + S_2$	0.0015	45.5
$5CO_2 + CS_2 = 2SO_2 + 6CO$	10^{-15}	141.7
$3S_2(g) = S_6(g)$	0.89	−64.4
$4S_2(g) = S_8(g)$	0.193	−95.5

*References (29, 39).

(a) What is this system's composition at total equilibrium at 900°K and 1 atm pressure?
(b) How much heat must be supplied to or removed from the reactor in Btu/lb mole of COS produced, assuming that the feed and product gases are at 900°K?
(c) How could this energy be added or removed?

PROBLEM 2-1

An evacuated box having a constant volume of 200 cu ft and containing 10 lb of solid carbon is maintained at 1600°F. Into the box is now injected exactly 1 lb mole of CO_2. What is the total pressure in the box after equilibrium is attained at 1600°F? Repeat assuming the solid carbon initially present weights 1 lb.

Note:

$$C + CO_2 = 2CO; \qquad K_p = 2.5 \text{ atm at } 1600°F.$$

The volume of the solid carbon in the box is negligible.

PROBLEM 2-2

A gas analyzing 10% H_2, 70% CH_4, and the rest nitrogen is to be preheated before being burned in a high temperature furnace. Operations are at 2 atm total pressure. Over what temperature range, if any, can this heater be operated without running the risk of carbon deposition on the walls? What is the equilibrium gas composition at 2 atm and 900°F?

Note:

$$C + 2H_2 = CH_4 \qquad \log K_p = 0.4837 \text{ at } 900°F$$

$$\log K_p = -1.1581 \text{ at } 1400°F$$

PROBLEM 2-3

In an experimental process for powdered iron, powdered FeO is mixed with an excess of a high grade carbon powder and a trace of starch paste to serve as binder, pressed into 2-in. pellets and dried. These pellets are next charged to an externally fired retort, suitably vented so that gas can escape when the internal pressure exceeds 1 atm. The retort is evacuated before heating. Free energy data for the relevant reactions are as follows (22):

$$C + O_2 = CO_2 \qquad \Delta G^\circ = -94{,}200 - 0.2T$$

$$2C + O_2 = 2CO \qquad \Delta G^\circ = -53{,}400 - 41.9T$$

$$2Fe + O_2 = 2FeO \qquad \Delta G^\circ = -126{,}400 + 30.9T$$

These equations apply from 298°K to over 1600°K. The standard free energy change, namely ΔG°, is in cal/g mole.

(a) For operations at 1 atm pressure, what is the lowest temperature to which the charge must be heated for reaction to proceed to completion, if the briquettes contained 25% by weight of carbon?

(b) What is the equilibrium off-gas composition? How much heat must be supplied, in Btu/lb of iron produced, assuming operations to be isothermal throughout?

(c) In your opinion, must the retorts be evacuated before heating? Discuss.

PROBLEM 2-4

Sodium bicarbonate decomposes thermally by the following reaction:

$$2NaHCO_3 = Na_2CO_3 + H_2O + CO_2$$

The total decomposition pressure π of CO_2 plus H_2O over sodium bicarbonate–carbonate mixtures is reported in Reference (61), Vol. 7, 305 to be

$$\log \pi = 8.9377 - \frac{3340}{T}$$

where π is in atm and T is in °K.

A batch of 84 lb of pure $NaHCO_3$ is to be fluidized* at 1 atm pressure, using a gas feed analyzing 80% of CO_2 and 20% of N_2. Such fluidized beds* of solids characteristically show excellent temperature uniformity due to vigorous solids mixing. Electrical heaters are to be submerged in the bed of the proposed operation, with control such that the bed temperature will be uniform at all times at 77°C. The off-gases will be at equilibrium with the bed solids with respect to the above reaction.

(a) If gas and solids both enter and leave at 77°C, how many moles of feed gas must be supplied to carry out this decomposition?

(b) How many kw-hr of electrical energy per lb of soda ash produced are required by this decomposition?

(c) It is now decided to supply the necessary heat as sensible heat of the

*See Chap. 9 for description of fluidization.

feed gases, rather than electrically. The off-gases and the bed itself are to be again at 77°C. What must the feed gas preheat temperature be?

(d) If operations were continuous rather than batch, would the answers to the foregoing questions be different?

Note: Assume for this problem only that MC_p for N_2, CO_2, and H_2O(g) are 7, 10.5, and 10.5 Btu/(lb mole)(°R), respectively.

PROBLEM 2-5

It is proposed to produce a special grade of $Ca(OH)_2$ from limestone, namely $CaCO_3$, by reaction with superheated steam. The product is to contain no CaO or $CaCO_3$ and, during reaction, it is considered important for CaO never to be allowed to form. Operations are to involve passing steam at high pressure and high temperature through an externally heated rotating drum in which a batch of $CaCO_3$ is to be tumbled. This drum cannot be operated above 10 atm pressure.

Since rates of reaction generally improve with temperature, this operation is to be carried out at the highest safe temperature. What is this temperature? Assuming equilibrium to be attained, how many lb of steam would be required per pound of solid product, if operations were kept isothermal at this highest possible temperature by suitable heating, and if the $CaCO_3$ were charged at this temperature?

Note: Dissociation pressures of CO_2 over $CaCO_3$ are 0.229 atm at 800°C and 1.00 atm at 894°C. Dissociation pressures of water over $Ca(OH)_2$ are 0.128 atm at 450°C and 1.0 atm at 541°C, Reference (61), Vol. 7, 294.

PROBLEM 2-6

Hydrodealkylation is the reaction of an alkylaromatic compound and hydrogen to form the parent aromatic compound and the corresponding alkane, as in the conversion of toluene to benzene in reaction (a) below. Reactions (b), (c), and (d) also tend to occur and are, of course, undesirable.

Reactions	log K_p(130)		
	800°F	1160°F	
$C_6H_5CH_3(g) + H_2 = C_6H_6(g) + CH_4$	3.17	2.36	(a)
$C_6H_6(g) + 3H_2 = \text{cyclo}C_6H_{12}(g)$	−4.26	−7.92	(b)
$CH_4 = C(c) + 2H_2$	−0.95	0.49	(c)
$\text{cyclo}C_6H_{12}(g) + 6H_2 = 6CH_4$	25.02	18.96	(d)

Commercial hydrodealkylations are often carried out without a catalyst at elevated temperatures. Operations are usually above 1000°F both to limit hydrogenation of benzene to cyclohexane by reaction (b), and to limit methane formation by reaction (d). Both these reactions are undesirable, since they increase hydrogen consumption, diminish benzene yields, and evolve heat. Methane formed by reactions (a) and (d) may decompose by reaction (c), but such cracking cannot occur in the presence of sufficient hydrogen.

Catalytic hydrodealkylations are also operated industrially, with typical mole feed ratios of hydrogen to toluene of about 0.6. As Asselin (4) points out, catalytic operations are in general worth considering when catalysts are available which increase reaction rates, thus reducing reactor size, lowering operating temperatures, or increasing conversion per pass. Catalytic operations may also be justified when reaction selectivity is enhanced, meaning a reduction in the extent of side reactions (b) and (d). At least one of these advantages must be great enough to offset catalyst investment, and to pay for its periodic replacement or reactivation.

(a) Reconcile the statement "cracking [as by reaction (c)] cannot occur in the presence of sufficient hydrogen" with the generalization made by one of your associates "regardless of the magnitude of its equilibrium constant, no reaction ever quite goes to completion."

(b) A gaseous mixture of toluene and hydrogen is to be reacted to make benzene. For an operation at 18 atm in which the product gases are at 1160°F, what should be the feed composition in order to avoid carbon deposition by reaction (c) and yet have a maximum yield of benzene? Assume equilibrium is attained by reaction (a), and that no cyclohexane forms because reaction (b) is uncatalyzed. Assume that the perfect gas laws apply.

PROBLEM 2-7

A certain porous ore contains iron and nickel orthosilicates (Fe_2SiO_4 and Ni_2SiO_4), which are reducible with CO, as follows:

$$\tfrac{1}{2}Fe_2SiO_4 + CO = Fe + \tfrac{1}{2}SiO_2 + CO_2 \qquad (1)$$

$$\tfrac{1}{2}Ni_2SiO_4 + CO = Ni + \tfrac{1}{2}SiO_2 + CO_2 \qquad (2)$$

The following reactions are also relevant:

$$FeO + CO = Fe + CO_2 \qquad (3)$$

$$NiO + CO = Ni + CO_2 \qquad (4)$$

$$C + CO_2 = 2CO \qquad (5)$$

The silicates in this ore are so intimately admixed that separation by physical means is not feasible. It is proposed to reduce the nickel selectively to metal, leaving the iron as the silicate. The nickel metal can then be separated from the partly reduced ore mass by leaching. To avoid losing porosity by fusion, temperature during reduction should be kept below 1000°C. Again, to maintain porosity, no solid carbon is to form in the reduced product.

Equilibrium constants (82) are as follows at around 900–1000°C.

$$\log K_p = \frac{58}{T} - 0.841 \tag{1}$$

$$\log K_p = \frac{1661}{T} - 0.0277 \tag{2}$$

$$\log K_p = \frac{950}{T} - 1.145 \tag{3}$$

$$\log K_p = \frac{2201}{T} + 0.0014 \tag{4}$$

$$\log K_p = -\frac{8950}{T} + 9.37 \tag{5}$$

In the foregoing expressions, K_p is in atm and T in °K.

(a) Plot $\log (p_{CO_2}/p_{CO})$ versus $1/T$ at a total pressure of 1 atm for each of the above reactions, including reaction (5), and show the range of conditions at 1 atm total pressure for a gas system containing only CO and CO_2 over which a nickel silicate-iron silicate mixture can be selectively reduced to metallic nickel without risking formation of metallic iron or the deposition of carbon.

(b) How could operations be conducted to keep the temperature at around 900°C, and the gas composition within the safe range? It will be assumed that there is a source of pure CO within the plant.

PROBLEM 2-8

A hot gas stream under 5 atm pressure and analyzing 10% CO, 3% CO_2, 12.6% H_2, 5% H_2O, and the rest nitrogen is to be passed over a suitable iron catalyst for hydrogen production by the water-gas shift reaction:

$$H_2O + CO = H_2 + CO_2 \tag{a}$$

Removal of CO_2 from the product gas from a shift reactor typically involves cooling the gas, then scrubbing with a selective solvent for CO_2 such as aqueous carbonate or an ethanolamine solution.

It is proposed that increased conversions could be attained by removing the CO_2 as fast as it forms in the reactor itself, thus forcing reaction (a) to the right. Such CO_2 removal could be accomplished by continuously injecting lime directly into the reactor, where $CaCO_3$ would then form

$$CaO + CO_2 = CaCO_3 \qquad \text{(b)}$$

Solid would be withdrawn at a suitable rate from the reactor, calcined to convert any $CaCO_3$ present back to CaO, and recycled to the reactor. To make reactions (a) and (b) occur simultaneously, a mixture of pellets of lime and of alumina carrying the iron catalyst are to be charged to a fluid bed reactor. Under fluidizing conditions in the reactors, these two types of pellets will be well mixed. A side stream of these pellets is withdrawn from the reactor and separated magnetically into pellets of $CaCO_3$ and catalyst. The $CaCO_3$ pellets are calcined to lime and recycled to the bed, while the catalyst pellets are returned directly to the bed without treatment. The catalyst pellets do not pass through the calcining step since it is believed that the high temperature of calcining would destroy their catalytic effectiveness.

This proposal seems interesting and deserves testing by you. The "wet" product gas is not to contain over 1% CO_2. It may be assumed that both reactions (a) and (b) proceed to equilibrium in the reactor. Calculate the composition of the product gases at 5 atm from the "mixed reactor" recognizing that the perfect gas laws are closely obeyed at these conditions in this mixture. What would be the gas composition under these conditions if no lime were present? Equilibrium expressions for these two reactions are as follows (103):

$$\text{reaction (a):} \qquad \ln K = \frac{4580}{T} - 4.33$$

$$\text{reaction (b):} \qquad \ln K = \frac{19630}{T} - 16.833$$

The pressures in these equilibrium constant expressions are in atm and T is in °K. Propose a flow sheet for this operation and discuss any features of the proposed operation which may involve difficulties or requires testing.

PROBLEM 2-9

A gas formed by burning elemental sulfur with dry air at atmospheric pressure analyzes 1.5% O_2, 1.0% SO_3, 18% SO_2, and the rest N_2. This burner gas, available at 1000°C, is to be used for the preparation of digester liquor in the paper industry. Desirably, both the SO_3 and the oxygen are to be eliminated from this hot gas before use. It is proposed that this might be done by passing the crude burner gas through a fixed bed, whose initial composition is that of magnetite (Fe_3O_4). When "exhausted," this bed will be regenerated by blowing through it a hot gas containing 4.75% sulfur vapor (which may be taken as S_2), SO_2, and N_2. Such a gas can be made, of course, by operating the sulfur burner on the "rich" side, rather than on the "lean" side.

In testing this proposal, it will be assumed that equilibrium is attained since all reactions are fast, and that operations are isothermal at 1000°C throughout.

Reactions	$\Delta F°$, kcal*			
	100°C	800°C	1000°C	
$3Fe_2O_3 + SO_2 = 2Fe_3O_4 + SO_3$	38.9	28.6	18.4	(1)
$2Fe_3O_4 + \frac{1}{2}O_2 = 3Fe_2O_3$	−35.0	−28.8	−22.6	(2)
$Fe_3O_4 + SO_2 = 3FeO + SO_3$	59.6	42.8	40.8	(3)
$\frac{1}{2}Fe_3O_4 + \frac{5}{4}S_2 = \frac{3}{2}FeS + SO_2$	−10.4	−10.6	−10.5	(4)
$6Fe_2O_3 + \frac{1}{2}S_2 = 4Fe_3O_4 + SO_2$	−1.2	−10.2	−19.0	(5)
$10Fe_2O_3 + FeS = 7Fe_3O_4 + SO_2$	5.0	−10.5	−24.7	(6)
$Fe_3O_4 = 3FeO + \frac{1}{2}O_2$	63.5	42.6	36.6	(7)

*Reference (85).

The off-gas from the bed is to be a mixture of SO_2 and N_2, with no more than a trace of SO_3, O_2, and S_2.

(a) Can a suitable gas be made in the "work" part of the cycle?
(b) Can the exhausted bed be regenerated as proposed? Will the product gases meet specifications?
(c) If one or both parts of this cycle are workable, present for each the heat requirements to keep the bed isothermal at 1000°C, in kcal/100 moles of gas fed.
(d) How would you propose to control the reactor, so that changeover from oxidizing to reducing conditions came at the right time?
(e) What important questions would require pilot plant answers?

PROBLEM 2-10

(a) A finely divided mixture of ZnO and carbon is briquetted and placed in a retort. After evacuating and sealing, this retort is heated externally, whereupon the following two reactions proceed to equilibrium:

$$C(c) + CO_2 = 2CO \tag{a}$$

$$ZnO(c) + CO = Zn(g) + CO_2 \tag{b}$$

Show that this system is monovariant. Prepare a plot of log pressure in atm versus $1/T$, where T is in °K, presenting curves for the equilibrium partial pressures of CO, CO_2, and zinc in this system, as well as the total pressure π. Now plot the vapor pressure of pure liquid zinc on this diagram. Within what temperature range, if any, will a liquid zinc phase form in the retort?

(b) This same operation is now carried out in a retort so designed that gas will be vented when the internal pressure exceeds 1 atm. The above two reactions again proceed rapidly, and equilibrium is substantially attained at all temperatures. At what minimum operating temperature will gas be evolved from the retort?

DATA*

Temperature, °C	K_a, atm	K_b, atm	Zinc Vapor Pressure, atm
600	0.10	1×10^{-5}	0.0145
800	7.3	1.23×10^{-3}	0.309
900	37	7.38×10^{-3}	0.948
1000	130	7.29×10^{-2}	2.41

*Reference (150).

PROBLEM 2-11

Magnesium metal is made by the melt electrolysis of magnesium chloride. This salt must be as nearly anhydrous as possible, since it reacts with water under cell conditions to form insoluble compounds like MgO and MgOHCl which do not electrolyze to metal. Moreover, these materials promote formation of stable metal droplets in the electrolyte, preventing formation of a metal layer in the cell and so interfering with metal harvesting. Preparation of suitably dry cell feed is made difficult by the presence of various hydrates as well as by the hydrolysis reactions, as listed:

$$MgCl_2 \cdot 6H_2O(c) = MgCl_2 \cdot 4H_2O(c) + 2H_2O(g) \qquad (1)$$

$$MgCl_2 \cdot 4H_2O(c) = MgCl_2 \cdot 2H_2O(c) + 2H_2O(g) \qquad (2)$$

$$MgCl_2 \cdot 2H_2O(c) = MgCl_2 \cdot H_2O(c) + H_2O(g) \qquad (3)$$

$$MgCl_2 \cdot H_2O(c) = MgCl_2(c) + H_2O(g) \qquad (4)$$

$$MgCl_2(c) + H_2O(g) = MgOH \cdot Cl(c) + HCl(g) \qquad (5)$$

$$MgCl_2 \cdot H_2O(c) = MgOHCl(c) + HCl(g) \qquad (6)$$

$$MgOHCl(c) = MgO(c) + HCl(g) \qquad (7)$$

$$MgCl_2(c) + H_2O(g) = MgO(c) + 2HCl(g) \qquad (8)$$

Equilibrium Pressures in Atm*

Reaction No.	1	2	3	4	5	6	7	8
	p_w	p_w	p_w	p_w	p_w***	p_{HCl}	p_{HCl}	p_w***
298°K	1×10^{-3}	6.15×10^{-4}	1.41×10^{-5}	5.64×10^{-8}	0.0078	7.71×10^{-7}	2.80×10^{-12}	1.0
350	3.2×10^{-2}	3.48×10^{-3}	9.77×10^{-4}	8.13×10^{-6}	0.125	6.44×10^{-5}		
389.9	2.47×10^{-1}**							
400		6.24×10^{-2}	2.00×10^{-2}	2.8×10^{-4}	0.172	1.51×17^{-3}	1.30×10^{-7}	0.999
450			2.06×10^{-1}	4.36×10^{-3}	0.218	1.75×10^{-2}		
454.7		6.03×10^{-1}**						
500			1.30	3.84×10^{-2}	0.259	1.20×10^{-1}	6.68×10^{-5}	0.986
600				9.42×10^{-1}	0.33	2.08	4.27×10^{-3}	0.913
700					0.38		7.36×10^{-2}	0.713
800							6.12×10^{-1}	0.410

*Reference (71).

**Temperature at which a saturated solution phase forms.

***The associated value of p_{HCl} is such that $(p_w + p_{HCl})$ equals 1 atm at all times.

Values of p_w, the equilibrium decomposition pressures of water over reactions (1) through (4), and of p_{HCl}, the decomposition pressure of HCl for reactions (6) and (7), are presented below at various temperatures. Also presented are the equilibrium pressures of water over the heterogeneous reactions (5) and (8), assuming that the total pressure is 1 atm and that the only other gas phase component is HCl.

Hexahydrate decomposes to tetrahydrate and saturated aqueous solution at 390°K, where $p_w = 0.247$ atm. Tetrahydrate decomposes to the dihydrate and saturated aqueous solution at 454.7°K, where p_w is 0.603 atm. The dihydrate decomposes to the monohydrate at temperatures such that p_w is above 1 atm. The same is true for the decomposition of the monohydrate to anhydrous salt. The anhydrous chloride melts at 985°K.

Vapor pressure data of water for saturated solutions of magnesium chloride, as well as for water at equal temperatures, are presented below:

SATURATED MAGNESIUM CHLORIDE SOLUTIONS

T, °K	p, atm	Conc. g/1000 g Water	Solid Phase	$p°$, Pure Water, atm
373	0.218	730	$MgCl_2 \cdot 6H_2O$	1.0
389.9	0.231*	855	$MgO_2 \cdot 6H_2O + MgO_2 \cdot 4H_2O$	1.78
393	0.253	864	$MgCl_2 \cdot 4H_2O$	1.89
413	0.416	918	$MgCl_2 \cdot 4H_2O$	3.57
413	0.602	1001	$MgCl_2 \cdot 4H_2O$	6.10

*This pressure is in reasonable agreement with that reported over solid $MgCl_2 \cdot 6H_2O$ at this temperature.

(a) Prepare a graph having as ordinate the partial pressure of water, namely p_w, and the temperature in °K as abscissa. The range for p_w is from 0 to 1.0 and that for T is from 500°K to 750°K. Plot the equilibrium values of p_w for reactions 1 through 8, recognizing that the gas phase contains HCl and H_2O vapor with a total pressure of 1 atm. Sketch the location of the vapor pressure curve for the saturated solution.

(b) Indicate the composition of the equilibrium solid phases which exist in the various zones of this diagram, assuming no melting to occur within this range.

(c) Propose a process for atmospheric pressure operation whereby anhydrous $MgCl_2$ could be made from $MgCl_2 \cdot 6H_2O$, in which no liquid phase occurs at any time.

PROBLEM 2-12

A plant contains three identical fixed bed kilns with No. 1 kiln initially filled with 6 lb moles of Fe_2O_3, No. 2 initially filled with 4 lb moles of Fe_3O_4, and No. 3 initially filled with 12 lb moles FeO. All kilns operate at atmospheric pressure and are kept isothermal at all times at 1500°F. A mixture of 50% H_2 and 50% N_2 is blown into the bottom of kiln 3. The product gas from the top of kiln 3 enters the bottom of kiln 2, the top gas from kiln 2 enters the bottom of kiln 1, and the top gas from kiln 1 is vented. It may be assumed that gas "hold up" within the kiln system is negligible. Gas flow is uniform across the beds at all times, consequently "piston flow" is achieved. All reaction rates are great, and equilibrium is attained at all points. The iron oxides in the kilns are present in the form of small uniform porous pellets, hence it is believed that reactions take place within relatively narrow horizontal zones which move up through the kilns as reactions proceed.

(a) What solid phases and how much of each phase (in lb moles) is present in each of the vessels after exactly 1 lb mole of gas mixture is passed through the system?

(b) As operations proceed, the amount of Fe_3O_4 present in kiln 2 diminishes and finally becomes zero. At the moment when Fe_3O_4 vanishes in kiln 2, what is the composition and amount of each solid phase present in kiln 2? In kiln 1? In kiln 3?

Data for 1500°F:

$$3Fe_2O_3 + H_2 = 2Fe_3O_4 + H_2O \qquad K_p = 98.0$$

$$Fe_3O_4 + H_2 = 3FeO + H_2O \qquad K_p = 1.86$$

$$FeO + H_2 = Fe + H_2O \qquad K_p = 0.37$$

PROBLEM 2-13

Oxidations of metal sulfides are vigorously exothermal, and almost invariably show very favorable equilibrium constants even at elevated temperatures. Designating the sulfide of a divalent metal as MS, then these oxidation reactions can be written:

$$MS + \tfrac{3}{2}O_2 = MO + SO_2(g) \tag{a}$$

Depending on the case, the metal oxide formed is either solid or molten.

The following partial oxidation reactions of metal sulfides also have very favorable equilibrium constants:

$$MS + O_2 = M + SO_2(g) \tag{b}$$

Reactions of type (b) are more attractive industrially than of type (a) because metal is produced directly, avoiding the subsequent step of reducing the metal oxide to metal.

(a) It is maintained that, by continuously blowing lead sulfide and air into a large pool of molten lead, when maintaining the mole ratio of PbS/O_2 at exactly unity, then only reaction (b) can occur if equilibrium is attained (98). In other words, any accidentally formed lead oxide will somehow react further to form metallic lead. Is this correct? If so, at what minimum temperature would you recommend operation, if at all?

(b) It is further maintained that, regardless of the ratio of air to metal sulfide, FeS would always react in accordance with reaction (a) rather than reaction (b) in the equilibrium case. In other words, unlike lead, metallic iron cannot be produced in a partial combustion operation. State and defend your position on this possibility.

MOLAL ENTHALPIES AND FREE ENERGIES OF FORMATION FROM THE ELEMENTS, KCALS

Compounds	$\Delta H^\circ_{298°K}$	$\Delta H^\circ_{1200°K}$	$\Delta F^\circ_{298°K}$	$\Delta F^\circ_{1200°K}$
PbO	−53.05	−44.1	−45.05	−24.45
PbS*	−37.7	−36.8	−31.9	−14.6
SO_2	−86.28	−86.39	−81.21	−65.45
FeS*	−37.8	−35.9	−32.64	−21.08
FeO	−63.5	−63.3	−58.15	−44.45

*Standard state for sulfur in these tables is S_2(g).

PROBLEM 3-1

The equilibrium diagram for the system sodium nitrate—sodium sulfate-water at 25°C is presented in Fig. 3-3.

(a) If 1 lb of darapskite is mixed with 0.5 lb of $Na_2SO_4 \cdot 10H_2O$ (Glauber's salt), what are the compositions and quantities of the phases formed after equilibrium is reached at 25°C?

(b) What is the composition of the saturated solution which (upon evaporation isothermally at 25°C until the liquid phase disappears) will yield pure darapskite? Will the composition of the solid change during this evaporation? Is there danger of contamination of the solid product? How much water must be evaporated in this process, expressed as lb/lb of solid?

(c) Supplies of $NaNO_3$ and Glauber's salt are available. Pure darapskite is desired, and it is considered inadvisable to depend upon the complete transformation of one solid such as Na_2SO_4 to darapskite during the precipitation process. Assuming all operations must be at 25°C, how would you proceed to produce uncontaminated darapskite?

PROBLEM 3-2

Relatively pure $NaNO_3$ and pure KNO_3 are to be separated from an intimate mixture of these two salts. A solution-precipitation procedure based upon solubility differences of these two salts is studied. An older engineer makes the familiar point "alternative methods should always be considered before making a final process selection." He suggests that fine grinding followed by a sink-float procedure may be effective for separation. Write a memorandum discussing this suggestion and considering these questions: What would be suitable liquids to use, what laboratory data would be required and how could they be obtained, and what would be a suitable flow sheet?

PROBLEM 3-3

A brine containing 20% $MgCl_2$ and 6% NaCl by weight becomes available. Solid magnesium chloride hexahydrate crystals are to be produced, containing less than 0.2% NaCl. Propose a possible operating procedure using batch operation. Propose a possible continuous operation. What sort of a development program would be required before undertaking plant design? The phase equilibrium data below are from Reference (61), Vol. VIII, 310.

	25°C		105°C	
Solid Phases	$MgCl_2$, %	NaCl, %	$MgCl_2$, %	NaCl, %
$MgCl_2 \cdot 6H_2O$	35.47	0	42.78	0
$MgCl_2 \cdot 6H_2O$ + NaCl	35.80	0.21	42.78	trace
NaCl	29.92	1.12	37.14	0.84
NaCl	24.40	4.11	34.55	1.16
NaCl	20.01	6.01	27.82	3.74
NaCl	16.84	8.79	26.69	4.17
NaCl	10.22	15.01	22.51	6.67
NaCl	5.95	19.5	13.09	14.44
NaCl	2.75	23.24	6.06	21.83
NaCl	0	26.27	1.66	26.35
NaCl			0	28.29

PROBLEM 3-4

In an experimental study at 55°C water, solid NaOH, and NaCN were mixed in such proportions that undissolved solids were always present in the equilibrium system obtained after prolonged agitation. In each run, samples were taken simultaneously of the clear saturated liquid-phase and of the associated "wet residues," namely the solids present together with some adhering mother liquor. Compositions are as reported below. The wet residue compositions presented are the average compositions of the solids and liquids in each sample.

COMPOSITIONS IN WEIGHT % AT 55°C*

Saturated Liquid		Wet Residues	
NaCN	NaOH	NaCN	NaOH
45.2	0	—	—
38.1	7.6	81.4	2.4
30.2	16.1	76.0	5.5
17.1	45.9	62.0	20.8
17.2	46.8	41.8	32.8
16.5	47.5	24.0	46.6
15.8	48.5	34.0	46.5
14.8	49.7	34.8	47.3
12.6	52.6	32.0	48.0
10.3	56.0	25.7	52.3
9.3	58.3	23.9	52.6
9.0	58.5	16.1	56.5
8.6	58.2	6.6	61.0
6.7	57.2	3.3	63.0
3.5	57.9	2.8	60.4

*Reference (108).

Construct the phase diagram for this system at 55°C on triangular coordinates. What composition of saturated solutions, upon evaporation, would yield the maximum amount of dry compound NaCN·NaOH without risking contamination with any other solids? What happens as increasing amounts of pure NaCN are added to a solution saturated with NaOH at this temperature? What happens as increasing amounts of pure NaOH are added to a solution saturated with NaCN? When NaOH·H_2O is added?

PROBLEM 3-5

Sodium nitrate is to be converted to KNO_3 by suitable addition of KCl. Solution-crystallization procedures are studied by your associates for carrying out this operation. An older engineer again makes the point "alternative methods should always be considered before any final process selection is made." He suggests that one alternative might involve the use of ion exchange resins. Write a memorandum discussing this suggestion, proposing possible flow sheets. Could energy consumption be kept at a low level?

PROBLEM 3-6

Pure KNO_3 is to be produced from a crude solid mixture of KNO_3, $NaNO_3$ and various insoluble compounds. The weight ratio of KNO_3 to total nitrate, namely, KNO_3 plus $NaNO_3$, is 23%. Operations are to be in a vessel equipped with an agitator, jacketed so that the vessel contents can be heated with steam and cooled with cooling water. Start-up is as follows: add to this vessel a batch of crude salt containing 100 lb of soluble nitrates, along with just enough water so that all of the KNO_3 but as little of the $NaNO_3$ as possible will be dissolved at 100°C. Filter this slurry at 100°C, and then add just enough water so that, upon cooling to 25°C, a maximum crop of pure solid KNO_3 is obtained at this temperature. Filter out these KNO_3 crystals at 25°C, and use this mother liquor (now identical with the mother liquor from step IV below) for the following cycle:

I. A fresh batch of crude salt containing 100 lb of soluble nitrates is mixed with an appropriate amount of mother liquor from step IV and heated to 100°C so that, at this temperature all of the KNO_3 is brought into solution while leaving as much $NaNO_3$ undissolved as possible.

II. Filter at 100°C, rejecting solids.

III. Add such an amount of water to the filtrate that, upon cooling to 25°C, a maximum yield of pure KNO_3 is obtained.

IV. Filter this slurry at 25°C. The filter cake represents the desired KNO_3 product. The mother liquor is in part recycled to step I, in part discarded as waste. This cycle is now repeated.

In a steady-state operation, how many lb of pure KNO_3 are produced per 100 lb of soluble nitrates fed? It will be assumed that the filter cakes produced can be sufficiently freed of mother liquor so that the wash waters quantities are negligible. Solubility data at 25°C are presented in Fig. 3-2.

COMPOSITION OF SATURATED AQUEOUS SOLUTIONS IN WEIGHT* % AT 100°C

$NaNO_3$, %	KNO_3, %	Equilibrium Solid Phase
63.7	0	$NaNO_3$
49.5	25.3	$NaNO_3$
41.1	40.0	$NaNO_3$
36.7	48.1	$NaNO_3 + KNO_3$
29.7	51.4	KNO_3
18.7	57.7	KNO_3
0	70.9	KNO_3

*Reference (61).

PROBLEM 3-7

Recovery of pure KNO_3 from crude salt (the solubles are 23% by weight KNO_3, the rest $NaNO_3$) was described in the foregoing Prob. 3-6. This cycle was unsatisfactory because its yield of pure solid KNO_3 is rather poor. It is desired to obtain all of the KNO_3 in the crude salt in the form of pure crystals. A new improved cycle is therefore proposed, as follows:

Start-up. Enough pure salts are at hand for making a solution saturated at 100°C with both KNO_3 and $NaNO_3$, but free from suspended solids.

I. With 100 lb of the start-up solution just described as a basis, add that quantity of water which, upon cooling this diluted liquor to 25°C, causes a maximum crop per cycle of pure KNO_3 crystals to form. Remove these crystals by filtering.

II. Heat the filtrate from step I, and evaporate off the proper amount of water.

III. Add such an amount of crude salt to the mother liquor at 100°C that all the KNO_3 contained in this crude dissolves to form a solution saturated with both nitrates.

IV. Filter at 100°C, rejecting solids, sending the hot filtrate to step I.

In this cycle, the liquor from step IV must be identical in quantity and composition from one cycle to the next. Solubility data at 100°C are presented in Prob. 3-6, and at 25°C in Fig. 3-2.

Will this cycle "work," producing a 100% yield of pure KNO_3 crystals in step II? How much water is added in step III, in lb/lb KNO_3 produced?

PROBLEM 3-8

Modify the cycle described in the foregoing problem so that a 100% yield is obtained, not only of pure crystalline KNO_3, but of pure crystalline $NaNO_3$. Solubility data at 100°C are presented in Prob. 3-6 and at 25°C in Fig. 3-2.

PROBLEM 3-9

The two salts in a very intimate solid mixture analyzing 90% by weight KNO_3 and the rest $NaNO_3$, are to be separated in a cyclic solution-precipitation procedure. As in any cycle, the net addition of each component (KNO_3, $NaNO_3$, and H_2O) to the system is zero around each cycle.

Start-up. Add water to a batch of the crude salt mixture at 100°C, in such quantity that all of the $NaNO_3$, but as little as possible of the KNO_3, dissolves. Filter off the undissolved KNO_3 at 100°C. The mother liquor produced here has the same composition as that from step IV of the cycle below.

I. A fresh batch of crude salt is added to an appropriate amount of mother liquor filtrate from step IV, all at 100°C. The system is next cooled to 25°C after adding enough water to dissolve all of the $NaNO_3$, but as little as possible of the KNO_3 present at 25°C. This solution is now saturated with both salts at this temperature.

II. After equilibrium is attained at 25°C, filter the slurry at this temperature.

III. Reheat the filtrate of step II and evaporate a suitable amount of water.

IV. Filter the slurry from step III at 100°C and reuse the filtrate in step I of the cycle.

(a) Show the cycle on a triangular diagram (solubility data for this system are presented in Prob. 3-6 and Fig. 3-2).
(b) What is the maximum attainable yield of solid KNO_3?
(c) How much water is evaporated in step III per lb KNO_3 produced?

PROBLEM 3-10

A large deposit analyzing 70% NaCl, 20% Na_2SO_4, and the rest silica is discovered in the Far West. This analysis is based upon samples taken from "many drill holes." The mass is in the form of very tiny, tightly interlocked, and interwoven anhydrous crystals. The deposit is 2000 ft below the surface, where it exists as a bed about 20 ft thick, very slightly sloped downward toward the northwest. The deposit is evidently very large in extent.

It is suggested that solution mining may be possible for this deposit. That is, water or an aqueous solution would be pumped into the deposit at one point, and a saturated solution removed at another point separated by a suitable distance vertically or horizontally from the first point. It is proposed to produce sodium sulfate decahydrate, using crystallization-evaporation procedures or any other suitable process. Comment on this plan, propose whatever scheme seems best suited, and indicate probable heat, power, and water needs. What by-products are produced, if any? Does the proposal deserve further study?

Would an orthodox mining procedure, bringing the dry crude salt to the surface for separation, simplify the separation problem?

Part of the solubility data available on this aqueous system in Reference (61), Vol. IV, 287 are as follows:

SATURATED SOLUTION COMPOSITION, WEIGHT %

Solid Phase	0°C		25°C		105°C	
	NaCl	Na_2SO_4	NaCl	Na_2SO_4	NaCl	Na_2SO_4
NaCl	26.26	0	26.42	0	28.28	0
NaCl + $Na_2SO_4 \cdot 10H_2O$	25.54	1.30	*	*	*	*
Na_2SO_4 + $Na_2SO_4 \cdot 10H_2O$	*	*	13.53	15.42	*	*
NaCl + Na_2SO_4	*	*	22.71	8.06	26.14	4.41
$Na_2SO_4 \cdot 10H_2O$	10.70	1.50	9.74	15.65	0	29.49
$Na_2SO_4 \cdot 10H_2O$	0	4.65	4.21	18.25		
$Na_2SO_4 \cdot 10H_2O$			0	21.79		

*Sodium sulfate decahydrate is unstable above 33°C; no anhydrous sodium sulfate exists in contact with NaCl solutions at 0°C.

Additional solubility data at 25°C are as follows:

Solid Phase	$Na_2SO_4 \cdot 10H_2O$		Na_2SO_4		NaCl	
	NaCl, %	Na_2SO_4, %	NaCl, %	Na_2SO_4, %	NaCl, %	Na_2SO_4, %
	0	22.11	16.57	11.62	23.92	4.72
	2.27	20.44	19.15	9.11	24.91	2.44
	4.42	18.59			26.26	0.0
	6.35	16.23				
	9.94	14.55				
	11.61	14.42				

PROBLEM 3-11

(a) Referring to Fig. 3-4 and operating always at 25°C, 1 mole of dry KCl is added to a saturated, aqueous solution containing 1 mole of $NaNO_3$. What are the amounts and compositions of the phases formed?

(b) Operating always at 50°C, as in Fig. 3-7, a solution saturated with $NaNO_3$, KNO_3, and NaCl is filtered and evaporated to dryness. What is the composition of the last drop of liquid? Repeat for a solution originally saturated with KCl, KNO_3, and NaCl.

(c) Solid $NaNO_3$ is added to a solution saturated with KNO_3 and KCl. If operations are at 50°C as in Fig. 3-7, and if the solution initially contains 1 mole of total dissolved salts, how much $NaNO_3$ must be added before $NaNO_3$ exists as a solid-phase?

PROBLEM 3-12

A solid mixture analyzing 40 mole % NaCl, 20 mole % $NaNO_3$, and 40 mole % KNO_3 is processed as follows to obtain pure KNO_3.

Start-up. Operating at 25°C, add just enough water to 100 lb of mixed salt so that the mother liquor composition is at point *j* on Fig. 3-4. How much water must be added here? Filter off the solid-phase (what is it?) at 25°C and proceed with step 1.

I. Operating at 50°C with the clear mother liquor from point *j* on Fig. 3-4 obtained either by start-up as just described or from step IV below, evaporate or add water, then add an appropriate amount of fresh crude salts.

II. Filter off pure NaCl at 50°C (see Fig. 3-7 for data).

III. Cool the filtrate to 25°C, add or evaporate water, and filter off pure KNO_3.

IV. If necessary, reject part of the filtrate (whose composition is at point *j* on Fig. 3-4) and recycle the rest to step I of the cycle.

What is the minimum amount of water which must be added or evaporated in step I of this cycle per mole of KNO_3 produced? Equilibrium data are presented in Figs. 3-4 and 3-7.

PROBLEM 3-13

(a) A saturated solution at 125°C which is exactly equimolal in KNO_3 and NaCl but contains no solids is evaporated isothermally at 125°C until 10% of the water has been removed. Describe the remaining system, using Fig. P-1 for data.

(b) The system is further evaporated at 125°C until 90% of the water has been evaporated. Describe the remaining system.

(c) A saturated solution initially at 125°C lies in Fig. P-1 at a value for Y of 0.8 and for X of 0.2. No solids are present initially in this solution. Show the history of the evaporation of this solution on the figure.

(d) A saturated solution at 125°C lying at point k in diagram P-1 is cooled to 25°C, with water added or removed as required. What is the maximum attainable yield of KNO_3? Data for 25°C are presented in Fig. 3-4. It should be noted that abscissas and ordinates are not identical in Figs. 3-4 and P-1.

PROBLEM 3-14

Figure P-2 shows equilibrium aqueous solution compositions at 15°C for the reciprocal salt pair NaCl-NH_4HCO_3 under a pressure of 1 atm of CO_2. This salt system is the basis of the Solvay process for making sodium bicarbonate and finally sodium carbonate from common salt. The constant

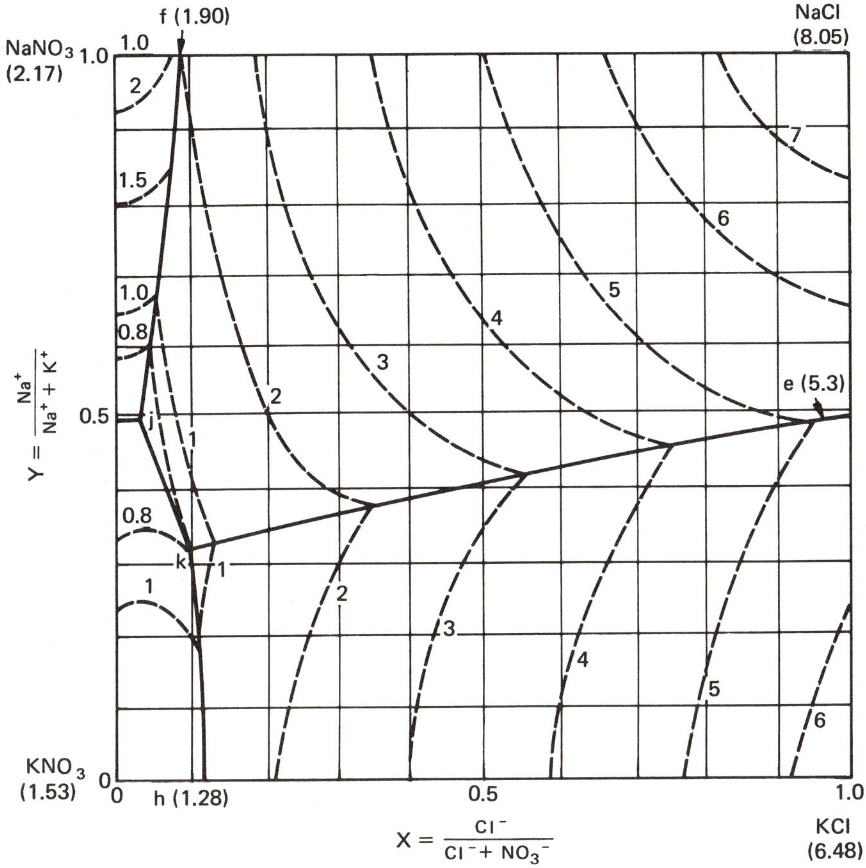

j = 0.56 moles water per mol dissolved salt
k = 0.80 moles water per mol dissolved salt

Fig. P-1. The system K^+, Na^+, Cl^-, NO_3^-, H_2O at 125°C. The dotted contour lines show moles of water per mole dissolved salt in saturated solutions (151). Moles of water at various points are shown in parentheses.

water lines on the diagram represent grams of water per gram equivalent of chloride ion in solution, this being a convenient basis for analysis of the Solvay operation. The constant water lines are dotted to indicate that these are approximations, since limited solubility data are published.

(a) To show details in the upper part of the diagram, the scale of the ordinate of Fig. P-2 is larger than for the abscissa. Calculate by

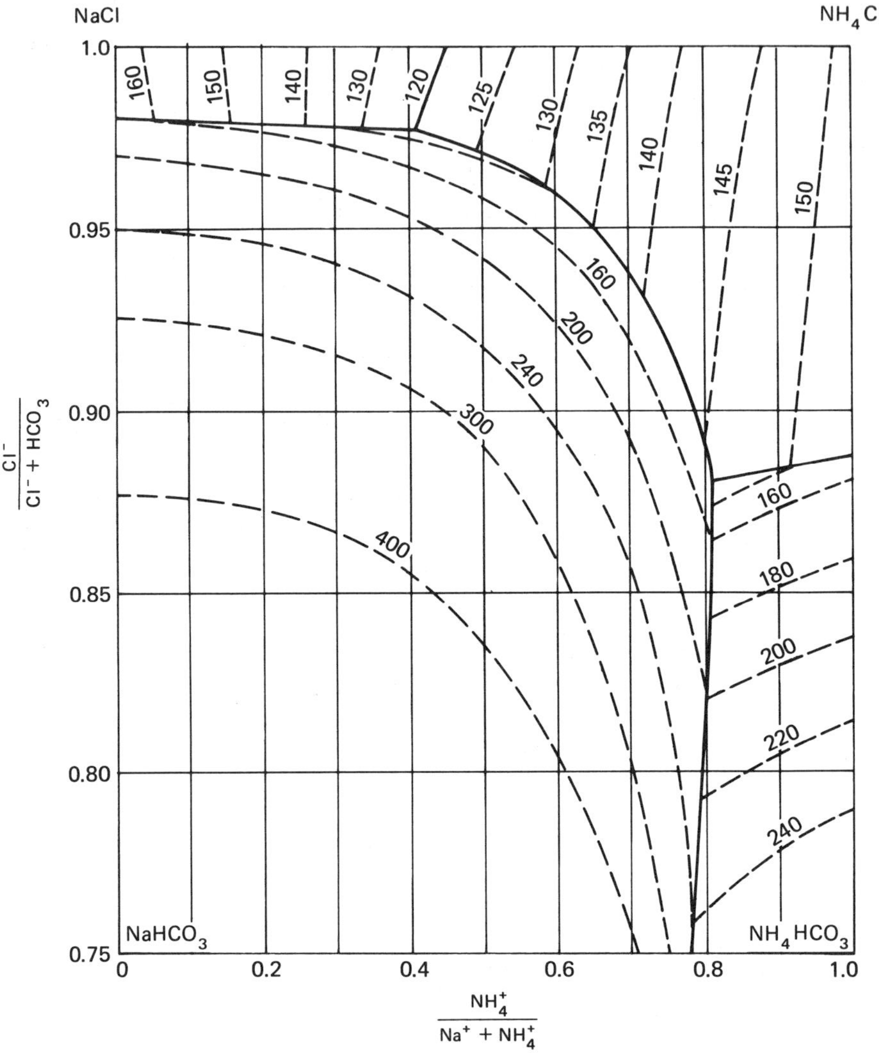

Fig. P-2. Janecke diagram for the reciprocal salt pair, NaCl–NH_4HCO_3 under a total pressure for CO_2 of 1 atm at 15°C (123). The numbered lines are the grams of water per gram equivalent chloride ion.

stoichiometry the composition of a mixture formed from a solution containing 100 moles of salt having a value for Y of 0.95 and for X of 0.1, and a solution containing 50 moles of salt with Y and X values of 0.75 and 0.2, respectively. Now repeat graphically in Fig. P-2, assuming the Inverse Lever law to apply. Do the two methods give the same answer? Does the Inverse Lever law apply to this "distorted" diagram?

(b) Sketch the path followed on the diagram as solid NH_4HCO_3 is added to a solution initially saturated with NaCl.

(c) Repeat (a) for the addition of NH_4HCO_3 to a solution containing 200 g of water per g mole of NaCl.

(d) What should be the initial concentration of a NaCl solution in order that maximum percentage conversion of sodium to $NaHCO_3$ be obtained by the addition of dry NH_4HCO_3? Trace the path followed on the diagram.

PROBLEM 3-15

A plant produces a waste solution of the following analysis:

$$\frac{NH_4^+}{NH_4^+ + Na^+} = 0.78$$

$$\frac{Cl^-}{Cl^- + HCO_3^-} = 0.80$$

This solution is saturated with both $NaHCO_3$ and NH_4HCO_3, but contains no solids. It is desired to produce solid pure sodium bicarbonate from this solution in a maximum yield, by operations at 15°C without evaporation or addition of water. A supply of pure solid NaCl is on hand which may be utilized to any desired degree in the operation. Equilibrium data are presented in Fig. P-2.

(a) Describe a procedure for obtaining such a maximum yield of sodium bicarbonate uncontaminated with any other solid.

(b) What is theoretically the maximum fraction of bicarbonate, originally in solution which is recoverable as pure sodium bicarbonate (uncontaminated by any other solid) in the operation?

PROBLEM 4-1

Crude aluminum can be made by the carbon reduction of alumina. It is proposed to recover pure metal (m.p. at 659°C) from this crude by the following reaction:

$$AlCl_3(g) + 2Al(l) = 3AlCl(g) \qquad K_p = p^3_{AlCl}/_{AlCl_3}$$

From 600–1200°C, the equilibrium constant K (in atm^2) is related to T in °K as follows (51):

$$-2.3RT \log K = 92940 - 61.2T$$

Continuous operation at atmospheric pressure is to involve charging impure aluminum and recycled $AlCl_3$ vapor, both at 1350°K, to reactor I. The gases leaving reactor I are at equilibrium with metallic aluminum at this reactor's temperature of 1450°K. These product gases flow to reactor II where they are cooled to 1350°K, reversing the above reaction. The gases from reactor II are recycled to I, while the molten aluminum formed in II is drawn off continuously as product.

Heating and cooling in each reactor is by passing the gas through a shower of aluminum droplets thrown off by rapidly rotating wheels partly submerged in the pools of molten metal in the bottoms of the reactors. Reactor I is heated electrically by induction. Heat is removed from reactor II by water-cooled tubes submerged in molten salt floating on the aluminum in a pocket in the reactor, thus avoiding direct contact of the cooling tubes with the molten aluminum, in which most metals have a high solubility. For this orientational analysis, assume that no purge is needed in the recycle loop. If the off-gases from each reactor are at equilibrium,

(a) How many lb moles of gas leave reactor I per lb atom of refined metal produced? Assume the concentration of impurities in solution in the impure aluminum to be negligible.
(b) How much heat must be supplied to reaction I per lb atom of refined aluminum produced?
(c) Materials of construction are a very serious problem in this operation because of high temperature and the corrosive atmosphere. What alternatives are there? How would you suggest moving the recycling gases? How could heat be added and removed from this high temperature system?

For impure liquid aluminum feed, assume the heat capacity to equal 5.0 cal/°K for every gram atom of aluminum metal present. For gaseous $AlCl_3$, MC_p equals 20 cal/(°K)(g mole) over the temperature range in question. The vapor pressure of aluminum metal is negligible at these temperatures.

PROBLEM 4-2

Nitric acid is made by bubbling a gas [analyzing (dry basis) 18% N_2O_4, 2% NO, 10% O_2, and the rest nitrogen] up through the liquor on a single bubble plate operating at 25°C and 1 atm pressure. It is assumed that no oxidation of NO to NO_2 and N_2O_4 occurs during the brief time while the gas bubbles move up through the liquor on the plate, but that equilibrium between gas and liquid is attained with respect to the heterogeneous reaction:

$$\tfrac{3}{2}N_2O_4(g) + H_2O(l) = 2HNO_3(l) + NO(g)$$

The equilibrium partial pressure (in atm) of NO and N_2O_4 for this reaction are related to w, the weight fraction of HNO_3 in the aqueous liquid phase, as follows (19):

$$\log (p_{NO}/p_{N_2O_4}^{1.5}) = 7.412 - 20.28921\,w + 32.47322\,w^2 - 30.87\,w^3$$

This equation is stated to be independent of temperature in the range of 0–75°C.

(a) The bubble plate in question is initially filled with water, after which no liquor is added or removed. What is the analysis of the acid on the plate after a longer time operation at 25°C, assuming evaporational losses of water and acid to be negligible? Could such evaporational losses be in fact made negligible?

(b) An aqueous nitric acid solution of unknown concentration is next fed continuously to the plate. Liquor is simultaneously withdrawn from the plate at a rate of 28.3 lb for each mole of mixed gas feed, and after a steady-state is attained, this liquor analyzes 51.3% HNO_3 by weight. If the off-gases are at equilibrium with the liquor on the plate (assumed thoroughly mixed), what is the analysis of the liquid feed to the plate? The water content of the gas phase may be neglected.

(c) Reasoning from the above equation, how much heat must be removed from the plate of part (b) per lb of acid liquor produced, if all streams enter and leave at 25°C? Discuss how this answer would differ if the plate operations were adiabatic, with all streams entering at 25°C.

PROBLEM 4-3

The hydrogen chloride by-product formed in the direct gas phase chlorination of saturated hydrocarbons represents a disposal problem in many locations. Such hydrogen chloride production is avoided by use of a suitable catalyst and the addition of air to the reacting gases. The reactions occurring are illustrated below with methane:

$$CH_4 + Cl_2 = CH_3Cl + HCl \qquad \Delta H^\circ_{298^\circ K} = -23.8 \text{ kcal} \qquad (a)$$

$$HCl + \tfrac{1}{4}O_2 = \tfrac{1}{2}H_2O + \tfrac{1}{2}Cl_2 \qquad \Delta H^\circ_{298^\circ K} = -6.9 \text{ kcal} \qquad (b)$$

$$CH_4 + \tfrac{1}{2}Cl_2 + \tfrac{1}{4}O_2 = CH_3Cl + \tfrac{1}{2}H_2O \qquad \Delta H^\circ_{298^\circ K} = -30.7 \text{ kcal} \qquad (c)$$

Inspection shows that HCl from reaction (a) is converted to chlorine in reaction (b), namely the Deacon reaction, and that the so-called oxychlorination reaction (c) is the sum of reactions (a) and (b). The CH_3Cl formed in reaction (c) is in turn oxychlorinated to CH_2Cl_2, $CHCl_3$, and CCl_4, with overall reactions as follows:

$$CH_3Cl + \tfrac{1}{2}Cl_2 + \tfrac{1}{4}O_2 = CH_2Cl_2 + \tfrac{1}{2}H_2O \qquad \Delta H°_{298°K} = -40.8 \text{ kcal} \quad (d)$$

$$CH_2Cl_2 + \tfrac{1}{2}Cl_2 + \tfrac{1}{4}O_2 = CHCl_3 + \tfrac{1}{2}H_2O \qquad \Delta H°_{298°K} = -18.9 \text{ kcal} \quad (e)$$

$$CHCl_3 + \tfrac{1}{2}Cl_2 + \tfrac{1}{4}O_2 = CCl_4 + \tfrac{1}{2}H_2O \qquad \Delta H°_{298°K} = -33.0 \text{ kcal} \quad (f)$$

While the equilibrium constant of reaction (b) is only 2.5 at 450° and 1.8 at 500°C, the equilibrium constants for the chlorination reactions like reaction (a) are very large, consequently the equilibrium constants of oxychlorination reactions (c), (d), (e), and (f) are also large. All these reactions, again excepting reaction (b), are vigorously exothermal.

In an experimental study, a mixture of methane, chlorine, and air was passed through a heated tube containing a fixed bed of catalyst, 12 in. long (97). The catalyst consisted of $\frac{1}{8}$-in. porous granules of pumice soaked in strong copper chloride solution and dried at 120°C. During operation, the temperature profile was determined by passing a thermocouple through an axial well in the tube. With fresh catalyst, a "hot spot" which moved slowly, through the tube over a period of days, was observed at the tube entrance during the course of continued operation. The tube was kept in an electric furnace at 500°C, while the "hot spot" was 100–200° higher in temperature. The chlorinated product was always found to contain all four chlorinated methanes, in proportions which depended on the ratio of Cl_2 to CH_4 in the feed (97). Despite favorable equilibria, small amounts of unconverted HCl and oxygen were often found in the off-gases, though no elemental chlorine ever appeared. Some oxidation of the hydrocarbons present always occurred.

(a) What is the probable cause of the hot spot and of its movement? Could operations be modified to avoid its formation? Would you expect catalyst deterioration in consequence of the formation of this hot spot?

(b) The methane fed is in part chlorinated to a mixture of the chlorinated methanes, whose average composition may be represented as $CH_{4-x}Cl_x$. The off-gases are scrubbed with water to remove HCl, and with solvents to remove the chlorinated hydrocarbons. In one of the runs made, the final off-gases analyzed 4.3% CO_2, 0.9% CO, 1.8% CH_4, 6.2% O_2, the rest N_2. No free chlorine, HCl, carbon, or oxygenated hydrocarbons were found in this run, while the average "X" was 2.15. What fraction of the methane fed was burned? What fraction was chlorinated?

(c) The disproportion reactions, whereby CH_4 and CH_2Cl_2 react to form CH_3Cl, CH_3Cl and $CHCl_3$ react to form CH_2Cl_2, and $C_2H_2Cl_2$ and CCl_4 react to form $CHCl_3$, apparently do not occur at any significant rate over the usual oxychlorination catalysts. That is, feeding a mixture of CH_3Cl and $CHCl_3$ (with or without air) to the reactor shows no CH_2Cl_2 to be produced. Nevertheless, the off-gases from a methane oxychlorination reactor always contain the four chlorinated methanes in proportions which would have existed, had these disproportionation reactions occurred and proceeded to equilibrium, with an equilibrium constant for each reaction of unity. The implications of these findings regarding the kinetics of these reactions are discussed elsewhere (97) (see also Prob. 7-15).

For purposes of product control, the following plot is convenient: The abscissa is X, namely the ratio (moles of chlorine per mole of methane) in the feed, ranging from zero to four. The ordinate is the fraction of the methane fed which reappears in the off-gas as methane, mono, di, tri, or tetrachloride, and ranges from 0 to unity. Assume that all the chlorine fed reacts to form the chlorinated methanes plus HCl, that no methane oxidation occurs and that the disproportionation reactions are all at equilibrium. Draw curves showing how the fraction of methane in the off-gas varies with X. Draw similar curves for mono-, di-, tri-, and tetrachloromethane.

PROBLEM 4-4

Ethyl chloride is a clear liquid boiling at 12.2°C, with vapor pressures of 1.33 atm at 20°C and 11.5 atm at 100°C. One of several methods of making ethyl chloride is by direct addition of dry HCl to ethylene (137):

$$C_2H_4 + HCl = C_2H_5Cl \qquad K_{p,\,500^\circ K} = 0.84$$

In the so-called vapor phase process, feed plus recycle gas pass through the catalyst-packed tubes of a cooled reactor, then through the main condenser, with the uncondensed gases being recycled. Liquid ethyl chloride from the main condenser contains dissolved gases, and so is sent to a stripping column, whose top off-gases pass through a partial condenser, and are then combined with gases from the main condenser for recycle. The ethyl chloride bottoms from the stripper are sent to storage as product.

The plant under consideration is to operate at 10 atm total pressure, and the gases leaving the reactor are at 227°C. The catalyst used, namely copper chloride supported on porous silica, is effective enough so that the value of the term $p_{C_2H_5Cl}/(p_{C_2H_4})(p_{HCl})$ in these gases equals 95% of the true equilibrium constant value for these conditions. The combined gases from the main condenser and partial condenser are at 20°C and 10 atm pressure, saturated with ethyl chloride, and are partially purged and recycled.

(a) The fresh feed to the plant is an equimolal mixture of pure HCl and pure ethylene. A purge quantity equal to 5 moles % of the fresh feed gas is to be taken. How many moles of total gas, namely, fresh feed plus recycle, enter the reactor per mole of liquid ethyl chloride product from the stripper?

(b) The fresh feed now contains 50% HCl, 49% C_2H_4, and 1% N_2. If operating temperatures and pressures are as before, and if the purge is again 5% of the fresh feed gas, what are now the moles of total gas entering the reactor for mole of final ethyl chloride product?

PROBLEM 4-5

Because of low vapor pressures, crystals of some semiconductor materials like gallium arsenide (GaAs) cannot be grown by simple sublimation and condensation. Preparation of such crystals from a melt is difficult because of incongruent melting. Gallium arsenide can, however, be deposited from the following reversible reaction (34):

$$GaI_3(g) + 2GaAs(c) = 3GaI(g) + \tfrac{1}{2}As_4(g)$$

A stream of gas analyzing 10 mole % GaI_3, the rest N_2, is passed over crude GaAs maintained at T_1, causing the above reaction to proceed to the right. This gas stream is then passed over a strip of GaAs seed crystal maintained at T_2, whereupon the above reaction reverses and the seed crystal grows. Deposition rates on solid surfaces other than gallium arsenide are to be assumed low. The system is at 0.2 atm, and equilibrium is attained by the gas in contact with the crude solid feed at T_1 and over the seed crystal at T_2.

(a) If 10% of the GaI_3 in the gas stream passing over the crude GaAs solid is to react by the reaction above, what is T_1?

(b) If 20% of the GaAs gasified at T_1 is precipitated on the seed crystal, estimate T_2.

(c) It is important that the purities of the reactants be high. Growth rates are slow, and crude GaAs is not inexpensive. Propose a possible apparatus for this crystal growing operation. Contamination with dust, air, water, grease, etc., is to be avoided.

Note: Solid GaI_3, GaI, and arsenic develop sublimation pressures of 1 atm at 345°C, 400°C, and 615°C, respectively. The values of $\Delta F°$ and $\Delta H°$ for the above reaction at 1000°K are 17.5 and 107.5 kcal.

PROBLEM 4-6

The gaseous reduction of granular magnetite (Fe_3O_4) to metallic iron is carried out continuously and without melting at 800°C and 5 atm pressure in a two-stage countercurrent system as shown in Fig. P-3. The magnetite fed is

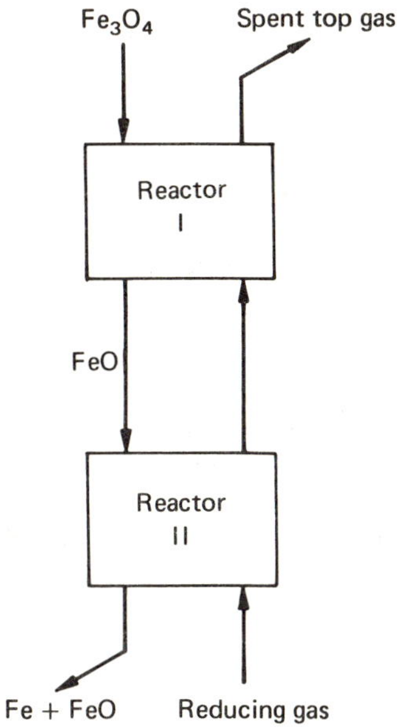

Fig. P-3. Countercurrent reduction of Fe_3O_4 in two stages, as in Prob. 4-6.

reduced in reactor I to FeO, and this material is then dropped into reactor II where it is further reduced to a product in which 95% of the iron is present as metal, the rest as FeO. The reducing gases entering the bottom of reactor II analyze 33.0% H_2, 1% H_2O, the rest N_2. The gases from reactor II flow up through reactor I, as shown in Fig. P-3. The solids in each reactor form a fluidized bed in which the mixing of solids is rapid. Reaction rates for the ore used are also to be assumed very rapid, hence the gas and solid streams leaving each reactor are in equilibrium at 800°C. The reactions occurring, together with equilibrium constants, are

$$Fe_3O_4 + H_2 = 3FeO + H_2O \qquad K_{p,\,800°F} = 3.0 \qquad (a)$$

$$FeO + H_2 = Fe + H_2O \qquad K_{p,\,800°F} = 0.55 \qquad (b)$$

(a) For each mole of FeO entering reactor II, what is the minimum number of moles of fresh reducing gas which must enter this reactor, if the off-gases from reactor II are at equilibrium with the bed solids? Is this gas in equilibrium with the FeO feed to this reactor?

(b) The off-gases from reactor II enter reactor I. There is now some disagreement as to the composition of the gases leaving reactor I. One engineer states that since the solids are well mixed and reaction rates are high, equilibrium is attained with respect to reaction (a).

Another engineer disagrees, stating that too little H_2 is available in the gas fed to reactor I "for equilibrium to be attained in this heterogeneous reaction." Still another engineer claims that "there is too much H_2 in the gas fed to reactor I" for this heterogeneous equilibrium to be attained. What in fact is the composition of the off-gas from reactor I? Comment on the positions taken by your associates.

(c) Operations are changed so that 12 mole of gas (34.7% H_2, the rest N_2) are fed to Reactor I, for every $\frac{1}{3}$ mole of Fe_3O_4 fed to Reactor II. What are now the compositions of the streams from Reactor II? From Reactor I?

(d) The reduction of Fe_3O_4 to a solid product containing 95% of the iron as metal, the rest as FeO, is again to be carried out continuously at 800°C, but this time in a single reactor. The gases and the solid product leaving the reactor are again to be in equilibrium. Calculate the minimum number of moles of reducing gas needed per lb mole of Fe produced.

PROBLEM 4-7

A bituminous coal is to be "gasified" continuously at 1050 psig to produce a gas of high methane content, so that a heating value of at least 900 Btu per standard cu ft can be attained. Operations involve four stages (40):

Stage I. React coal with oxygen and steam in a "transport reactor," which is a duct through which the finely divided coal travels as a dust cloud suspended uniformly in the moving hot gas stream, reacting as it proceeds. The ash and partly reacted coal particles are removed in cyclones at the end of the reactor, after which the gas travels to stage II. In pilot studies, the gas from the stage I cyclone is at 1700°F, analyzes 15.4% CO_2, 25.4% H_2, 38.8% CO, 17.5% CH_4, and the rest N_2, on a dry basis. The reactions occurring are presumably the "water-gas shift" and the methane formation reactions:

$$CO + H_2O = CO_2 + H_2 \qquad K_{p,\,1700^\circ F} = 0.7$$

$$2C + 2H_2O = CO_2 + CH_4$$

$$C(\text{graphite}) + 2H_2 = CH_4 \qquad K_{p,\,1700^\circ F} = 0.016$$

Stage II. Cool the gas from stage I and pass it through a catalytic converter where more hydrogen is made by the water-gas shift reaction.

Stage III. Cool the gas from stage II to about 180°F and pass it through a scrubbing tower, where most of the CO_2 present is removed by adsorption in aqueous sodium carbonate solution. Some water is also removed here by condensation.

Stage IV, Reheat the gas, add steam, and pass the mixture through a cooled catalytic reactor called the "Methanator", where water reacts with CO and CO_2 to from methane. The pilot plant off-gases from stage IV analyze 2.8% CO_2, 5.1% H_2, 0.1% CO, 90.1% CH_4, and the rest N_2, on a dry basis. The small nitrogen content of these gases, being obtained by difference, is considered too unreliable to be used as a "basis" in calculations.

(a) How many tons of oxygen are fed to the gasifier of stage I per ton of coal fed?

(b) If equilibrium is attained in stage I with respect to the water-gas reaction at 1700°F, how many tons of steam are used per ton of coal fed?

(c) In travelling from stage I through stage IV, CO_2 is removed from the gas in stage III, while water is added, condensed, and reacted in several places in the flow scheme. How many moles of dry product gas leave stage IV per ton of coal fired?

Note: The bituminous coal used may be taken to contain 8% ash, negligible nitrogen and sulfur, and has the typical empirical formula on a dry ash-free basis: $CH_{0.65}O_{0.03}$. The perfect gas laws may be used for gases encountered here at the temperature and pressures indicated. The oxygen fed contains 3% N_2; the solid residue may be considered to carry negligible carbon.

PROBLEM 4-8

Ethylene oxide (a toxic liquid boiling at 10.75°C) is made by the partial oxidation, at low temperature, of pure ethylene with an excess of air or oxygen over a silver catalyst. A high concentration ratio of oxygen to ethylene improves ethylene yields and conversions per pass. To avoid forming explosive mixtures, ethylene concentrations in these mixtures are kept below 3.4%. In one process, catalyst granules are packed in tubes externally cooled by recirculating oil, with tube temperatures slowly increased from about 260–300°C in operations over a period of months to offset loss of catalyst activity with age. Operating pressures of 5 atm or so are used to reduce equipment size and simplify recovery. These high pressures effect little improvement in reaction rates in this operation (138).

One method of preparing the granular catalyst is by precipitation of silver oxide from a silver nitrate solution with caustic soda, followed by mixing the washed oxide paste into porous pumice particles, drying, and reducing with H_2. Small amounts of "anticatalysts" such as ethylene dibromide, ethylene chlorohydrin, etc., when added to the feed gases, help suppress the oxidation of ethylene all the way to CO_2 and water. Such complete oxidation is undesirable both because of the ethylene yield loss and the large heat release, which introduces serious reactor design problems:

$$C_2H_4(g) + \tfrac{1}{2}O_2 = C_2H_4O(g) \qquad \Delta H^\circ = -31 \text{ kcal}$$

$$C_2H_4(g) + 3O_2 = 2CO_2 + 2H_2O(g) \qquad \Delta H^\circ = -315 \text{ kcal}$$

Materials of construction must be carefully chosen, since many oxides, such as iron oxide will catalyze the total combustion of ethylene to CO_2 and H_2O. Paraffins like ethane are more prone to total oxidation than ethylene, consequently, very pure ethylene is used as raw material in ethylene oxide manufacture.

A flow sheet of the old Zweckel plant for ethylene oxide using a silver catalyst packed into water-cooled tubes is presented in Fig. P-4. As is typical with silver catalysts, no oxygenated materials other than ethylene oxide, CO_2, and H_2O were found in the reactor product gases. The feed rates of crude ethylene and air were reported as 130,800 and 1,488,000 cu ft/day, respectively, measured at standard conditions, while the volume of gases leaving the second converter and the recycled gas after the purge were 2,976,000 and 1,488,000 standard cu ft/day, respectively, all on a water-free basis. The product gas from the second converter analyzed 2.2% ethylene oxide, 0.9% ethylene, and 3.6% CO_2 on a dry basis. These off-gases were scrubbed with water containing about 1% sulfuric acid, as shown. As a result, the recovered ethylene oxide promptly hydrolyzed in the scrubbing liquors of the absorption towers to form ethylene glycol. Scrubbing was very

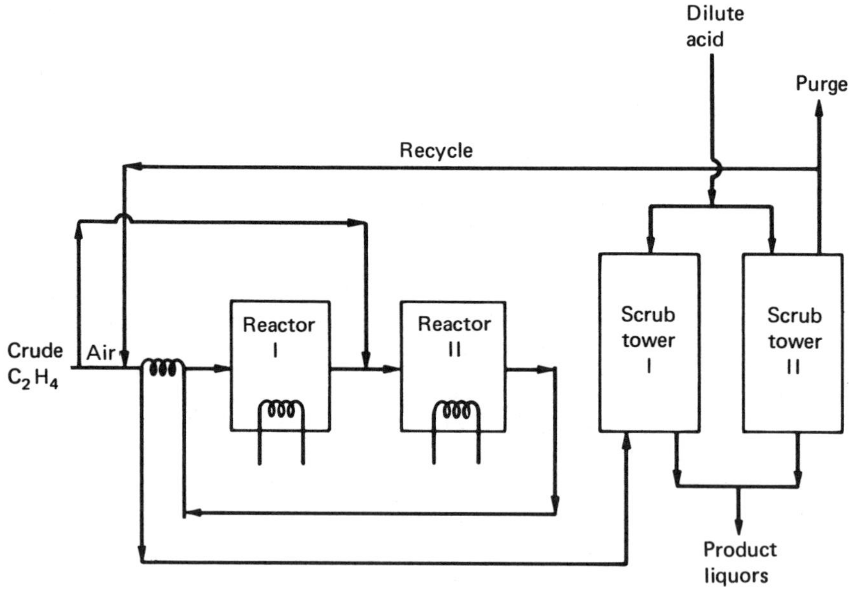

Fig. P-4. Flow sheet of the Zweckel plant for ethylene oxide (137), described in Prob. 4-8.

efficient, and losses of ethylene oxide and glycol in the scrubber off-gases were negligible. Glycol was recovered from the scrubber liquors by evaporation and distillation operations not shown on this flow sheet. Assume that no CO_2 is lost in scrubbers.

(a) How many lb of ethylene oxide were made per hour?
(b) How many standard cu ft of purge gas were produced per day?
(c) What was the mole % of ethylene in the crude ethylene feed to the process?
(d) What was the molecular oxygen concentration in recycle stream?
(e) What was the concentration of ethylene in the gas entering each converter? Assume each converter made the same amount of C_2H_4O, burned the same amount of ethylene per hour, and that the crude ethylene feed stream was split equally between the two reactors. Why was the crude ethylene feed divided like this between the two reactors?
(f) If the gas into and out of the first converter was at 200°C, how much heat was removed from this converter in Btu/day?

PROBLEM 4-9

Removal of CO_2 from gas mixtures is usually by preferential solution in solvents like water, or else by chemical reaction with alkaline scrubbing agents like aqueous ethanolamine or carbonate solutions. The Bureau of Mines (37, 149) studied removal of CO_2 from N_2-H_2-CO mixtures by scrubbing with aqueous K_2CO_3, for which the following reaction occurs:

$$CO_3^{--} + CO_2 + H_2O = 2HCO_3^-$$

For each mole of original carbonate, the moles of CO_2 reacted to form bicarbonate are represented by X, where

$$X = \frac{KHCO_3}{KHCO_3 + 2K_2CO_3}$$

The partial pressures in psia of CO_2 over solutions which initially analyzed 40% by weight K_2CO_3 are reasonably approximated as follows:

$$\text{at } 120°\text{C:} \qquad p_{CO_2} = \frac{19X^2}{(1 - X)}$$

$$\text{at } 70°\text{C:} \qquad p_{CO_2} = \frac{3.8X^2}{(1 - X)}$$

These equations do not apply at X values above about 0.9. The partial pressure of water at 120°C may, for purposes of this problem, be assumed constant at 18.5 psia over these solutions initially, analyzing 40% by weight K_2CO_3, regardless of X values, (actual variation from 21.9–15.2 psia for X values from 0 to 0.8) and at 3.6 psia at 70°C.

Gas analyzing 15% CO_2, 1.2% water vapor, and the rest inerts is passed up through a scrubbing column countercurrent to aqueous 40% K_2CO_3 solution. This scrubber is operated at 300 psia and 70°C, and serves to reduce the CO_2 content of the gas to 3%, on a wet basis. The hot solution from the bottom of this tower, now in part converted to bicarbonate, is to be passed through a reducing valve discharging at about 2 atm pressure, heated to 120°C, and fed to the top of a stripper packed with ceramic rings. This stripper is operated on "open steam" passed up through the reboiler. The stripper, like the reboiler, may be assumed adiabatic.

(a) Plot Y, defined as the ratio (p_{CO_2}/p_{H_2O}), versus X as defined above, at 70°C and at 120°C.

(b) What is the minimum attainable steam consumption in the two-tower system described, expressed as lb of steam per lb of CO_2 removed, assuming infinite towers? What is the corresponding liquor rate?

(c) What is minimum liquor recirculation rate and the corresponding steam rate?

Note: It may be assumed as a first approximation that the partial pressure of H_2O remains constant at 18.5 psia over all solutions in the stripper and at 3.6 psia in the scrubber. Neglect water volatilization from the liquor. Each tower is substantially isothermal in operation, since heat effects are small.

PROBLEM 4-10

Acetaldehyde is a colorless liquid boiling at 20.8°C and miscible with water in all proportions. One of several important commercial methods of acetaldehyde production is by hydration of acetylene at 60–95°C using an aqueous solution of mercuric salts as catalyst:

$$C_2H_2(g) + H_2O(l) = CH_3CHO(g) \qquad \Delta H^\circ = -27 \text{ kcal}$$

To suppress the slow reduction of the catalyst during operation, ferric salts are dissolved in the catalyst solution to reoxidize the mercurous ions as they form. Even with ferric salt addition, it is still necessary from time to time to remove a sludge of reduced catalyst and tars from the reactor for reactivation. The catalyst solution typically analyzes 25% sulfuric acid, 0.2–2% mercury, and 4–6% iron as sulfates, the rest being water and a little acetaldehyde.

The flow sheet of an acetaldehyde plant operating the so-called German process is shown in Fig. P-5. The reactor is a cylindrical tower typically 30

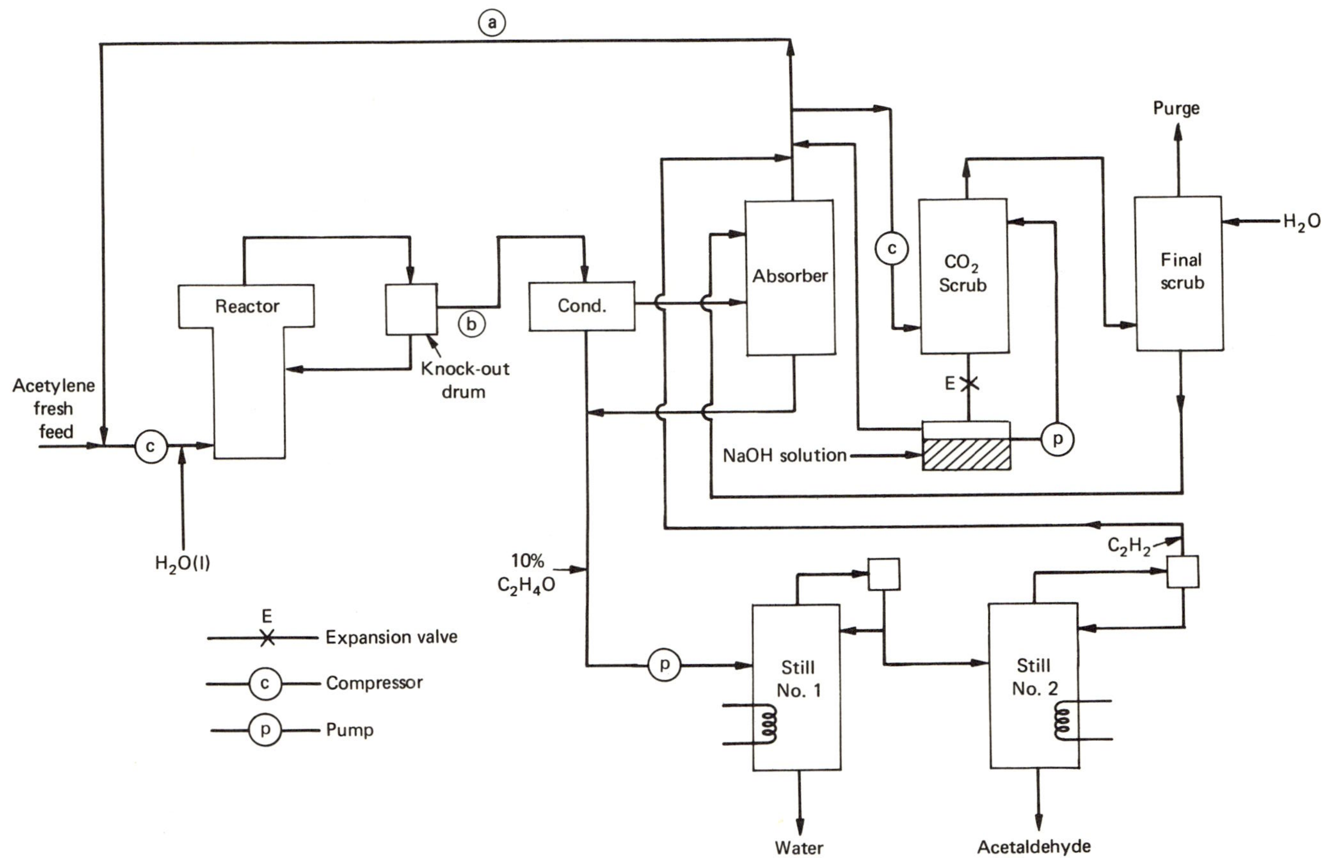

Fig. P-5. Acetaldehyde from acetylene by the German process as in Prob. 4-10.

ft high and 4 ft in diameter, filled with catalyst solution and capped with an empty expansion section for foam breaking. The acetylene-containing gases at 2 atm enter at the bottom of the tower and then bubble up through the vessel which operates adiabatically and continuously at about 95°C. Because of the excellent liquid mixing in bubble reactors, this tower operates isothermally, with the heat of reaction being removed by vaporization of water and acetaldehyde (111).

The product gases from the reactor pass through a knockout drum, a partial condenser, and a water absorber and are then recycled. The combined aqueous liquors from the partial condenser and absorber containing 10% by weight acetaldehyde, are pumped into fractionating towers operating at 2.5 atm to avoid the use of refrigeration in the condensers. A side stream is withdrawn from the absorber off-gases, compressed to 2.5 atm, and scrubbed with caustic soda to remove the small amounts of CO_2 formed by oxidation in the reactor. These gases from the caustic tower are finally scrubbed with water before being sent to the stack (30).

(a) The fresh gas feed to the plant analyzes 97% C_2H_2, 3% N_2, while the purge is 10% C_2H_2, 90% N_2, containing no CH_3CHO or CO_2. Caustic consumption is 1 lb mole of NaOH per 50 moles of fresh feed gas to the plant, while the spent liquor from the caustic scrubber contains only $CO_3^=$ as anions. While some acetic acid, crotonaldehyde, tar, etc., is produced, such by-product production will be neglected in this preliminary analysis. If the acetylene in the feed is either hydrated to CH_3CHO, oxidized to CO_2, or purged as such, how many moles of CH_3CHO are produced per 100 moles of dry feed gas?

(b) What is the mole ratio CO_2/N_2 in the recycle gas at point (*a*) in Fig. P-5? Clearly show your reasoning.

(c) The volumetric flow rate past point (*a*) is known to equal that of the feed, both on a dry basis. Based upon pilot plant findings it is believed that the mole ratio of CH_3CHO to C_2H_2 at point (*b*) is unity, while the gas at (*a*) contains no acetaldehyde. What is the composition of the gas at point (*b*)? At point (*a*)?

PROBLEM 4-11

With reference to the flow sheet of Fig. P-5, Prob. 4-10, comment on the advantages and disadvantages of each of the following possible flow sheet changes:

(a) The final water scrubber is to precede rather than to follow the CO_2 scrubber.

(b) The purge gas stream to the CO_2 scrubber is to be connected at point (a) rather than at its present location.
(c) The return gas stream from still no. 2 connected at point (b), rather than at its present location.
(d) The caustic scrubber is at point (b), rather than at its present location.
(e) A young engineer suggests that condensate and absorber bottoms should not be mixed before feeding to still no. 1, but should be fed to separate plates. Comment.
(f) It is suggested that still no. 1 should take water and acetaldehyde as bottoms, with the acetylene overhead being recycle. Still no. 2 would then separate acetaldehyde and water. Comment.

Volumes of gas (0°C, 760 mm) dissolved in one volume of water under a partial pressure for the gas at 1 atm are (138)

Temperature, °C	Acetylene	CO_2
10	1.31	1.194
30	0.85	0.665
50	0.61	0.436

PROBLEM 4-12

The flow sheet for the Chisso process (111) for acetaldehyde is shown in Fig. P-6. Acetylene at 1.5 atm passes up through a reactor, which is a ceramic ring-packed tower down through which catalyst solution flows counter to the rising gases. Composition of this catalyst liquor is much like that for the German process (Prob. 4-10). Most of the acetylene fed to the tower is hydrated, and the acetaldehyde formed leaves the tower in solution in the catalyst liquors. As shown, makeup water for hydration flows to the reactor from the scrubber, where it is first used to remove unreacted acetylene and acetaldehyde from the tail gases. The reactor liquors are at 70°C, and pass to the adiabatic flash chamber operating at 66°C and 0.3 atm, where 60% of the dissolved C_2H_4O is vaporized. The partly stripped catalyst liquor leaves this flash chamber at 66°C and is returned to the reactor, while the overhead vapor (50% C_2H_4O and 50% H_2O by weight) passes to a partial condenser where some water is condensed, along with small amounts of the acetic acid and crotonaldehyde formed. The vapors leave this partial condenser at 35°C, and are 85% by weight acetaldehyde, the rest water. These vapors are compressed to 2.5 atm in a turbo compressor, and are enriched in the final column. This column's overhead product is 99.5% acetaldehyde at the boiling temperature corresponding to 2 atm pressure, namely, 43°C. The acetylene efficiencies of the Chisso process and the German process are roughly equal.

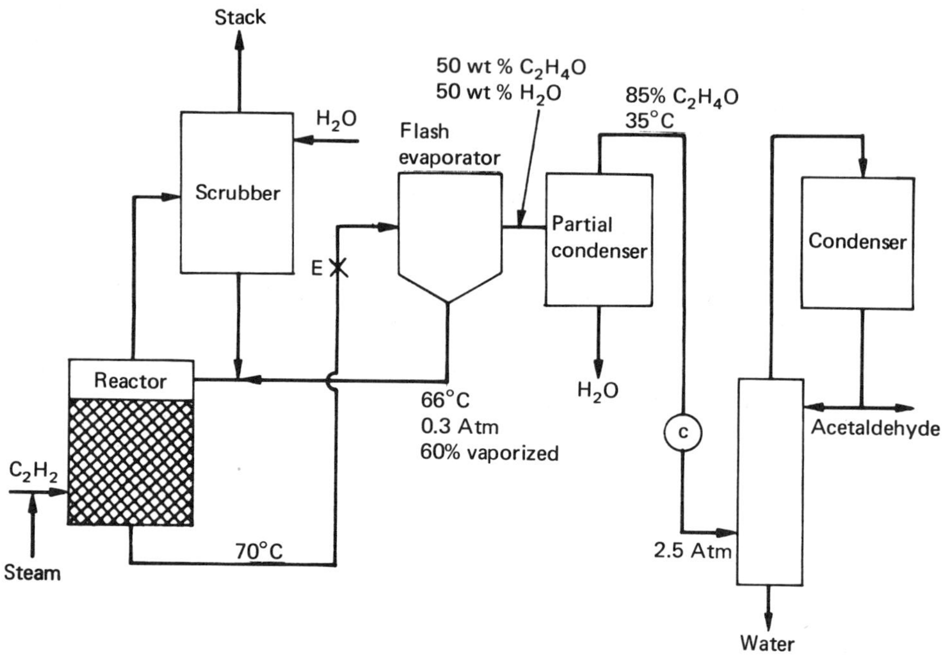

Fig. P-6. Chisso process for acetaldehyde for Prob. 4-12.

(a) Estimate the concentration of acetaldehyde in the catalyst liquor entering the flash chamber of the Chisso process.

(b) Compare the Chisso process with the German process of Prob. 4-12, mentioning especially heat requirements, cooling water needs, and power costs for gas compression.

Notes: $C_2H_2(g) + H_2O(1) = C_2H_4O(g) \qquad \Delta H^\circ = -27$ kcal

Approximate heats of vaporization of water and acetaldehyde from the catalyst solution are 540 kcal/kg and 135 kcal/kg, respectively. The heat capacities of the catalyst liquors may be taken as 1 cal/(g)(°C), regardless of acetaldehyde content, and their density is 1.1 g/cc. See preceding problem for gas solubilities.

PROBLEM 5-1

An adiabatic countercurrent heat exchanger is to be used to convert 70,000 lb of water, initially at 60°F, to steam at 400°F. The entire operation is to be at atmospheric pressure. This heating will be accomplished with hot furnace gases, available at 1000°F. What is the theoretical minimum quantity of these gases (lb/hr) required? At this minimum flow rate, what is the temperature of the gases leaving the exchanger? Assume heat capacities for $H_2O(l)$, $H_2O(g)$, and the hot combustion gases to be constant at 1, 0.45, and 0.32 Btu/(lb)(°F), respectively, and the heat of vaporization of water to be 970 Btu/lb at 212°F.

PROBLEM 5-2

$CaCO_3$ is to be burned completely to CaO in a shaft kiln by combustion of CO with air at atmospheric pressure. Additional shaft kilns as well as gas-gas heat exchangers are available (together with conveyors for moving hot and cold solids) and these shaft kilns may be used as heat exchangers between solids and gases, if needed. The only fuel available is pure CO, which is very costly in this region, so that every effort is to be made to minimize fuel consumption.

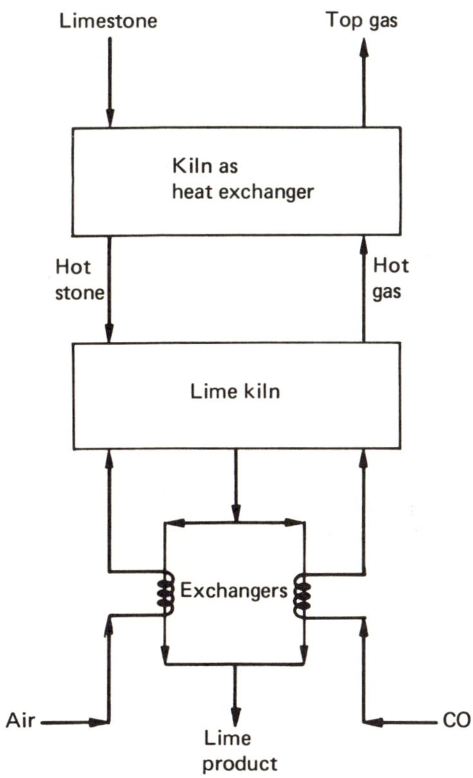

Fig. P-7. Flow sheet of lime burning for Prob. 5-2.

(a) An idealized flow sheet is proposed in Fig. P-7. It is maintained that this operation would require the absolute minimum of fuel. Do you agree? Propose a better scheme, if one exists.

(b) What is the minimum volume (standard cu ft) of CO required as fuel for the best possible flow sheet?

Note: Assume that all kilns, heat exchangers, pipe lines, and conveyors operate adiabatically. The formation of much calcium hydroxide is to be avoided. Assume MC_p for CO, O_2, N_2, CO_2, CaO, and $CaCO_3$ are 7, 7, 7, 10.5, 10, and 20.5 Btu/(lb mole)(°F), respectively.

$$CO + \tfrac{1}{2}O_2 = CO_2 \qquad \Delta H^\circ = -67.5 \text{ kcal}$$

$$CaCO_3 = CaO + CO_2 \qquad \Delta H^\circ = +42.8 \text{ kcal} \qquad K_p = 1 \text{ at } 900^\circ\text{C}$$

$CaCO_3$ decomposes to CaO and CO_2 at a finite rate in operations at atmospheric pressure only at 900°C and above. The $CaCO_3$, CO, and air feeds are available at the surroundings temperature of 100°F.

PROBLEM 5-3

Methane, CO_2, and steam, each at 30 psig pressure and at the temperatures shown in Fig. P-8, are to be preheated, mixed, and passed through a nickel catalyst bed. This catalyst operates isothermally at 1500°F and, based on pilot studies, the off-gas composition is estimated at 2% CH_4, 4% H_2O, 2% CO_2, 28% CO, and 64% H_2.

(a) What is the composition of the mixed feed gas before reaction occurs?

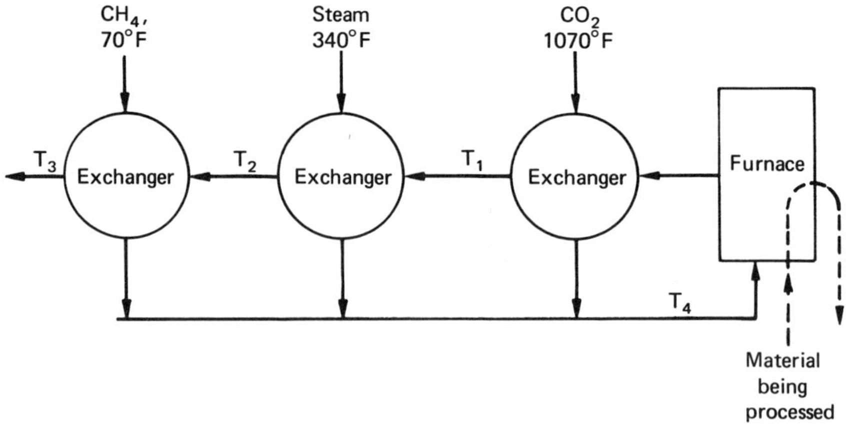

Fig. P-8. Flow sheet for synthesis gas manufacture for Prob. 5-3.

(b) Assuming infinite exchangers to be used, calculate T_1, T_2, T_3, and T_4. Should these be cocurrent or countercurrent?

(c) What is the amount of heat supplied to the reactor, in Btu/100 moles of product gas?

(d) There is extensive need for high temperature heat in the plant, consequently the gas stream or streams leaving this unit should be at as high a temperature as possible. Using infinite exchangers, propose a flow sheet wherein the heat added to the reactor is reduced to a minimum (the temperatures and pressures of the CH_4, H_2O, and CO_2 entering the system and of the catalyst bed are to be unchanged). What is now the heat required, in Btu/100 moles of product gas?

Notes: For purposes of this problem only, assume that MC_p for H_2, CO, CO_2, H_2O, and CH_4 are 7, 7, 11, 11, and 17 Btu/(lb mole)(°F), respectively. Approximate enthalpies of reaction are

$$CO + H_2O = CO_2 + H_2 \qquad \Delta H^\circ = -15000 \text{ Btu/lb mole}$$

$$CH_4 + H_2O = CO + 3H_2 \qquad \Delta H^\circ = +97000 \text{ Btu/lb mole}$$

PROBLEM 5-4

A waste gas at 220°C contains 5% HCN, the rest is N_2. This gas is to be disposed of by reaction with 150% of the air needed for complete combustion, with the HCN burning to CO_2, H_2O, and N_2. Before mixing with the waste gas, the air is preheated in a countercurrent heat exchanger by the hot off-gases from the adiabatic reactor, which is filled with pieces of broken ceramic to catalyze combustion. The reactor, exchanger, and ducts are all well insulated. Air enters the exchanger at 25°C.

(a) The gas leaving the reactor is to be at 1100°C. How many sq ft of heat exchange surface must be installed in the exchanger? The overall air-gas coefficient of heat transfer in the exchanger may be taken as 10 Btu/(hr)(sq ft)(°F).

(b) All gas rates through the system are doubled. What is now the reactor temperature, if the overall coefficient of heat transfer in the exchanger typically varies as the mass flow to the 0.8 power?

(c) Keeping all other operating conditions the same, if an exchanger of infinite area were used, what temperature would be reached in the reactor? If heat transfer surfaces in the exchanger actually used became fouled, what would happen?

Data: Complete combustion may be assumed in the reactor. For purposes of this problem only, assume MC_p values are constant, as follows: for H_2,

N_2, O_2, $H_2O(g)$, CO_2, and HCN(g), 7, 7, 7, 10.5, 10.5, and 10.5 cal/(g mole) (°C), respectively. Enthalpies of formation at 25°C for CO, CO_2, $H_2O(g)$, and HCN(g) are −26.8, −94, −58, and +31.2 kcal/g mole (29), respectively. Equilibrium constants of the combustion reactions are all large. One hundred lb moles of waste gas are made per hour.

PROBLEM 5-5

The three methylamines are made simultaneously by the catalyzed gas phase combination of ammonia and methanol at around 400°C with reactions presented in Prob. 1-10. The disproportionation reactions (monoamine reacting with triamine to form dimethlamine, and NH_3 reacting with di- to form monomethylamine) also occur over the same catalyst used for the main reactions. Operating pressures are usually at least several atmospheres, since elevated pressures benefit rates and help suppress a tendency of the reaction partners to decompose into their elements. Obviously materials of construction for the reactor must be chosen with care in order not to catalyze such decomposition.

The catalyzed reactions for amine formation are relatively rapid and proceed to equilibrium in a well-designed reactor. Since equilibrium constants lie not very far from unity, these reactions do not go to completion, and so are sometimes called "equilibrium reactions." Over industrially interesting ratios of methanol to NH_3 in the feed, all three amines together with some unreacted NH_3 are, therefore, present in the off-gases.

Separation of the three methylamines at atmospheric pressure is difficult because of azeotrope formation: 78% mono- in the mono-tri-azeotrope boiling at −5°C; 26% di- in the di-tri-azeotrope boiling at 3°C; and 73% NH_3 in the NH_3-tri-azeotrope boiling at −33.5°C. All these azeotropes disappear at above 26 atm [Reference (152), Vol. 12, 420].

(a) Propose a flow sheet for an operating scheme for converting methanol and NH_3 to dimethylamine with no net production of the mono- and trimethylamines. What is the composition of the feed to this plant?

(b) The claim is made that total power requirements of a plant for making dimethylamine at elevated pressure are less than for atmospheric pressure operation. Discuss and explain this remark.

Notes:

VAPOR PRESSURES VERSUS TEMPERATURE

	NH_3	Mono-	Di-	Tri-
100 mm Hg	−68.4°C	−43.7°C	−32.6°C	−40.3°C
1 atm	−33.6°C	−6.3°C	+7.4°C	+2.9°C
30 atm	66.1°C	106.3°C	132.2°C	—

PROBLEM 5-6

An adiabatic converter in a sulfuric acid plant is fed with a burner gas at 400°C analyzing 10% SO_2, 11% O_2, and the rest N_2. The gas leaves this converter at equilibrium, passes through a cooler where the temperature is reduced to 400°C, and then enters a second adiabatic converter, from which the gas again emerges at equilibrium. The maximum allowable gas flow into either converter is 100 lb moles/hr. Instead of operating in series as at present, it is proposed to operate these two converters in parallel, recycling part of the product gas back to the converter feed gas as needed to obtain the same yield of SO_3 as with the two converters in series. The mixed feed gas is again to enter both converters at 400°C. What is the production of moles of SO_3 per unit time attainable by this new parallel arrangement relative to that of the series operation? The total pressure may be assumed atmospheric at all times.

Notes: Approximate heat capacities and heats of reaction are as in example 5-4, while equilibrium data are presented in Fig. 5-4.

PROBLEM 5-7

Two adiabatic converters are operated in series with interstage cooling to produce SO_3 from sulfur burner gas. The gases enter the first converter at 400°C, analyze 10% SO_2, 11% O_2, and the rest nitrogen. The product gases from the first converter are at equilibrium, are cooled to 400°C and then enter the second converter, which they again leave at equilibrium.

Because of a malfunction, the interstage cooler is to be taken off stream to be repaired. It is proposed to continue operations by mixing cold dry air (25°C) with the hot gas from the first converter to make a mixed gas at 400°C. This mixed gas is then to be fed to the second converter, which it will again leave at equilibrium. The maximum allowable mass flow into either converter is 100 lb mole/hr.

(a) How many moles of air must be added between converters per 100 moles of burner gas fed to the first converter?
(b) What is the temperature of the gas leaving the second converter when operating at maximum capacity with the added air?
(c) Contrast the output of SO_3 obtained in these two types of operation, expressed as moles of SO_3/hr. What are the per cent conversions of SO_2 to SO_3 in these two cases.

Notes: Use the heat capacities, enthalpies of reaction, and equilibrium chart of example 5-4.

PROBLEM 5-8

An exothermal reaction is carried out in an adiabatic reactor containing a bed of granular catalyst. The catalyst for this system is highly effective, and equilibrium is attained in the off-gases. After several months of operation the pressure drop through the reactor starts to increase unexpectedly. Upon taking the reactor off stream for inspection, it is found that part of the catalyst particles have broken up extensively and others have changed color. This color change is stated to indicate overheating, which would mean the attainment of temperature significantly higher than the adiabatic exit temperature corresponding to the feed conditions. Your young engineering assistant says he cannot see how such temperatures could possibly develop, and feels there must be another explanation since "the laws of thermodynamics cannot be violated." Moreover, the records show that the temperature, composition, and rate of gas feed to the adiabatic reactor remained constant at "safe levels." Another engineer states he sees no inconsistencies with thermodynamics in these findings.

You must decide upon the possibility of a "hot spot" existing in this reactor. Prepare a statement explaining your position.

PROBLEM 5-9

A flow sheet for an experimental sulfuric acid plant is shown in Fig. P-9. The gases fed to the reactor (which is a fluid catalyst bed operation) analyze 10% SO_2, 11% O_2, and 79% N_2, and are at 300°C and 1 atm pressure. The catalyst bed itself is isothermal at 480°C. Cooling of this reactor is accomplished by recycling the solid catalyst between the reactor and a second fluid bed in the air heater, where the catalyst exchanges heat with cold air entering at 20°C. Air from this fluid bed air heater goes to the sulfur burner, which is also fed with solid sulfur at 20°C. The sulfur burns to SO_2 in this unit, and the hot burner gases are cooled in a waste heat boiler, then enter the catalytic reactor as shown. The waste heat boiler generates saturated steam at 200°C.

(a) What is the temperature of the fluid bed in the air preheater?

(b) How much catalyst is recycled, in lb/100 lb moles of gas entering the reactor?

(c) How much steam (lb/lb of sulfur fed to unit) is generated, if the heat of vaporization of water at 200°C is 950 Btu/lb?

Notes and Data: The catalyst used is in good condition, so that equilibrium of the off-gases is attained. Fluidization is excellent, hence the solids and gases leaving each fluid bed reach temperature equality. Each unit in this system is well insulated, so that heat losses to the surroundings may be

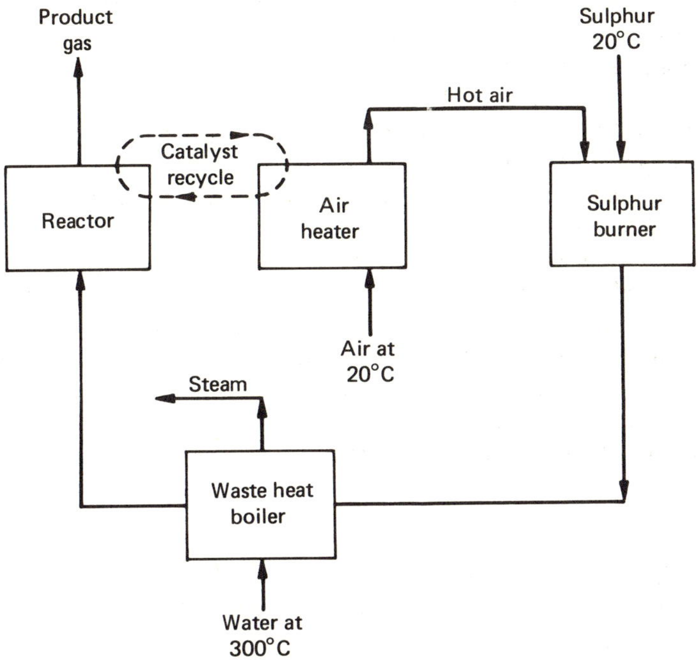

Fig. P-9. Flow sheet of an experimental sulfuric acid plant for Prob. 5-9.

neglected. The heat capacities for the gases and the heat of SO_2 oxidation are as in example 5-4. The heat capacity of the catalyst is 0.1 Btu/(lb)(°F), and the enthalpy of formation of SO_2 at 20°C from solid sulfur and oxygen is −71 kcal/g mole. The water enters the waste heat boiler at 200°C.

PROBLEM 5-10

Formaldehyde is made by the catalyzed partial combustion of methanol with air at atmospheric pressure, with recovery of the product by water scrubbing. The feed gas composition must lie outside the explosive limits of 6.7–30 volume % of methanol in air. When using silver wire gauze or silver crystals as catalyst, the feed gas generally contains over 35% methanol in air, and enters the converter at around 75°C. Only a thin catalyst bed is required because silver is an effective catalyst for both of the following reactions:

$$CH_3OH(g) + \tfrac{1}{2}O_2 = CH_2O(g) + H_2O \qquad \Delta H^\circ = -37 \text{ kcal}$$

$$CH_3OH(g) = CH_2O(g) + H_2 \qquad \Delta H^\circ = +21 \text{ kcal}$$

The oxidation reaction is so fast that molecular oxygen disappears within the first few millimeters of the catalyst bed, while the slower dehydrogenation reaction continues in the later sections of the bed. Typically, the product gas contains significant amounts of unreacted methanol, which is recovered with the formaldehyde in the scrubber and then separated by distillation. Silver has little catalytic effect on the decomposition of either methanol or formaldehyde to CO and H_2, which would otherwise be extensive at these operating conditions. Yields of formaldehyde are therefore high, running well over 97% on the methanol reacted.

The plant equipment for this process is relatively simple. The air used is first scrubbed with caustic soda to remove any SO_2 and CO_2 present, since these gases are catalyst poisons. The feed mixture is then filtered to remove dust, since even a small particle of iron lodged on the catalyst will cause methanol to carbonize, and so reduce the catalyst activity (158). The equilibrium constant for the oxidation reaction is very large at all temperatures, and for the dehydrogenation reaction is as follows (106):

$$T \log K_p = 6.47T - 4600$$

where pressure units are in atm and T is in °K.

In an early study of this operation, Homer (56) reports the product gas of one run on an experimental atmospheric-pressure converter to analyze 4.8% CO_2, 0.2% CO, 0.3% O_2, 20.2% H_2, and the rest N_2. In view of the date of this work, the analysis was probably by Orsat, and so is on a formaldehyde and methanol-free basis (why?). The feed to the converter contained 1 mole of methanol/mole of dry air.

(a) What yield was obtained (moles of CH_2O produced per mole of CH_3OH reacted)? What was the conversion per pass (moles of CH_3OH reacting per mole of CH_3OH fed)?
(b) If the off-gases from this experimental reactor had been at equilibrium with respect to the dehydrogenation reaction above, what would their temperature have been? Actual off-gas temperature was probably 620°C.
(c) If the off-gases from the reactor are at 620°C, what is the temperature of the gas stream entering the converter? If preheating is necessary, would you preheat the mixed gas stream?

Standard enthalpies of formation at 1 atm at 25°C, in kcal, are as follows: −94.05 for CO_2; −26.42 for CO; −57.8 for H_2O(g); −68.3 for H_2O(l); −17.8 for CH_4. Heat of combustion in kcals/g mole at 20°C and 1 atm for CH_2O(g) to gaseous CO_2 and liquid water is −134.1, and for CH_3OH(g) is −182.7. For purposes of this problem only, assume constant MC_p values for H_2, O_2, N_2, CO, CO_2, CH_4, H_2O(g), CH_2O(g), and CH_3OH(g) for 7, 7, 7, 7, 10.5, 14, 10.5, 14, and 21 cal/(g mole)(°C), respectively.

PROBLEM 5-11

The following two reactions have been proposed for extracting oxgyen from air:

$$Ag_2O(c) = 2Ag(c) + \tfrac{1}{2}O_2 \qquad \text{(a)}$$

$$Ag_2CO_3(c) = Ag_2O(c) + CO_2 \qquad \text{(b)}$$

Equilibrium constants (all pressures in atm) for these two reactions are as follows (73, 104):

$$\log K_a = -\frac{1430}{T°\text{K}} + 3.1426 \text{ for reaction} \qquad \text{(a)}$$

$$\log K_b = -\frac{3855.9}{T°\text{K}} + 7.9040 \text{ for reaction} \qquad \text{(b)}$$

Porous silver oxide granules have been prepared in the laboratory by soaking $\frac{1}{4}$-in. pumice particles with a concentrated aqueous solution of silver nitrate, then drying, and heating to decompose the nitrate. These freshly prepared granules contain 30% by weight of Ag, and show a packing density of 50 lb/cu ft. With these granules, rates of reactions (a) and (b) are large only at above about 480°K, hence it is agreed that operations of the reactors will be at this temperature or higher.

It is suggested that pure oxygen (free of N_2 and CO_2) can be made as follows:

I. Dry air (21% O_2, 0.5% CO_2, the rest N_2) at 480°K and a pressure of 12 atm is passed through a fixed bed of the granules just described. The bed is kept isothermal at 480°K by appropriate heating and/or cooling. The off-gases may be assumed at equilibrium with this bed.

II. When the bed is "saturated," air flow is stopped and pure oxygen at 480°K and a pressure of π_2 is passed through the bed to recover oxygen. The bed is again isothermal at 480°K, and equilibrium between the bed and its off-gases is again attained.

For these orientational calculations, it is agreed to neglect purge losses of oxygen and contamination with other gases which occur during change-overs between stages I and II. It is also assumed that only half of the silver present in the granules participates chemically in these cyclic operations. The bed used is 10 ft deep and superficial hot gas velocities are 3 ft/sec. It is pointed out by an older associated that this process appears of only very limited interest because of obvious high investment requirements. The following questions are nevertheless to be asked:

(a) What is the off-gas composition during stage I?

(b) What is the highest value (if any) of the total pressure at which stage II of this process can be operated?
(c) Assuming equilibrium in the off-gases, how many minutes are required for stage I of any cycle?
(d) It is proposed that this operation could readily be made continuous. Discuss and propose a flow sheet. If two reactors are to be used between which solid is to circulate, propose suitable apparatus. Could the heat evolved in one reactor be utilized in the other?

PROBLEM 5-12

In the so-called *Hypro process* (117), methane is cracked to hydrogen and carbon in a catalytic fluid bed unit, called the *cracker*. The methane feed enters into the base of the cracker at 600°C. Off-gases from the bed (operating at 900°C) analyze 10% CH_4 and 90% H_2. The carbon formed adheres to the solid fluidized catalyst granules and travels with this catalyst while it is recycled between the cracker and a second unit called the regenerator. The regenerator is also a fluid bed, and operates at 1050°C. Fluidization here is with air, which burns off some of the carbon, so that the regenerator product gases contain only N_2 and CO. The regenerated catalyst returned from the regenerator to the cracker carries 0.3 lb of carbon per lb of catalyst, while the catalyst entering the regenerator carries more carbon. Operation is continuous and in steady-state, and the entire system is well insulated.

(a) How many moles of CO are generated per mole of CH_4 fed to the unit?
(b) How many pounds of catalyst (carbon-free) are recycled per lb mole of H_2 produced in the product gas?
(c) How much carbon (lb/lb of catalyst) is carried by the catalyst leaving the cracking chamber?
(d) At what temperature does the air enter the regenerator chamber? How could air preheat needs be reduced?

Notes: For purposes of this problem, assume MC_p values as follows: 7, 7, 7, 7, 17.5, and 3.5 Btu/(lb mole)(°F) for H_2, N_2, O_2, CO, CH_4, and solid carbon, respectively. The heat capacity per lb of catalyst is 0.2 Btu/(lb)(°F). The enthalpies of formation of CH_4 and CO from the elements at 25°C are −17.9 and −26.9 kcal/g mole, respectively. Note that fluidized solids can be circulated between reactor vessels very rapidly, and with minimized gas carry-over from one vessel to another. Equilibrium is not attained in the cracker off-gases. The fluidized solids in the cracker are well mixed, as are the fluidized solids in the regenerator.

PROBLEM 5-13

When Cu_2S reacts with oxygen, the following reaction occurs:

$$Cu_2S + \tfrac{3}{2}O_2 = Cu_2O + SO_2 \qquad \Delta H^\circ = -93 \text{ kcal} \tag{a}$$

The equilibrium constant of this reaction is very large at all temperatures. When a limited amount of oxygen is passed through a bed of Cu_2S, then a correspondingly limited amount of Cu_2O forms by reaction (a), and this Cu_2O tends to react further with nearby Cu_2S. The reaction and its equilibrium constant (124) are as follows:

$$Cu_2S + 2Cu_2O = 6Cu + SO_2 \qquad \log K_p = -\frac{5070}{T^\circ K} + 5.02 \tag{b}$$

It is suggested that the metallic copper so produced could be used to remove H_2S from a hot, reducing gas stream. The reaction and its associated equilibrium constant (13) are as follows:

$$2Cu + H_2S = Cu_2S + H_2 \qquad \log K_p = +\frac{2275}{T^\circ K} + 0.86 \tag{c}$$

The copper sulfide from reaction (c) could then be recycled to reaction (a) for reuse.

A copper-containing material suited to the foregoing reaction scheme was made by impregnating granular silica gel with a strong copper acetate solution, drying, heating, and reducing with hydrogen. A batch of these granules was placed in a vertical externally heated tube and fluidized with hot hydrogen containing 5% H_2S. Fluidization was good at a superficial velocity for the hydrogen of 3 ft/sec, with the bed solids showing a density of 60 lb/cu ft. With a 6-in. deep bed kept at 800°C, and an entering crude gas temperature of 600°C, the off-gas composition was found to remain at equilibrium with respect to reaction (c) until 50% of the copper had reacted. As still more copper reacted, H_2S removal became poorer. When this bed was then regenerated by fluidizing with air entering at 25°C, keeping the bed itself at 850°C, an off-gas formed which analyzed 79% of N_2 and 21% of SO_2 as long as at least 5% of the copper was present as the sulfide.

A gas stream in a plant analyzes 5% H_2S, 5% N_2, the rest hydrogen, and is available at 600°C and 2 atm pressure at a rate of 10^6 standard cu ft daily. It is proposed simultaneously to desulfurize and heat this gas to 800°C for use in a metallurgical operation as follows: pass the crude hydrogen at 600°C up through a fluid bed of silica granules carrying 5 lb of copper per 95 lb of silica, prepared as described above. This bed, called the desulfurizer, is to be at 800°C, consequently the purified hydrogen off-gases will be at this temperature. Solids will be continuously drawn off from this bed and cycled through a second fluidized bed for controlled oxidation and heating, then returned to the first bed. This second bed, called the oxidizer, is fluidized with air under 2 atm pressure, with only SO_2 and N_2 in the off-gases. Any heat required or removed in the desulfurizer is to be carried as sensible heat in the recycled hot solids from the oxidizer. The temperature of the oxidizer and of the air entering this oxidizer remain to be established. For safety, not more than $\frac{1}{3}$ of the copper in the desulfurizer, nor less than $\frac{1}{10}$ of the copper in the oxidizer, is to be present as Cu_2S. Solids mixing in these fluid beds is again excellent.

(a) What fraction of the H_2S in the crude hydrogen entering the desulfurizer remains in the gas?
(b) What is the rate of the solids circulation, in lb/hr fed to the desulfurizer?
(c) What is the heater temperature?
(d) It is stated by a capable associate that, due to equilibrium limitations, there exist operating temperatures for both the oxidizer and the desulfurizer below which the system cannot function. Discuss. What are these temperatures?
(e) The same associate maintains that a life test must be made before large sums of money are committed to this process. Discuss.

Note: For purposes of this problem, assume MC_p values for N_2, O_2, H_2, H_2S, and SO_2, of 7, 7, 7, 10.5, and 10.5 cal/(g mole)(°C), respectively. For SiO_2, Cu, and Cu_2S, heat capacities are 0.25, 0.11, and 0.115 cal/(g)(°C), respectively.

PROBLEM 5-14

A furnace operating at atmospheric pressure is to be heated by partial combustion of methane with air. The methane is so proportioned to the air fed that the mole ratio of H_2O/H_2 in the furnace off-gases is unity, since a higher ratio results in oxidation of the furnace charge. The heat generated must be available above 1800°F to be useful, consequently the furnace off-gases must leave at this temperature. Three alternative methods of operation are proposed:

(a) Methane and air, each at 70°F and 1 atm, are fed directly to the furnace burner. The off-gases contain only N_2, H_2, H_2O, CO, and CO_2 and leave the furnace at equilibrium with respect to the water-gas reaction at 1800°F for use elsewhere in the plant:

$$H_2O + CO = H_2 + CO_2 \qquad K_{p,\,1800°F} = 0.60$$

(b) Operations are as in part (a), except that the fuel gas and air are preheated by heat exchange with the hot flue gases. To avoid cracking in the exchanger, the natural gas is preheated to only 600°F, while the air is preheated to only 1200°F to avoid excessive corrosion of the exchanger, which is made of ordinary steel.

(c) The natural gas (80a) is preheated to 600°F, mixed with an equal volume of steam at 600°F, and passed at atmospheric pressure through a catalytic reformer, yielding a gas at 1800°F analyzing (wet basis) 6.2% CH_4, 58.7% H_2, 16.2% H_2O, 16.2% CO, and 2.7% CO_2. This hot gas is injected into the furnace along with air preheated to 1200°F. As before, the furnace flue gases leave at 1800°F, contain no CH_4, have a ratio of H_2O/H_2 of unity and are in equilibrium with the water-gas reaction at this temperature. The off-gases are used to (a) preheat the air and natural gas, (b) generate the steam consumed in the reformer, and (c) supply the heat needed by the endothermal reformer reactions. For the latter purpose, these hot flue gases are burned in the furnace space around the catalyst-packed reformer tubes, using theoretical air preheated to 1200°F.

Determine the Btu released in the furnace above 1800°F per mole of CH_4 fed, for each of the three schemes presented above. Compare and contrast the advantages of these three schemes.

Notes: For purposes of this problem only, assume MC_p for H_2, N_2, O_2, CO, CO_2, H_2O, and CH_4 to be 7, 7, 7, 7, 10.5, 10.5, and 17.5 cal/(g mole)(°F), respectively. Enthalpies of formation of CO, CO_2, H_2O(g), and CH_4 are −28, −94, −58, and −18 cal/g mole, respectively. The heat of vaporization of H_2O is 10 kcal/g mole, and the value of MC_p for liquid water is 18 kcal/(g mole)(°C). It may be noted in passing that the reformer off-gases are not in equilibrium with the reforming reactions.

PROBLEM 6-1

After much trial, a small-scale pilot furnace has been developed for making TiO_2 pigment by the direct combustion of titanium tetrachloride with air at elevated temperatures. The $TiCl_4$ vapor flows through a suitable duct into a large and relatively stagnant furnace space, which is filled with a hot mixture of air and products of combustion. The pigment-forming reaction is exothermal and proceeds as follows:

$$TiCl_4(g) + O_2 = TiO_2(c) + 2Cl_2 \qquad \Delta H^\circ = -46 \text{ kcal}$$

The TiO_2 is produced as extremely fine particles, whose size distribution is a sensitive function of operating conditions, namely flame residence, times, ratio of $TiCl_4$ to air blown into the furnace, etc. Particle size and size distribution must be held within narrow limits to meet pigment specifications.

(a) What factors must be considered in scaling up this burner?

(b) A severe problem has been the rapid formation of incrustations on the burner pipe tip and on the walls, which interfere with mixing and reduce quality. Have you any suggestions?

PROBLEM 6-2

Certain large metal castings are reheated in a special furnace. Long flames are formed by propane issuing from ports along the furnace wall, and burning within the furnace itself. After much trial and error the burner settings as now used were arrived at, and the resultant flame length and heat release are considered to be optimum.

Natural gas becomes available at a low price, and so conversion of the furnace from propane to natural gas is proposed. What recommendations have you to make for minimum disturbance of production? Justify your position.

PROBLEM 6-3

Tetrafluoroethylene, the monomer of "Teflon," is made by the pyrolysis of gaseous monochlorodifluoromethane ($CHClF_2$) in an "empty" heated tube, as follows:

$$2CHClF_2 = C_2F_4 + 2HCl \qquad \Delta H^\circ = 18 \text{ kcal}$$

The equilibrium constant for this reaction is very large at 700°C. Park et al. (118) fed $CHClF_2$ vapor at 1 atm pressure to a reaction tube in an electric "tube" furnace 12 in. long. Temperatures were measured by a single thermocouple located in the tube wall in the middle of the heated section. Park estimated that the gas was "at the indicated temperature" over 8 in. of this tube, implying that preheating occurred earlier in the tube.

Off-gases were analyzed for HCl (by water scrubbing and titration), and for C_2F_4 (by condensation and fractionation—no chromatographic techniques were available at the time of these studies). Small amounts of by-products were found, including members of the homologous series $H(CF_2)_nCl$. Traces of unstable and "quite toxic" materials formed unexpectedly, illustrating again the care with which pioneering studies should be undertaken.

(a) Per cent conversion was defined as the ratio (100) (moles HCl in product gas per moles $CHClF_2$ in feed). $CHClF_2$ vapor was fed at atmospheric pressure with the reactor tube at 700°C. When feeding at 120 g/hr, 25% conversion was obtained with a "contact time" of 0.35 sec. What is the tube's approximate inside radius? Assuming no side reactions to occur, how much heat must be supplied, in Btu/(sq ft of internal tube area)(hr), on the average?

(b) Park indicates that when operating at 700°C and 1 atm pressure with two different feed rates, conversions of 14 and 35% were obtained in 0.23 and 0.51 sec, respectively. Would these findings be reasonably consistent with first order kinetics typically found in pyrolysis reactions? Are they more consistent with second order kinetics? With zero order?

(c) For a given contact time, Park reports that percent conversion for a given contact time decreases somewhat with increasing pressure. Is this consistent with the answer to part (b) above?

(d) The run described in question (a) is repeated with the same linear velocity and wall temperature in a tube of four times the inside diameter. What is now the expected conversion in 0.35 sec contact time?

PROBLEM 6-4

Ethylene is made by pyrolysis of hydrocarbons, that is, by severe high temperature cracking. Ethylene is unstable at elevated temperatures, and so reactor residence times should be as short as possible to avoid product loss. Ethane, an important ethylene raw material, is usually cracked to over 60% conversion per pass with the uncracked portion recycled. This cracking is strictly thermal, since no catalysts have as yet been developed which do not rapidly become inactivated by formation of a carbon coating. Pyrolysis is carried out in a tube furnace, with short throughput times (resulting from high flow velocities) to minimize undesired side reactions. Steam is mixed with the ethane feed to reduce residence time, thus helping to avoid close approach to equilibrium. That is, near equilibrium, secondary reactions involving ethylene or its free radicals proceed at increasing rates, leading to the formation of heavy liquid fractions, tar, and coke. Therefore, a well-designed coil is one in which the reactants flowing through the system encounter a continuously rising temperature gradient, so chosen that equilibrium is never closely approached (136, 164). The primary reaction, of course, is

$$C_2H_6 = C_2H_4 + H_2 \qquad K_p = 0.74 \text{ atm at } 1500°\text{F} \tag{1}$$

A pipe reactor is to be designed. For orientation, assume continuous isothermal operation at 1500°F and 2 atm absolute, with a feed of 50 mole %

H_2O and 50 mole % C_2H_6. The feed rate is 8 lb/(sec)(sq ft), and piston flow may be assumed. The pipe is 3 in. in I.D.

(a) What is the equilibrium conversion to be expected under these conditions?

(b) A reaction tube is to be designed in which the fraction of ethylene reacted is 50% of the equilibrium conversion. How long must this tube be? The back reaction is to be neglected.

(c) What would be the residence time?

(d) Now recognize that the back reaction does in fact occur. How long must the tube be?

Notes: The rate equation for the forward reaction (1) is of the form

$$-\frac{dn_e}{d\theta} = k_e n_e$$

where k_e is 1.0 reciprocal sec at 1500°F, and n_e is moles of C_2H_6 in the system. For the backward reaction,

$$\frac{dn_e}{d\theta} = k_B n_H \frac{n_{et}}{V}$$

where n_H is moles of hydrogen, n_{et} is moles of ethylene, V is the volume of the system, and k_B is the backward reaction's rate constant. It may be assumed that the foregoing rate equations apply even in the vicinity of equilibrium.

PROBLEM 6-5

The thermal decomposition of nitric acid vapor in the range of 100–500°C occurs as follows:

$$2HNO_3(g) = H_2O(g) + 2NO_2 + \tfrac{1}{2}O_2 \qquad \Delta H^\circ = +22.6 \text{ kcal}$$

The equilibrium constant for this reaction at these temperatures is large. This gas phase decomposition was studied in glass vessels and was found to be homogeneous above 400°C, with the simultaneous wall reaction becoming noticeable below 400°C in small vessels (66). For partial pressures of HNO_3 of 25 mm and below, with inert gases added to increase the total pressure to 1 atm, the homogeneous decomposition reaction is first order in nitric acid, with the rate independent of concentrations of H_2O, NO_2, and O_2. The homogeneous reaction rate constant in reciprocal sec was found to be 0.26 at 450°C and 0.018 at 400°C.

(a) Values of the logarithm of the rate constant k versus reciprocal temperature fall on a reasonably straight line. What is the energy of activation?

(b) In the decomposition of certain inorganic nitrates, a gas stream leaves the reactor at 425°C and 1 atm total pressure, and contains 50 mole % of HNO_3 vapor, the rest being inert substances. This gas flows through a large, well-insulated duct to a condenser. Estimate the partial pressure of HNO_3 at the end of this duct, if the residence time in the duct is 4 sec, and if piston flow may be assumed. The heterogeneous reaction is to be taken as negligible in this operation.

Notes: For this problem, assume MC_p values in cal/(g mole)(°C) for HNO_3, H_2O, NO_2, and O_2 of 16, 10.5, 9, and 7, respectively. The inerts in the gas stream have an average heat capacity of 8.

PROBLEM 6-6

"Most homogeneous gas reactions are so rapid that, when they go at all, the rate of mixing rather than kinetics controls," according to Chilton (20). He points out that the gas phase oxidation of NO with molecular oxygen to form NO_2 is unusual, being one of the very few known third-order reactions with a rather slow rate which diminishes with increasing temperature. This reaction and its rate as found by Bodenstein (10) for a batch reactor of constant volume are

$$NO(g) + \tfrac{1}{2}O_2 = NO_2 \qquad K_{p150°C} = 1.5 \times 10^3;\ \Delta H° = -13.5 \text{ kcal}$$

$$-\frac{dp_{NO}}{d\theta} = kp_{NO}^2 p_{O_2}$$

where k has values of 6 and 2 (atm^{-2}) (sec^{-1}) at 150 and 350°C, respectively.

A gas stream from a certain nitrate decomposition analyzes 70% N_2, 18.5% O_2, 2.5% NO_2, and the rest NO. This gas flows continuously through a chamber whose volume equals 5 sec flow of product gas measured at the chamber's operating conditions of 3 atm and 150°C. Mixing within this chamber is excellent because of the jet action of the entering gas mixture.

(a) Comment, based on your experience, on Chilton's generalization.
(b) What is the composition of the gas leaving this chamber?
(c) Assuming the reactor to be adiabatic, what was the temperature of the feed? Values of MC_p in cal/(g mole)(°C) for N_2, NO, O_2, and NO_2 are 7, 7, 7, and 10.5, respectively.

PROBLEM 6-7

Acetylene can be produced by partial combustion of methane with less oxygen than required for complete combustion, followed by quenching after a brief residence time at elevated temperature. In one study (60), natural gas and oxygen were mixed at room temperature and then fed into a cylindrical combustion chamber, 1.6 in. in diameter, through a perforated plate acting as a cap on one end of the cylinder. Reaction occurred immediately and the product gases were sampled at various points downstream from the perforated plate. A small water jacket served to protect the sampling tube as well as to quench the gas samples withdrawn. Composition of the gases at various distances downstream from the perforated plate were as follows:

	Product Gas*			
Downstream, in.	5	7	9	11
C_2H_2, mole %	3.0	5.7	6.5	6.0
H_2	25	45	52	53
CO_2	3.5	4.5	4.7	4.7
CO	18.5	31	32	33
CH_4	30	9	4	3
O_2	20	5	1	0.3

*Solid carbon formation was negligible.

The natural gas feed analyzed 93.3% CH_4, 3.8% C_2H_6, 1.1% C_3H_6, 1.5% CO_2, 0.2% C_4H_{10}, and 0.1% C_5H_{12} and was fed at the rate of 15.1 lb/hr.

(a) How many lb of acetylene were formed in the gas at the 9-in. point per 1000 cu ft of natural gas fed to the furnace at 68°F and 1 atm? What percentage of carbon in the feed was converted to acetylene?

(b) How many pounds of oxygen were used per lb of natural gas?

(c) The reactor temperature, measured with a properly shielded thermocouple, was 1903°K at a distance of 9 in. from the perforated plate. Is this gas at equilibrium with respect to any reaction occurring? The standard free energies of formation at 1900°K for C_2H_2, CO_2, H_2O, CO, and CH_4 are 51.95, −94.24, −33.7, −66.50, and 28.15 kcal/g mole, respectively.

(d) In view of these results, prepare a brief list of the requirements which a reactor system must meet for this method of acetylene manufa cture.

PROBLEM 6-8

Ethane is being cracked to ethylene in the experimental plant whose flow sheet is shown in Fig. P-10. Water vapor at 2000°F and preheated ethane at T_i are injected into an adiabatic reactor so designed that its contents are well mixed at all times. The hot reactor off-gases pass through one leg of the exchanger for preheating ethane, as shown, and leave the plant at 1104°F.

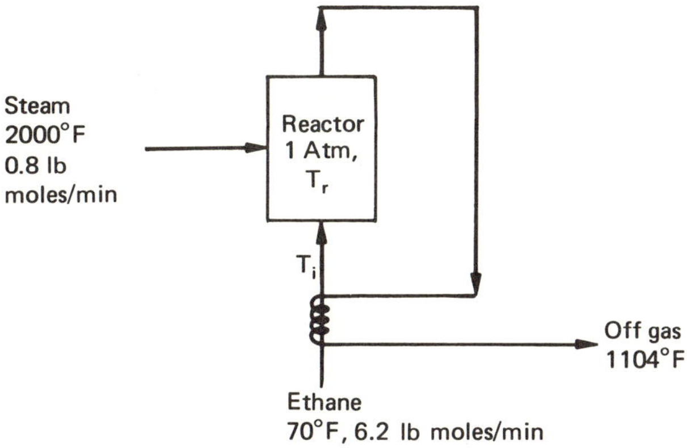

Fig. P-10. Ethane cracking operation of Prob. 6-8.

Operations are at atmospheric pressure throughout. The feed to the adiabatic reactor is 0.2 lb moles of ethane and 0.8 lb moles steam per min. The ethane enters the exchanger at 70°F. It may be assumed that the only reaction occurring is the following:

$$C_2H_6 = C_2H_4 + H_2; \qquad \Delta H = 60{,}000 \text{ Btu/lb mole}$$

The exchanger has a heat transfer area of 29 sq ft and an overall coefficient of 10 Btu/(hr) (sq ft) (°F).

(a) What is the composition of the product gas? Average values of MC_p for C_2H_6, C_2H_4, H_2, and $H_2O(g)$ are 24, 17, 7, and 10.5 Btu/(hr) (°F), respectively.

(b) The rate for the above reaction may be expressed as

$$-\frac{dn_e}{d\theta} = kn_e$$

where k is 10 and 0.1 reciprocal seconds at 1670°F and 1360°F, respectively (136). What is the reactor volume, in cu ft?

PROBLEM 6-9

The gas phase demethylation of toluene to benzene proceeds "thermally and homogeneously," i.e., without catalysts, at atm pressure above 700°C. The following reaction sequence has been proposed (144):

$$\begin{array}{rcl} H_2 & = & 2H\cdot \\ H\cdot + C_6H_5CH_3 & = & C_6H_5\cdot + CH_4 \\ C_6H_5\cdot + H_2 & = & C_6H_6 + H\cdot \\ \hline H_2 + C_6H_5CH_3 & = & C_6H_6 + CH_4 \end{array}$$

The first of these reactions has a very small equilibrium constant but is assumed to be very fast, hence atomic hydrogen is always present at its minute equilibrium concentration. It is stated that this reaction sequence leads to the rate expression

$$-\frac{dc_{\mathrm{T}}}{d\theta} = k_T c_{\mathrm{T}} c_{\mathrm{H}}^{1/2}$$

where c_{T} is toluene concentration, c_{H} concentration of molecular hydrogen, and θ is time.

This system was studied (166) by passing toluene-hydrogen mixtures through a preheater, flow reactor, and condenser. The Vicor preheater was a tube 11 ft long and of small diameter for relatively short residence times, while the Vicor reactor tube was 2 ft long and had a volume of 2.5 liters. Reactor residence times were from 1.8–5 sec.

It was found that the experimental data were in good agreement with the foregoing rate expression. Rates of toluene disappearance and benzene appearance were equal, indicating that no complicating side reactions occur.

(a) Show that the reaction sequence presented does in fact lead to the above rate expression.

(b) The rate constant was evaluated (166) from the equation

$$k_T = \frac{6 \times 10^4}{V} \cdot F\left(\frac{TG}{P}\right)^{1.5} \int_0^x \frac{dx}{(1-x)(b-x)^{1/2}}$$

where k_T is in $(\text{moles})^{1/2}/(\text{l})^{1/2}(\text{sec})$, T is reactor temperature in °K, P is reactor pressure in in. of Hg absolute, x is moles of toluene decomposed per mole of toluene fed, b is moles of hydrogen per mole of toluene in the feed, while by some oversight G is undefined. Concentrations are all in g mole/liter, F is ml of liquid toluene fed per min (the toluene feed has a density of 0.87 g/ml), while V is reactor volume in liters. Derive this integral equation from the rate expression indicating and justifying the assumptions made, and define G. Is there a printing error in the constant 6×10^4?

PROBLEM 6-10

The diester of adipic acid and 2-ethyl hexanol (an isomeric octyl alcohol) was prepared as part of a study of synthetic lubricating oils. Operations involved dissolving the solid acid in the alcohol, then heating the mixture to the temperature to be studied. The run was considered to start when the esterification catalyst, namely sulfuric acid, was added. The extent of reaction was followed by continuously bubbling nitrogen gas up through the reacting liquid, and condensing out the vapors of alcohol and water so removed. This condensate passed continuously through a settler and decanter, where the alcohol layer was separated from the water layer and returned to the reactor, while the water layer was measured and rejected.

(a) In the run under consideration, made at 110°C, 146.1 g adipic acid, 390.0 g 2-ethyl hexanol, and 2.92 g sulfuric acid were mixed, yielding liquid phase having an initial volume of 0.520 l. It may be assumed that volume changes of the reacting system were negligible. What is the theoretical yield of grams water from this system, assuming complete esterification to be attainable? Would complete reaction be possible in this system, recognizing that esterifications are all typically "equilibrium limited"?

(b) The monoester presumably forms first in this esterification, followed by diester formation from the monoester. The "forward" esterification reaction rates at constant temperature are respectively proportional to the concentration of acid and alcohol, and to the monoester and alcohol. The "backward" reaction rates would be proportional to the water and to the ester concentrations. The cumulative water formation versus time was as follows:

Time, min	1	2	4	6	8	10	11	12
Water formed, cc	3.5	8.5	16.0	22.5	27.5	30	31.5	32.5
Time, min	13	14	15	16	17	18	19	20
Water formed, cc	33.0	33.5	34.0	34.1	34.3	34.5	34.7	34.9

Estimate the time required for 99% and 99.9% of the theoretical water yield, assuming all water formed to be removed by the nitrogen.

(c) Calculate the daily output of diester from a reactor containing 500 lb of charge at the outset of each cycle, when operating batchwise to attain 99.5% of the theoretical water yield. Assume a half hour between batches for dumping and filling, and 24 hr a day of operation.

(d) This well-stirred reactor is now to be operated continuously and in steady state, with 500 lb of material always present. The feed contains catalyst, alcohol, and adipic acid in proportions as before, while 99.5% of the theoretical water is to form, and to be carried out of the kettle in a gas purge. What is now the daily production of the diester?

(e) A visiting academic physical chemist claims that the laboratory rate experiments as performed "would never be carried out this way in a proper kinetic study." What is he concerned about, and how would you answer him?

PROBLEM 6-11

Acetic anhydride can be made in several ways, one involving the direct reaction of ketene (CH_2:C:O) with acetic acid. Ketene in turn is made by pyrolysis of acetone at 600–700°C in externally heated alloy tubes:

$$CH_3COCH_3 = CH_2CO + CH_4 \tag{1}$$

The ketene made in this operation is in part lost by pyrolysis, as follows:

$$2CH_2CO = C_2H_4 + 2CO \tag{2}$$

Some cracking of ketene and of ethylene also occurs, resulting in formation of coke, hydrogen, and (in the case of ketene) carbon monoxide. Such cracking is reduced but not eliminated by addition of a little CS_2 to the feed.

(a) After removal of "condensibles," usually ketene and unconverted acetone, a typical off-gas analysis obtained when pyrolyzing acetone vapor at atmospheric pressure in glass tubes at 667°C is 82.2% CH_4, 11.2% CO, 5.1% C_2H_4, 0.9% H_2, and the rest nitrogen. Estimate the per cent of acetone disappearing which was converted to ketene. What assumptions were made in this calculation? Where did the H_2 probably come from? The nitrogen?

(b) Reaction (1) is known to be first order. That is, if n_A is the moles of acetone in the system and θ is time, then

$$\frac{dn_A}{d\theta} = -k_1 n_A$$

It is proposed that reaction (2) is also first order. That is, when feeding pure ketene to a hot tube, the rate of decomposition (where n_k is moles of ketene) by reaction (2) would be as follows:

$$\frac{dn_k}{d\theta} = -k_2 n_k$$

Recognizing that reactions (1) and (2) occur sequentially, set up an expression relating X, the moles of acetone surviving per mole of acetone fed, to Y, the moles of ketene surviving per mole of acetone fed. Assume that only reactions (1) and (2) occur.

(c) It is reported (23) that an 85% ketene yield is attained at 10% acetone disappearance, and a 70% yield at 25% disappearance. Are these two consistent with the assumption in part (b) that these two reactions are sequential and first order? Do these data "prove" both reactions to be first order?

(d) A reactor having 40 ft tubes is currently operating at 600°C at design capacity, and the conversion, namely the moles of ketene produced per mole of acetone fed, is to be a maximum. If the ratio k_1/k_2 is as in part (c), what is this conversion?

(e) The plant is to be designed for operations at 600°C. Sufficient flexibility is to be included so that efficient operations at two-thirds capacity are possible. How should this be accomplished?

PROBLEM 6-12

Hexamethylenetetramine, $(CH_2)_6N_4$, often called "hexamine," is a colorless, odorless solid whose saturated aqueous solutions contain 44.8 and 46.5% by weight of hexamine, respectively, at 12 and 25°C. Hexamine must be recovered from aqueous solution by vacuum evaporation to avoid decomposition, which is noticeable above 60°C.

Hexamine is formed by mixing aqueous solutions of formaldehyde and ammonia:

$$6CH_2O + 4NH_3 = (CH_2)_6N_4 + 6H_2O$$

The heat evolved when all partners remain in solution is about 55 kcal. To aid in conversion and minimize by-product formation, a slight excess of NH_3 is used in industrial operations (95).

A study of this reaction was made in a steady-flow, stirred tank reactor (163). The reaction mechanism is complex; but an overall expresssion which reproduces the rate from about 10–50°C, with molal CH_2O/NH_3 ratios of from 1.5–2 and ammonia concentrations of from 0.2–4.3 g-moles/liter, is as follows:

$$r_H = kc_A c_F^2$$

where r_H = rate of hexamine formation g-moles/(liter)(sec), k_H = rate constant, (liters)^2/(g-mole)^2 sec, found to equal $1.126 \times 10^9 \exp(-14122/RT)$, where T is °K and R is in cal/(°K)(g mole), c_A and c_F are ammonia and formaldehyde concentrations in g moles/liter.

(a) How much greater is the rate constant at 60°C than at 20°C? Would the same answer be obtained when assuming the rate doubles for every 10°C.

(b) How long would it take to attain 99.5% conversion of the ammonia in a batch operation at 20°C, wherein a 30% formaldehyde solution is mixed with 27% ammonia in proportions such that the mole ratio of CH_2O/NH_3 is 1.5? Assume the above rate expression applies down to the low concentrations equivalent to 99.5% conversion. Assume also that densities of all solutions are the same as for water.

(c) When operating batchwise, the kettle is emptied when the desired concentration is attained, and immediately refilled with fresh reagents. The emptying and filling time required, for the size of the operation contemplated, is 10% of the total reaction time. This same kettle could be operated continuously, with the product withdrawn at 99.5% conversion. Compare the production rates for this continuous operation (lb/hr) with that attained in batch operation, assuming the plant to operate "around the clock."

PROBLEM 6-13

A manufacturer has been producing hexamethylenetetramine (see preceding problem for background) batchwise in jacketed kettles. A need arises for a new larger plant. It is desired to develop a process which economizes on power, labor, cooling, and equipment costs without sacrificing yield. These factors are to be considered:

(a) Bubbling gaseous ammonia directly into more concentrated formaldehyde solutions resulted in such violent absorption that shocks developed which endangered building and equipment. A young chemical engineer, however, maintains that no such shocks would develop in a packed tower when scrubbing ammonia out of a gaseous air-ammonia mixture.

(b) For each unit of ammonia content, the tank car price of pure liquified ammonia is significantly lower than that of aqueous ammonia solutions.

(c) It is believed that "Hexa" can be crystallized pure from a solution which contains unreacted NH_3 and CH_2O. To avoid side reactions it is necessary to maintain the mole ratio of CH_2O/NH_3 reasonably close to 1.5.

(d) The reaction rate, as pointed out in the preceding problem, increases very rapidly with temperature.

Propose an improved process for making hexamine. Support your contentions that this is an improved process. What points require testing in the laboratory? Suggest a development program.

Notes: For purposes of this problem only, assume that the density of all solutions is 1 g/cc. The heats of solution in water are approximately 8.3 kcal/mole of NH_3 and 15 kcal/mole of CH_2O. The heat of evaporation of water is 9.8 kcal/mole. Atmospheric boiling temperatures of formaldehyde solutions are 100.9°C for 20% and 101.8°C for 40% by weight. For ammonia solutions:

Weight fractions of NH_3	0	0.1	0.2	0.3
Boiling Point, °F	212	156	126	82

PROBLEM 7-1

Feed gas analyzing 10% SO_2, 11% O_2, the rest nitrogen, as in example 7-1, is again fed to the adiabatic reactor of this example. The gas enters the reactor bed at 437°C and 1 atm pressure as before, but with a superficial approach velocity of 7 ft/sec rather than the velocity used in example 7-1 of 5 ft/sec. The purpose of this increased velocity is, of course, to increase capacity. What is the exit gas composition from this bed, which was calculated in example 7-1 to be 0.47 ft deep? By what factor has the output of SO_3 from this converter been increased?

PROBLEM 7-2

The hydrogenation of ethylene to ethane is of no commercial significance, but is sometimes used to determine the characteristics of an experimental reactor. In the hydrogenation of ethylene over a nickel catalyst in a fixed bed, it was found that the rate of hydrogenation can be represented as follows, when a large excess of ethylene over the stoichiometric concentration is present (43):

$$-\frac{dn_{H_2}}{d\theta} = kmp_{H_2}$$

where n_{H_2} is g moles of H_2, θ is time in sec, k is the rate constant g moles/(sec)(g cat)(cm Hg), m is the weight of catalyst in g, and p_{H_2} is the partial pressure of H_2 in cm Hg. The temperature is constant at 100°C, the total pressure is 76 cm Hg, and the inlet gas analyzes 88.75% ethylene and the rest hydrogen.

(a) For a given bed weight, plot the percentage hydrogen conversions versus the ratio: (flow rates in a piston flow reactor)/(flow rates in a stirred reactor to attain these hydrogen conversions). Assume equal catalyst activity in these reactors.

(b) For equal conversions and flow rates, plot the ratio of the catalyst weights required against conversion in these two reactors.

(c) For the concentrations, temperature, and pressure indicated above, gas was first passed down through a packed bed of nickel-coated beads of mass m. Did the results presented in the following table justify the assumption of piston flow?

(d) With gas upflow, the catalyst beads fluidized, with results shown in the table. What conclusions are indicated?

Gas Downflow		Gas Upflow	
Flow rate (g moles feed gas/sec)	H_2, Conversion %	Flow rate	H_2, Conversion %
2.71×10^{-4}	97.5	1.35×10^{-4}	98.3
4.47×10^{-4}	89.1	2.15×10^{-4}	84.2
6.35×10^{-4}	79.9	2.71×10^{-4}	68.1
7.45×10^{-4}	74.1	4.93×10^{-4}	50.1
9.52×10^{-4}	64.5	8.11×10^{-4}	36.8

PROBLEM 7-3

At elevated temperatures, methane cracks on porous iron particles (made by the hydrogen reduction of iron ore) as follows:

$$CH_4 = C + 2H_2 \qquad \Delta H^\circ = 18.2 \text{ kcal} \qquad K_{p,\,1375^\circ F} = 13.5$$

The solid carbon formed adheres to the iron catalyst, and so little carbon dust appears in the product gas stream. The rate of disappearance of methane in lb moles/(min)(lb catalyst), namely r, is proportional to the partial pressure of methane in the system (154). A relationship which appears to hold to near equilibrium in fluidized beds of iron particles is

$$r = -k(p - p_e)$$

where p and p_e are the methane partial pressures in atm in the reactor and at equilibrium for the reactor conditions, respectively. At 1375°F, k was found to be 3.2×10^{-4} in the units indicated for particles of 60 to 150 mesh for the particular Lake Superior ore used, independent of the amount of carbon on the catalyst. It was also found that log k could be plotted as a straight line versus reciprocal absolute temperature, and that the energy of activation was 85,000 Btu/lb mole.

A gas analyzing 90% CH_4 and 10% N_2 is to be fed to a reactor containing a fluidized bed of these iron particles. Operations are to be at 1375°F, atmospheric pressure, 3 ft/sec superficial velocity, and the off-gases are to contain 70% H_2. A fluidized bed of this type contains 100 lb of iron/cu ft. How many pounds of catalyst are required per sq ft of bed cross section, if the reactor contents are perfectly mixed? If the gas is in piston flow? The solids are always well mixed. How deep would the bed be in each case? How much heat would have to be removed or supplied per sq ft of bed cross section per hr? How could this heat be supplied or removed?

PROBLEM 7-4

With reference to the preceding problem, it was found that the carbon adhering to the iron catalyst granules can be burned off with air, yielding an off-gas which contains CO and N_2, and practically no CO_2 at 1300–1600°F. The iron in these particles is not oxidized during this regeneration until the atom ratio of carbon to iron falls below 0.05. This regenerated catalyst can be reused for methane cracking, and shows no lessening in activity after three cycles. Propose alternative flow sheets for making pure hydrogen from methane by cracking methane on iron, followed by regeneration of the iron catalyst. What factors would require pilot study?

Notes: Large amounts of the Lake Superior ore used are available, at typical iron ore prices. No information on the suitability of other ores is at hand. Reactions are as follows (22):

$$CH_4 = C + 2H_2 \qquad \Delta F^\circ = +20{,}920 - 25.5T$$

$$2C + O_2 = 2CO \qquad \Delta F^\circ = -26{,}700 - 20.98T$$

PROBLEM 7-5

The water-gas shift reaction proceeds as follows:

$$H_2O + CO = H_2 + CO_2$$

This reaction is catalyzed by a chromia-promoted iron oxide catalyst, and is carried out by feeding "synthesis gas" to an adiabatic fixed bed converter. The following rate equation is proposed for the $\frac{1}{4} \times \frac{1}{4}$ in. cylindrical catalyst pellets to be used:

$$r' = k\left(ab - \frac{cd}{K}\right) \tag{1}$$

where the equilibrium constant K and the rate constant k for these pellets are related to the absolute temperature T in °R as follows (103):

$$\ln K = \frac{8240}{T} - 4.33 \tag{2}$$

$$\ln k = -\frac{17500}{RT} + 16.88 \tag{3}$$

Here the gas constant R is in Btu/(lb mole) (°R), while terms a, b, c, and d are the mole fractions of H_2O, CO, H_2, and CO_2, respectively, and r' is the volume of CO_2 in cu ft produced (measured at 60°F and 1 atm) per hr per cu ft of catalyst. The small effect upon rate of total pressure at constant composition for this catalyst will be neglected.

The gas feed to an adiabatic, fixed bed atmospheric pressure reactor analyzes 5.35% CO, 3.4% H_2, 4.5% CO_2, and the rest steam. Superficial linear velocity of the feed gas stream entering the catalyst at 600°F is 1.5 ft/sec, and the gas will be assumed to move through the bed in piston flow. How deep must the catalyst bed be, if the off-gas is to contain 2% CO? What is the exit temperature? A solution based on graphical procedures would, of course, be acceptable.

Notes: For purposes of this problem, assume the following MC_p values for H_2, CO, CO_2, and H_2O of 7, 7, 11, and 11, cal/(g mole)(°C), respectively. The value of $\Delta H°$ can be calculated from Eq. (2).

PROBLEM 7-6

One route to vinyl chloride is by the direct reaction of HCl and acetylene over a catalyst made by impregnating granular activated carbon (typically 6- to 8-mesh) with an aqueous 10% mercuric chloride solution and then drying. The feed gas contains a 10% excess of HCl over the stoichiometric value to attain high acetylene yields, since acetylene is the costlier raw material. To avoid both corrosion and acetaldehyde formation, the acetylene and HCl are dried before use. The reaction is vigorously exothermal:

$$C_2H_2 + HCl = CH_2{:}CHCl(g) \qquad \Delta H° = -22.8 \text{ kcal}$$

with an equilibrium constant of over 10^6 atm^{-1} at 200°C.

It was originally believed that this reaction could not be operated in fluid beds, because of the catalyst's poor resistance to abrasion. Earlier operations were in fixed beds with acetylene yields of around 99%. Some recently constructed plants operate fluidized bed reactors.

A typical flow sheet involves mixing C_2H_2 and HCl, preheating to 200°F, and passing the mixture through externally cooled, catalyst-packed steel tubes 1 in. in I.D. Reactor cooling is carried out with an oil rather than with water to avoid HCl corrosion which would result from a leak. The reactor off-gases pass through a condenser, with the condensate entering a low temperature topping column to remove unconverted HCl and C_2H_2. The bottoms from this still pass to a low temperature column to separate the vinyl chloride from the small amounts of by-product 1, 1-dichlorethane and acetaldehyde which are produced. The usual precautions against acetylene explosions must be taken.

A rate expression for this reaction at 257°F is as follows for the catalyst studied (160):

$$r = \frac{2.7p_A p_H}{(1 + 2.9p_H + 3.8p_{VC})(1 + 0.4p_A)}$$

where r is in lb moles of vinyl chloride/(hr)(lb catalyst), p represents partial pressures in atm and the subscripts A, H, and VC refer to acetylene, HCl, and vinyl chloride, respectively. The catalyst has a void volume of 30% and a bulk density of 25 lb/cu ft.

Assume that reactor operations are isothermal at 257°F and about 7 psig. What is the heat which must be transferred expressed as Btu/(hr)(ft of bed length), when 10% of the theoretical conversion to vinyl chloride has been attained? When 95% of theoretical conversion has been attained? The feed gas contains 10% excess HCl over the stoichiometric value and enters the catalyst bed at 257°F at 1 ft/sec. Propose and discuss a flow sheet for this operation.

PROBLEM 7-7

Styrene is produced by the gas phase dehydrogenation of ethylbenzene at atmospheric pressure, using as catalyst a mixture of copper and zinc oxides deposited on 4-mesh porous alumina pellets:

$$C_6H_5 \cdot C_2H_5(g) \longrightarrow C_6H_5 \cdot C_2H_3(g) + H_2 \qquad \Delta H^\circ = 29.8 \text{ kcal} \qquad (1)$$

with equilibrium constants of 0.025, 0.23, and 1.4 at 500°, 600°, and 700°C, respectively. It has been proposed (159) that r, expressed as lb moles of styrene/(hr)(lb catalyst), is related to p_E, p_S, and p_{H_2}, the partial pressures (in atm) of ethylbenzene, styrene, and hydrogen as follows:

$$r = k\left(p_E - \frac{p_S p_{H_2}}{K}\right)$$

where k is the rate constant and K is the equilibrium constant. The relation of k to temperature in °K is

$$\log k = -\frac{11{,}370}{4.575\,T} + 0.883$$

For orientation's sake, assume here that no other reactions occur (in actual operation, small amounts of benzene, toluene, and carbon also form). Operations are to be at 1 atm total pressure.

(a) Plot quantity r as ordinate versus temperature as abscissa for several conversions of ethylbenzene to styrene, namely 0%, 10%, 20%, 30%, 40%, 50%, and 60%.

(b) It is desired to convert 50% of a pure ethylbenzene feed to styrene in an experimental tubular reactor to which pure ethylbenzene is fed. Operations are to be isothermal but operating temperatures should not exceed 650°C. At what temperature level is the quantity W/F, namely, catalyst weight divided by the feed in moles/hr a minimum? What is W/F at this condition? Construct a line in the figure of part (a) representing the history of the gas flowing through the tubes of this reactor.

(c) This reactor is now operated adiabatically, with ethylbenzene feed at 650°F. Again construct a line in the figure of part (a) representing the history of the gas through this reactor. What weight of bed, expressed as W/F, has been traversed when the gas temperature falls to 600°C? What is the conversion reached? Can 500°C be attained by the gas stream? How would the addition of steam as a diluent affect conversion?

Note: MC_p for ethylbenzene, styrene, and H_2 may be taken as 65, 58, and 7 cal/(g mole)(°C), respectively, for the purposes of this problem.

PROBLEM 7-8

An exothermal gas reaction ($A + B = C$; $\Delta H° = -24{,}000$ Btu/lb mole of C) is carried out continuously in a fluidized bed of granules of a suitable catalyst. The bed is cooled by a water jacket, through which cooling water flows so rapidly that inlet and outlet water temperatures are substantially equal at T_w. The sequential reaction ($2C = D$) also tends to occur, but is negligible providing not more than 20% of A in the feed has reacted. Under all reactor conditions, both reactions are substantially irreversible.

Over the range of compositions of interest, it may be assumed that the primary reaction is of zero order, hence the following rate expression applies at 175°F:

$$\frac{1}{W}\frac{dn_c}{d\theta} = 0.2 \text{ lb moles/(lb catalyst)(hr)}$$

where $dn_c/d\theta$ represents moles of "C" formed per unit time, and W is catalyst weight. Laboratory scale studies have shown that the rate of this reaction doubles every 18° in the vicinity of 175°F. The feed to the reactor is 50% A, 50% B. The feed temperature is the same as the cooling water temperature.

(a) If exactly 20% conversion of A is to be obtained, and if the reactor is maintained at 175°F, what should be the feed rate to the reactor in lb moles/hr?

(b) If the reactor is operated as just described, what must be the cooling water temperature?

(c) Plot q, the Btu's per hr of heat generated by the chemical reaction in the kettle, versus T, the kettle temperature, between 80 and 180°F. On the same graph, plot the sum of the total heat for cooling the reactor and for heating the feed (with a 95°F cooling water and feed temperature), versus the temperature of the fluid bed. Is the operation proposed in this problem "stable"?

(d) Plot the possible steady-state operating temperatures for the bed as ordinate versus the temperature of the cooling water (assumed always equal to the feed temperature). What are the regions of the stable and unstable operations on this chart?

Notes: The bed has a cooling surface area of 3 sq ft, for which there is an overall heat transfer coefficient between the cooling water and the bed contents of 500 Btu/(sq ft)(°F)(hr). Values of MC_p for A, B, and C are 15, 15, and 30 Btu/(lb mole)(°F), respectively. The fluid bed contains 50 lb of solid catalyst.

PROBLEM 7-9

Polymer sheets 12 ft × 12 ft × $\frac{1}{4}$ in. thick are to be piled, one on top of the other, for storage. Each pile contains n sheets, thus the pile has the dimensions 12 ft × 12 ft × $n/4$ in. It is found that these sheets weld together on storage, despite cooling by water spray on the top and bottom surfaces of the pile. This cooling water temperature is 70°F. It is believed that such welding is caused by temperatures within the pile of over 160°F since this is the temperature at which sticking first develops. These elevated temperatures are caused by slowly continuing polymerization, a reaction which may here be regarded as of zero order, generating 50 Btu/(cu ft)(hr) at 60°F, and doubling in rate every 20°F. There is enough unconverted monomer in the sheets for this reaction to continue for several days.

The heat transfer coefficient from the sheet surfaces on the bottom and top of the pile to the cooling water is 100 Btu/(hr)(sq ft)(°F). The sheet material itself has a thermal conductivity k of 0.1 Btu/(hr)(sq ft)(°F)/(ft).

If the sheets are to attain but not exceed 160°F, what is the allowable number of sheets per pile, and what is the associated cooling water temperature required? Repeat for 100°F. For a cooling water temperature of 70°F, indicate how to calculate the maximum number of sheets to be included in a pile without sticking.

PROBLEM 7-10

In a study of the kinetics of methanol oxidation to formaldehyde (5), methanol vapors and air were premixed in equimolal proportions, preheated to a suitable temperature, and passed through a thin, fixed catalyst bed made up of 3–4 mm granules of pumice carrying 36% by weight of silver. Oxygen was found to react in two ways:

$$\tfrac{1}{2}O_2 + CH_3OH = CH_2O(g) + H_2O(g) \tag{1}$$

$$O_2 + CH_2O = CO_2 + H_2O(g) \tag{2}$$

Methanol was also dehydrogenated:

$$CH_3OH = CH_2O(g) + H_2 \tag{3}$$

Rates expressions of reactions (1) and (3) were established separately and are reported, where n_x and n_f are moles of oxygen and of formaldehyde, c_x and c_f are the corresponding concentrations in g mole/cc, V is catalyst volume, θ is contact time, and k_1 and k_3 are 105 and 160 reciprocal sec, respectively, at 450°C:

$$-\frac{dn_x}{Vd\theta} = k_1 c_x$$

$$+\frac{dn_f}{Vd\theta} = k_3 c_x$$

The performance of an industrial tubular reactor with equimolal air-methanol feed is now to be anticipated, using this same catalyst. Assume in these orientational calculations that reaction (2) may be neglected. The superficial approach velocity of the mixed gases in the reactor tubes (i.e., velocity calculated for empty tubes) is to be 3 ft/sec at the proposed operating conditions of 450°C and 1 atm total pressure. The catalyst bed is 50% void space.

(a) What is the gas composition when 80% of the oxygen has reacted?
(b) How deep must the bed be for 80% oxygen to react. Piston flow may be assumed.

PROBLEM 7-11

A gaseous mixture of A, B, and N_2 is fed continuously and at a constant rate to a tube packed with a granular catalyst. This tube is kept isothermal by suitable heating or cooling. Chemical reactions involving reagents A and B

occur in the tube, producing M along with some other gaseous products. At present, the feed rate is fixed at F_0 lb moles/min, the catalyst temperature and total pressure are constant at T_0 and π_0, and the concentration of M in the product stream is found to be constant at c_0.

(a) Keeping F_0, π_0, and the feed composition unchanged, the catalyst temperature is increased by 20°C from T_0 to T. Could c, the exit concentration of M in the off-gases, increase? Decrease? Remain constant? If any change can occur, what are the possible reasons for it? If it is not possible, why not?

(b) Keeping π_0, T_0, and the feed composition unchanged, the flow rate is decreased by 20% from F_0 to F. Could c, the concentration of M in the product gas, increase? Remain constant? Decrease? Present possible reasons for each case.

(c) Repeat the foregoing when only the total pressure is changed.

Note: The coefficients of mass transfer between a fluid and a solid surface tend to vary as the mass velocity raised to the 0.8 power.

PROBLEM 7-12

Phosgene is a colorless, highly toxic and very reactive material boiling at 8.3°C, which finds use industrially as an intermediate. Phosgene is made by passing a mixture of CO and Cl_2 over activated carbon, which catalyzes this exothermal reaction:

$$CO + Cl_2 = COCl_2 \qquad \Delta H^\circ_{298^\circ K} = -25.7 \text{ kcal}$$

Values for log K_p at 298 and 400°K are 12.86 and 7.8, respectively. Operations are reportedly kept below 250°C to avoid phosgene disproportionation to CO_2 and CCl_4.

The feed gases to a phosgene operation are kept oxygen-free to avoid attack on the catalyst, while sulfur compounds are eliminated to avoid forming undesirable sulfur chlorides. Hydrogen should not be present, since it reacts with both chlorine and phosgene to form HCl. This HCl in turn dissolves in liquid phosgene, and unless removed, attacks the iron storage containers commonly used for phosgene to produce hydrogen gas, thus generating dangerous pressures. Water and hydrocarbons should not be present because they also form HCl in the system.

For safety, phosgene must be removed from all purge streams, as by scrubbing with a water slurry of lime, with which phosgene reacts to form CO_2 and calcium chloride. Phosgene is also destroyed by hydrolysis in hot water, though somewhat less rapidly than in a lime slurry. Hydrolysis in cold water is slow.

A simplified version of the rate equation of Potter and Baron (119) for a given catalytic carbon is as follows:

$$r = kp_{CO}p_{Cl_2}$$

where r is in (g moles)/(g catalyst)(hr), pressures are in atm, and k is unity at about 400°K and 2 at 435°K for the granular catalyst in question. It may be assumed that $\ln k$ varies linearly with $1/T$°K.

(a) Each of the tubes of the multitubular reactor to be used is to be packed with catalyst granules so that equal flow and equal production is attained in each tube. Why is this done and how can it be done?
(b) How would you recover phosgene from the reactor off-gases? The presence of inert substances like N_2 and CO_2 in the recovered phosgene are considered particularly undesirable here. Discuss.
(c) If operations are to be kept isothermal at 303°K in an atmospheric pressure operation by cooling the reactor tubes with cold brine, how much catalyst is required for the production of 1 ton/hr of phosgene from pure reagents, if (1) the feed mole ratio of CO/Cl_2 is unity, and (2) the feed ratio is 1.1. Assume 99% conversion of the chlorine attained in both cases. Could catalyst requirements be further reduced?
(d) It is maintained that it would be possible so to arrange the reactor system (using pure CO and Cl_2 feeds) that both the following could be accomplished simultaneously: the feed to the reactor itself shows a mole ratio of CO/Cl_2 of 1.1, but the fresh feed to the plant system is in a ratio of unity. Discuss, indicating when and how this would be possible and what the control problems might be. Would the presence of inerts in the feed affect your thinking?

Notes: The catalyst is in $\frac{1}{16}$ in. granules, weighs 40 lb/cu ft. The vapor pressure of phosgene, Reference (152), Vol. 13, 493, from 0–182°C is $\log p_{atm} = 4.7187 - 1326/T$°K.

PROBLEM 7-13

Phosgene is to be produced by passing an equimolal mixture of chlorine and carbon monoxide, initially at atmospheric pressure and 303°K, through a tube packed with granules of activated carbon to serve as catalyst. Data on kinetics, enthalpy of reaction, etc., are presented in the preceding problem. The superficial flow velocity of the entering gas is to be 1 ft/sec. The cooling brine surrounding these tubes is assumed to be supplied in such a fashion that its temperature is constant at T_c through the entire reactor.

(a) The reactor tubes are $\frac{1}{2}$ in. in diameter. Calculate the value of T_c

if the catalyst temperature in the first foot of tube lengths is not to exceed 303°K.

(b) Would the $\frac{1}{2}$ in. tube be unstable to a transient accidental increase in temperature of say 0.5°C? Of 5°C?

(c) The apparent radial conductivity of tube packing increases slowly with increases in gas velocity (91, page 292). What effect might operation at a reduced throughput rate have on reactor stability?

Notes: Assume the equivalent conductivity of the packing to average 0.3 Btu/(hr)(sq ft)(°F)/ft, and the inside coefficient between the catalyst and the tube wall to be 10 Btu/(hr)(sq ft)(°F). The resistance to heat flow through the tube wall to the cooling water may be assumed negligible. The sensible heat requirements of the reacting gases may also be assumed negligible in this problem.

PROBLEM 7-14

Aniline continues to be an important intermediate in the manufacture of dyes and rubber chemicals. One source of aniline is nitrobenzene, which is reduced in the liquid phase batchwise in kettles with iron borings and acidified water:

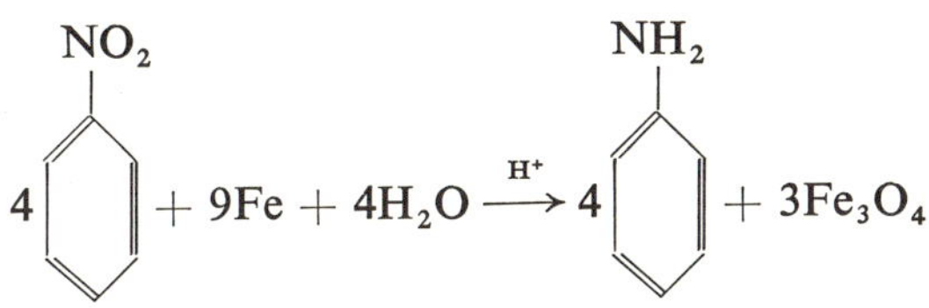

The heat of this vigorously exothermal reaction is removed by refluxing the vapors evolved from the reacting mass. This process requires much operating attention, involves a disposal problem, and consumes iron. Interest has recently focused on the direct vapor phase reduction of nitrobenzene, using copper and nickel supported on silica as catalysts:

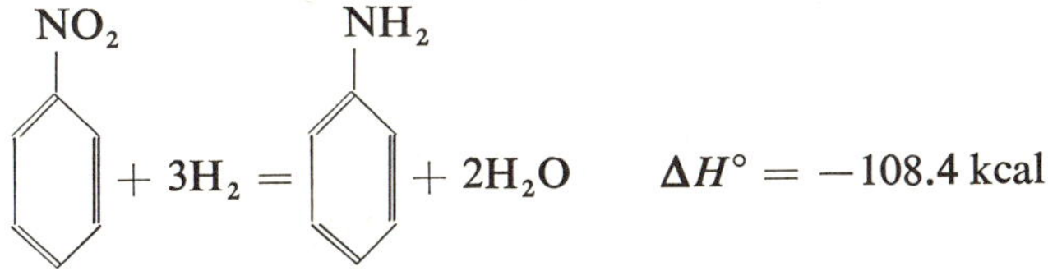

The equilibrium constant for this reaction is large below about 800°K. The following rate expression was found (121a) to apply to a particular granular catalyst:

$$r_A = \frac{dn_A}{W d\theta} = k p_N^{0.5} p_H^{0.5}$$

where r_A is the rate of aniline production in g mole/(hr)(g catalyst), p_N and p_H are the partial pressures of nitrobenzene and hydrogen in atm, T is in °K, and k in g mole of aniline/(hr)(g catalyst)(atm) is as follows:

$$k = 41{,}040 \exp\left(-\frac{8240}{T}\right)$$

A mixture of 80 mole % of N_2 (inert), 15% of H_2, and 5% of nitrobenzene is fed at 250°C to an atmospheric pressure reactor consisting of a tube packed with catalyst. The superficial approach velocity is 5 ft/sec, and the reactor is 1.6 ft long.

(a) Assuming piston (slug) flow of the gases in the reactor and adiabatic operation, what fraction of the entering nitrobenzene is converted to aniline? What is the exit temperature?

(b) The reactor tubes are immersed in a pool of liquid boiling at 250°C. Estimate the fractional conversion of nitrobenzene to aniline, assuming no radial temperature gradients through the catalyst bed. What is the off-gas temperature now?

Note: Assume the overall coefficient of heat transfer through the tube between the reacting gas and the coolant to be 12 Btu/(hr)(sq ft)(°F). The tubes used are 1 in. in I.D. The heat capacities of nitrobenzene (g), hydrogen, aniline (g), H_2O(g), and N_2 are 33, 7, 37, 8.5, and 7 cal/(g mole)(°C), respectively. The catalyst is 30% voids and has a heat capacity of 0.25 cal/(g)(°C), and a pellet density of 1.5 g/cc.

PROBLEM 7-15

Four sequential reactions appear to be involved (97) both in the chlorination and the hydrochlorination of methane (Prob. 4-3 presents further information):

$$CH_4 + Cl_2 = CH_3Cl + HCl$$

$$CH_3Cl + Cl_2 = CH_2Cl_2 + HCl, \text{ etc.}$$

Designating the moles CH_4, CH_3Cl, CH_2Cl_2, $CHCl_3$, and CCl_4 in the product gas per mole of methane feed as y_0, y_1, y_2, y_3, and y_4, then the net rates for forming each chlorinated product are presumably of the form:

$$\frac{dy_0}{d\theta} = -ky_0 xc^n$$

$$\frac{dy_1}{d\theta} = (ky_0 - ky_1)c^n$$

$$\frac{dy_2}{d\theta} = (ky_1 - ky_2)c^n, \text{ etc.}$$

where the constant k, though very large, is the same for each reaction at any given temperature, c is the chlorine concentration, and n is an exponent. After all the chlorine present has reacted, the product moles are stated to be related as follows (97):

$$y_1 = -y_0 \ln y_0$$

$$y_2 = -\frac{y_0}{2} \ln^2 y_0$$

$$y_3 = -\frac{y_0}{6} \ln^3 y_0$$

and

$$y_4 = 1 - y_0 - y_1 - y_2 - y_3$$

(a) Show that these equations can be derived from the foregoing differential rate expressions.
(b) Chlorine is well mixed with methane, and the mixture then heated to 400°C, whereupon all of the chlorine reacts to form the chlorinated methanes. Upon condensing these chlorinated products, this mixture is found to analyze 30 mole % CH_3Cl, the remainder being the three higher chlorinated methanes. What was the mole ratio of chlorine to methane in the feed gas?
(c) The three disproportionation reactions here are

$$CH_4 + CH_2Cl_2 = 2CH_3Cl$$
$$CH_3Cl + CHCl_3 = 2CH_2Cl_2$$
$$CH_2Cl_2 + CCl_4 = 2CHCl_3$$

It was stated earlier that chlorination reaction rates discussed in part (a) of this problem are such that each of these three reactions appears to be at equilibrium. That is, regardless of the ratio of chlorine to methane in the feed, the products form in such quantities that an equilibrium constant appears to be satisfied for each disproportionation. Moreover, these constants are all close to unity. Do you agree? This finding appears surprising and fortuitous, since these disproportionations do not seem to occur under the usual conditions of chlorination or hydrochlorination. That is, a feed of CH_3Cl and $CHCl_3$ does not appear to react to form CH_2Cl_2, etc.

PROBLEM 8-1

The adiabatic moving bed kiln of example 8-2 in the text is fed at the top with pure solid Cu_2O at 220°F. The gas fed to the bottom of the kiln at 220°F is a mixture of N_2 and H_2 in the mole ratio of $\frac{5}{1}$. Kiln operations are continuous and in a steady-state, and the following very fast reaction, having a very large equilibrium constant, occurs:

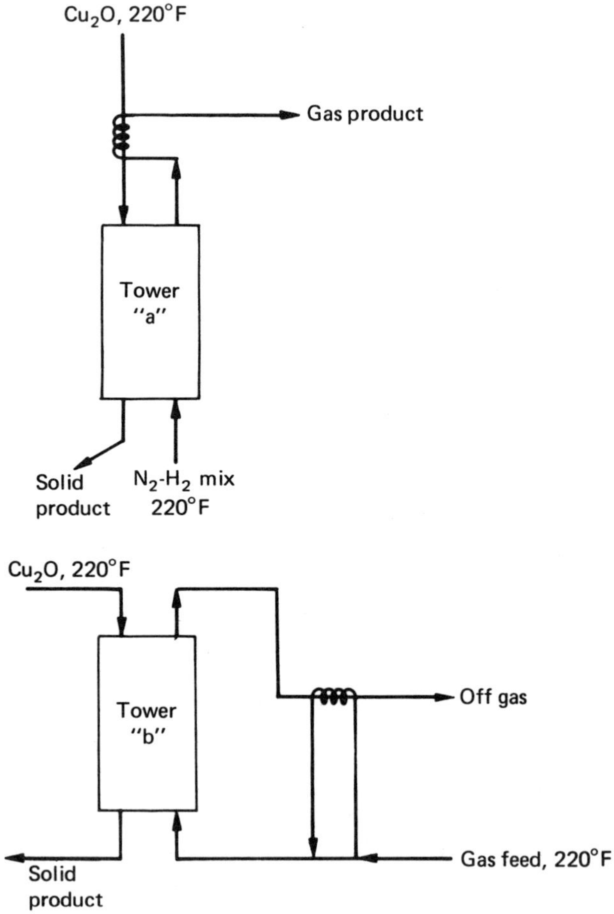

Fig. P-11. Flow sheets for Prob. 8-1.

$$Cu_2O + H_2 = 2Cu + H_2O \qquad \Delta H^\circ = -30{,}000 \text{ Btu/lb mole}$$

(a) The reaction zone is to be kept centered halfway up the kiln. What and where is the maximum temperature anywhere within the kiln system? *Note:* The moving bed of solids in the kiln is relatively deep, thus providing an extended interfacial area between gas and solid. Operations are at 1 atm pressure.

(b) The solid feed is now supplied at a rate of 1.5 lb moles Cu_2O/lb mole H_2 contained in the gas fed to the bottom of the tower. Where and what is the highest temperature within the kiln?

(c) Several proposals have been made for raising the kiln temperature to attain even higher reaction rates. One of these proposals is to insert an "infinite" exchanger into the flow sheet as shown in Fig. P-11a. What ratio of solids to gas feed is now required to keep this reaction zone midway up the tower? What and where is now the maximum temperature anywhere in this kiln-heat exchanger system, if the feed gas is N_2/H_2 in the ratio of $\frac{5}{1}$ at 220°F?

(d) An alternative proposal is the flow sheet in Fig. P-11b. Of the total gas fed to the system 10% flows through the infinite heat exchanger, and 90% bypasses this exchanger, as shown. Answer the questions of part (c) for this case.

Note: The values of MC_p of Cu, Cu_2O, H_2, N_2, and H_2O are 7, 17.5, 7, 7, and 10.5 Btu/(lb mole)(°F), respectively. Assume temperature equilibrium between gas and solid attained within the reaction zone.

PROBLEM 8-2

Acetylene black is a finely divided carbon made by pyrolysis of acetylene. Since acetylene is relatively expensive, this black costs more than black made from natural gas. Acetylene black particles have the unusual property of spontaneously clinging together in irregular chain-like structures which impart electrical conductivity to battery fillers, plastics, rubber, etc.

Acetylene black is to be made in a continuous atmospheric pressure operation by passing a mixture of 20% acetylene, the rest nitrogen (initially at 80°F) into an adiabatic retort, from which the carbon formed is withdrawn continuously as a suspension in the product gas stream, to be recovered after cooling by settling and filtration. Retort design and velocities of the entering gas are such that the contents of the retort will be well mixed. Acetylene decomposition is to be assumed to occur homogeneously in the gas phase, and to conform to the following isothermal rate expression (n is moles of acetylene in the system):

$$\frac{dn}{d\theta} = -kn$$

The half-life of acetylene may be assumed to be 1 min at 2200°F and 0.5 min at 2300°F. At what gas feed rate (moles/min) to an adiabatic retort having a volume of 10 cu ft will 90% of the acetylene in the feed be decomposed? The volume of solid carbon in the reactor is negligible at all times.

Note: Values of MC_p for C_2H_2, N_2 H_2, and solid carbon may be assumed for purposes of this problem only, as constant at 14, 7, 7, and 3.5 cal/(g mole) (°C), respectively. The enthalpy of formation of acetylene from the elements (solid carbon and gaseous hydrogen) at 2200°F is about +58 kcal/g mole, and log K_p for the formation reaction is −5.130 at this temperature.

Before undertaking work in this field, it is agreed that special care and precautions relating to safety must be exercised, due to possibilities of explosion.

PROBLEM 8-3

Caustic soda has been made in large quantities by reacting soda ash (sodium carbonate) in aqueous solution with slaked lime, whereupon the following double decomposition occurs:

$$Na_2CO_3(aq) + Ca(OH)_2(c) = 2NaOH(aq) + CaCO_3(c)$$

This operation, called *causticizing*, is carried out continuously in agitated tanks at about 95°C. Solid calcium carbonate is separated from the caustic soda by settling and filtering, after which the solid is washed to recover adhering mother liquor and these dilute liquors are used in subsequent cake washings and also to dissolve fresh soda ash for subsequent operations.

In a laboratory study of this causticizing reaction, calcium oxide was added to an appropriate amount of soda ash solution in an agitated vessel. This system was sampled from time to time, and the ratio of OH^- and $CO_3^=$ ions in the aqueous phase was determined (how?). The rate of reaction was found to decrease both with an increase in the particle size of the quicklime and an increase in the concentration of sodium carbonate solution used to slake this lime. At 95°C, all batch slaking tests came close to equilibrium in about 2 hr.

(a) The equilibrium constant for the foregoing reaction has been expressed as $(OH^-)^2/CO_3^=$, where the quantities in parentheses are molalities. Assuming all ionic activity coefficients to be unity, calculate the magnitude of this constant at 95°C from solubility data of the individual salts in water, and compare with the constants calculated from the following laboratory data for 95°C (109):

50% Excess CaO		5% Excess CaO	
Na_2CO_3 in Original Solution, Weight %	Conversion to Caustic Soda, % of Initial Na^+ Present	Na_2CO_3 in Original Solution, Weight %	Conversion to Caustic Soda, %
2.3	99.38	3.0	99.3
5.0	98.85	5.1	98.80
9.8	96.51	10.2	96.15
15.1	91.50	15.0	91.20
19.6	83.47	19.2	84.40

(b) An aqueous solution originally containing 15.1% soda ash is causticized at 95°C with 5% excess lime. The product slurry is filtered and evaporated until a 50% caustic soda solution remains. This solution is now cooled to 35°C, and the precipitated solids are removed by careful filtration from the now viscous caustic soda solution. What is the concentration of soda ash in the filtrate?

SOLUBILITIES IN WATER (% BY WEIGHT) AT 35°C

Na_2CO_3, %	NaOH, %	Solid Phase	Na_2CO_3, %	NaOH, %	Solid Phase
32.9	0	$s \cdot 7w$	0.5	34.2	$s \cdot w$
32	0.6	$s \cdot 7w + s \cdot w$	0.5	39.2	$s \cdot w + s$
25.3	4.9	$s \cdot w$	0.2	39.5	s
15.2	11.9	$s \cdot w$	0.2	44.8	s
7.5	18.7	$s \cdot w$	0.2	50.2	s
1.9	27.5	$s \cdot w$	0	55.4	$NaOH \cdot H_2O$

$s = Na_2CO_3$ $\quad s \cdot w = Na_2CO_3 \cdot H_2O$ $\quad s \cdot 7w = Na_2CO_3 \cdot 7H_2O$

(c) Propose a flow sheet for making a 50% caustic soda solution with high yield continuously from dry soda ash and lime.

Note: In saturated aqueous solution containing no CO_2, 1 liter of water dissolves 0.024 g of $CaCO_3$ at 95°C and 0.013 g of $CaCO_3$ at 25°C. Calcium hydroxide solubilities reported as CaO at 95° and 25°C are 1.29 g and 0.56 g/l, respectively.

PROBLEM 8-4

In bench-scale studies of $SiCl_4$ production from relatively pure metallic silicon, particles of 60- to 80-mesh size (0.025–0.012 cm) were fluidized in dry gas mixtues of HCl and N_2 (87). The bed was contained in a 1-in. I.D. silica tube located in an electric furnace. Off-gases from runs above 600°C, when analyzed chromatographically, were found to contain only N_2, H_2, $SiHCl_3$, $SiCl_4$, and never more than traces of HCl. All the chlorine in the feed, therefore, reappeared in the off-gases as volatile silicon compounds.

For run 3, which was at atmospheric pressure, the mole ratio (N_2/HCl) in the feed gas was 1.64, with a flow rate of 0.0513 g mole HCl/min. This run was continued until the initial charge of silicon (containing less than 0.5% Fe and 0.07% calcium as impurities) was exhausted. Particle loss by entrainment was small, except in the first 4 min. Selected data for run 3 are as follows:

Time, Min	Temp, °C	Bed Weight, g	$\frac{SiCl_4}{SiCl_4 + SiHCl_3}$	$\frac{HCl}{\Sigma\ Cl}$
4	611	41.2	0.436	0.014
13	631	37.7	0.46	0.015
43	645	25.7	0.714	0.015
70	642	15.8	0.753	0.02
110	649	0.3	0.433	0.008

The last column is the ratio of the atoms of chlorine as HCl to the atoms of chlorine in all forms in the off-gas. The next-to-last column is the mole ratio of silicon tetrachloride to tetra- plus trichlorosilane, and in run 3, this ratio varies more widely than in other runs reported. This ratio lies between 0.4–0.8 in runs between 400° and 1250°C, regardless of gas velocity and the ratio of HCl/N_2 in the feed. No mono- or dichlorosilane were found in the off-gases by the analytical procedures used (gas chromatography) and element balances in these runs were reasonably satisfactory.

Locker proposed that SiH_2Cl_2 was the primary reaction product, which then reacted homogeneously and very rapidly with HCl to form $SiHCl_3$; this in turn reacted more slowly to form $SiCl_4$. When a mixture of HCl and $SiHCl_3$ was passed through a heated tube containing ceramic packing, it was reported tersely that "silicon tetrachloride is produced."

(a) What was the superficial gas velocity at the 43-min point for the feed gas entering the bed? The product gas leaving the bed?

(b) Are these findings compatible with the assumption that chemical equilibrium for any of the possible reactions is attained?

(c) Propose a possible flow sheet of a plant making only $SiCl_4$ from silicon metal and HCl.

(d) Propose a possible flow sheet for making only trichlorosilane from silicon metal and HCl, supporting your proposal with equilibrium calculations, if required. What questions remain to be settled in bench-scale and pilot-scale work? Remember that corrosion may be severe.

Note: From the JANAF Tables (62) for formation from the elements at 900°K:

	$SiCl_4$	$SiHCl_3$	SiH_2Cl_2	HCl
log *K*, 900°K	31.244	21.602	13.446	5.812
log *K*, 1500°K	16.076	10.708	6.052	3.615
ΔH_f, kcal	−156.441	−112.326	−76.211	−22.505

Boiling points for $SiCl_4$ and $SiHCl_3$ are 57.6 and 33°C, respectively.

PROBLEM 8-5

Cylindrical 1 in. by 1 in. briquettes were made by compressing a mixture containing 50% by weight of zinc oxide, 35% by weight of bituminous coal, and 15% by weight of anthracite dust (150). The zinc oxide used was finer than 200-mesh, while the coals were ground to pass 100-mesh. The raw briquettes were first coked (heated) to 600°C in nitrogen to drive off coal volatiles, leaving behind a relatively strong porous briquette in which the ratio (atoms carbon/atoms ZnO) exceeded 3. One of these briquettes was then lowered into a retort kept at 950°C. This retort was initially filled with N_2 but suitably vented so that the internal pressure remained at atmospheric. It is fair to assume that only the following reaction occurred (CO_2 production is small):

$$ZnO(c) + C(c) = Zn(g) + CO(g) \qquad \Delta H^\circ = +152{,}000 \text{ Btu/lb mole}$$

This monovariant system develops a total decomposition pressure π of 1 atm at 910°C (π equals the partial pressures of zinc plus CO). The gas evolved from the briquettes quickly sweeps the N_2 out of the retort. The extent of reaction was followed by continuously weighing the briquette suspended in the retort.

(a) Assume as a first approximation that a cylindrical briquette is equivalent to a spherical briquette of the same mass. Assume that "zinc eliminations" from the briquette follows a shell and core model, with heat transfer from the briquette surface (950°C) and the shell-core interface (910°C) controlling. The external dimensions of the briquette remain unchanged as reaction proceeds. Calculate the average thermal conductivity of the briquette shell from the finding that 90% zinc elimination was attained in 4.5 hr. Assume (ρ/M) to be 1.4 lb moles of ZnO/cu ft, $(T_0 - T_c)$ to be constant at 40°C.

(b) Using this value of k, calculate the times required for 52 and 70% zinc removals, and compare with the observed times of 1 and 2 hr, respectively.

(c) At the temperatures studied, the thermal conductivity of the briquette shell is primarily due to radiation between particles, and hence can be calculated by the Damköhler equation [Reference (9), p. 290]:

$$k = 0.692\epsilon D_p\left(\frac{T}{100}\right)^3$$

where ϵ is fraction voids, D_p is particle diameter in feet, T is the local temperature in °R, and k is in Btu/(hr)(sq ft)(°F)/(ft). Assume ϵ in these briquettes to be 0.5, and particle diameter of the coke in the briquette shell to be 0.001 in. after reaction. Estimate the average thermal conductivity of the briquette shell by the Damköhler equation, and compare with the k value of part (a).

PROBLEM 8-6

A spherical particle initially at T_{p_0} is to be heated by immersion in a hot fluid at T_f. Relative fluid particle motion is to be negligible, as in stagnant suspensions of fine powders of low density in water or gas. The order of magnitude of θ, the necessary heating time is to be determined in the extreme cases: (I) the thermal conductivity of the fluid k_f is much larger than k_p, the thermal conductivity of the solid, and (II) the reverse.

Case I. The particle surface temperature rises immediately upon immersion to T_f, while temperatures within the particle increase only slowly as heat is conducted through the solid. Obviously the lowest temperature of the sphere is at its center, with symmetrical but constantly lessening temperature gradients between the center and the particle surface.

A generalized graphical solution for this case, showing temperatures at various points within a sphere versus heating times was developed by Gurney and Lurie (91). The following equation fits the line on the Gurney-Lurie plot showing the variation of the sphere's center temperature T_{p_c} in °F versus θ in hours:

$$\ln\frac{T_f - T_{p_c}}{T_f - T_{p_0}} = 0.43 - 4.4\frac{k_p\theta}{\rho c_p r_0^2}$$

where ρ and c_p are the density and heat capacity of the sphere, respectively, and r_0 is its external radius. Obviously, consistent units must be used. This equation applies for values of $(T_f - T_{p_c})/(T_f - T_{p_0})$ below 0.85.

Case II. The thermal conductivity of the fluid, k_f is small relative to k_p. Because of the high conductivity of the solid, the particle is isothermal at all times, with the center temperature always equal to the surface temperature. It is evident that the temperature difference between the bulk of the fluid and the sphere must be finite, but it diminishes with time. In other words, the temperature difference in this system is across the fluid film surrounding the particle, not within the particle.

(a) An associate claims that the following relation can be derived for Case II from Eq. (8-9) in the text:

$$\ln \frac{T_f - T_{p_0}}{T_f - T_{p_c}} = \frac{3k_f\theta}{\rho c_p r_0^2}$$

Derive this relation.

(b) The solid in a sphere 0.001 ft in diameter has conductivity k_p of 40 Btu/(hr)(sq ft)(°F)/ft, a density of ρ of 300 lb/cu ft and a heat capacity of 0.1 Btu/(lb)(°F). If k_f is 0.2, as for nitrogen, what is the time in sec needed to increase the sphere temperature from 70 to 1000°F when T_f is 1050°F?

(c) For this system, what is θ when the values of k_f and k_p are reversed?

(d) Repeat parts (b) and (c) assuming the sphere radius to be 0.1 ft.

PROBLEM 8-7

Finely divided coke particles are to be suspended in a stream of air, blown into a furnace chamber, and burned. This dust cloud system may be assumed to move in piston flow through the furnace reactor as it undergoes combustion. The coke used is pure carbon, the particles are nonporous and spherical at all times, and initial size distribution is such that 2.5% by weight of the particles are 20 μ in diameter, 27.5% are 50 μ, and 70% are 100 μ. These particles are uniformly dispersed in the air used, and a 15% excess of air beyond that needed for complete combustion is supplied. The total pressure is 1 atm.

It will be assumed that preheat and ignition of the air-particle system takes negligible time, and that combustion kinetics are extremely fast, so that diffusion controls during most of the combustion history of a particle. The g atoms carbon reacting per unit particle area per unit time are therefore represented by Eq. (8-23).

(a) What are the number fractions of the 20-, 50-, and 100-μ particles in the initial coke system? What assumptions were made in your calculations?

(b) Let x, y, and z represent the diameters at any time θ of the particles which are initially 20-, 50-, and 100-μ, respectively. What are the magnitudes of y and z when x first reaches zero? What is the magnitude of z when y first reaches zero?

(c) If the gas velocity is such that the largest particles are all consumed at a distance of 6 ft from the burner tip, estimate the distance at which the 50-μ particles disappear. In making this calculation, assume as a first approximation that k is constant and that only CO_2 forms as the product of combustion.

(d) An associate points out that, rigorously speaking, Eq. (8-23) cannot apply to the moment of particle disappearance. That is, this equation indicates an infinite rate for particles of radius approaching zero. Before this happens, Eq. (8-24) rather than Eq. (8-23) applies, making rate finite. Do you agree?

PROBLEM 8-8

Calcium sulfate occurs in nature extensively as the insoluble hydrate called gypsum, $CaSO_4 \cdot 2H_2O$. In times of sulfur shortage, gypsum has been used in sodium sulfate manufacture:

$$CaSO_4 \cdot 2H_2O(c) + Na_2CO_3(aq) = CaCO_3(c) + Na_2SO_4(aq)$$

In a recent study, groups of students made slurries of gypsum in sodium carbonate solutions of various strengths, sampling the liquid phase from time to time and determining the ratio of $(SO_4^=/CO_3^=)$ in solution (how?). Over prolonged time periods from 25°C to the boiling temperature, conversions of sodium carbonate to sodium sulfate of over 99.5% were attained when using a small excess of gypsum.

Samples from a quite pure surface deposit of gypsum in New Mexico were reacted with a sodium carbonate solution, with findings as shown in the table. This gypsum could probably be delivered in bulk form to a nearby plant for about 50¢/ton. In the study described here, the name "coarse gypsum" was given to material ground to pass through a standard 8-mesh screen. Of this coarse gypsum, 50% lay between 8- and 40-mesh, 21% between 40- and 100-mesh, 17% between 100- and 200-mesh, while the remainder was finer than 200-mesh. Corresponding numbers for more

CONVERSIONS AT 100°C

Run	Gypsum Used	S*	R**	% Na_2CO_3 Reacted 5 min	10 min	15 min	% Gypsum Reacted, 15 min
4	100- to 140-mesh	½	1.22	98.8	99	99.1	81
5	Coarse	½	1.04	70	74	75	72
6	Coarse	½	1.22	75	77	78	64
7	20- to 60-mesh	½	1.04	64	67.5	68	66
8	Fine	½	1.16	81.5	82	83	71
10	Fine	1.0	1.2	—	—	96	81

**S* represents the fractional saturation of the sodium carbonate feed solution used at 100°C. A saturated aqueous solution of sodium carbonate at this temperature contains 31.5% salt by weight.

***R* is the ratio (moles gypsum per moles sodium carbonate) in the feed. An excess of gypsum rather than soda ash was used since gypsum is the cheaper reagent.

extensively ground material, here called "fine gypsum," were 15.5, 30, 40, and 14.5%. The materials reported below as "100- to 140-mesh" and "20- to 60-mesh" were screened out of the "coarse gypsum." With all gypsum samples studied, the times required for any given degree of conversion decreased as temperatures rose.

In all of these runs, sodium carbonate conversion was rapid at the outset, then slowed down to a low rate after about 10 min. Thus in typical run 6, in addition to the data of the above table, conversion of sodium carbonate to sodium sulfate was 60% in 0.5 min, 68% in 1 min, 74% in 2.5 min, 78% in 16 min, 82% after 1 hour, with 99.5% attained after 24 hr.

Further study of "fine gypsum" was made for feed mole ratios of gypsum to sodium carbonate of 1.04. The gypsum was first separated into two fractions of roughly equal weight by screening, one containing particles coarser than 40-mesh, the other the finer particles. The coarser particle fraction was added to all of the sodium carbonate solution (half-saturated) and digested for 15 min at 100°C, whereupon 43% of the sodium carbonate and 89% of the gypsum was found to be reacted. The small particle fraction was now added to this reacting system, and after 15 min longer, 94% of the total gypsum added and 92% of the sodium bicarbonate was found to be reacted.

In another 15-min experiment at 100°C, 1.2 moles of fine gypsum was mixed with 1 mole of sodium carbonate, whereupon 81% of the gypsum and 96% of the sodium carbonate was found reacted. The washed filter cake from this run was mixed with 0.4 moles of sodium carbonate in a half-saturated solution. After 15 min, overall gypsum efficiency is reported at 99%.

Comment on the planning of experiments in this study. Propose a flow sheet for making a strong sodium sulfate solution containing less than 1 mole of $CO_3^=$/100 moles of $SO_4^=$ in solution. Of the various possible flow sheets for carrying out this reaction, discuss why your choice is to be preferred.

Note: At 100°C, solubility of sodium sulfate in pure water is 27.7 weight %. Solubilities of calcium carbonate and calcium sulfate in the aqueous systems involved here are vanishingly small.

PROBLEM 8-9

McKewan (94) made spherical pellets of Fe_2O_3 by rolling pure, finely divided Venezuelan iron ore into spheres, drying, and heating in air to 1370°C. The pellets so produced were very dense, free from pores, and very hard. These pellets were placed in a basket, and their change in weight determined while undergoing isothermal reduction in an atmosphere of H_2 and H_2O. The per cent reduction was calculated as the percentage of the initial oxygen present which was lost from the charge, and findings for pellets of different radii were as follows:

PURE HYDROGEN AT 811°C AND ATMOSPHERIC PRESSURE

Radius	0.290 cm	0.500 cm	0.841 cm
50% reduction	3.0 min	5.5 min	8 min
75% reduction	5	9.2	14
90% reduction	7.2	12.5	20.8

The rate of reaction, expressed as g of oxide reduced per sq cm of core surface per min, was proportional to the hydrogen partial pressure. The effect of gas velocity was not determined, but flow rates were such that the change in gas composition through the furnace was negligible. Pellets in various states of reduction were cut in half, and were always found to contain a dense spherical core of unreduced Fe_2O_3 surrounded by a porous shell of iron. Traces of layers of FeO and Fe_3O_4 were observed between the Fe_2O_3 core and the iron shell, but these appeared negligible in mass relative to the core and shell. The unreduced core remained spherical at all times, concentric with the pellet, and appeared to be quite impermeable to gas. The porous iron shell, on the other hand, was quite gas permeable.

(a) It is maintained that these data are compatible with the assumption that the gas-solid reduction of Fe_2O_3 to iron at the core surface is controlling, rather than gas diffusion or solid diffusion. Do you agree?

(b) Other studies have been made on Venezuelan ores finds without, however, first pelletizing or making the particles denser by heating (39, 99). It was found in these investigations that the fractional reduction attained in a given period of time under conditions of temperature, gas composition, etc., similar to those used by McKewan is independent of particle size below 30-mesh. These studies were made in fluid beds, operated batchwise. Can these findings be reconciled with those of McKewan's?

PROBLEM 8-10

Water and sulfur trioxide as either liquid or vapor combine immediately upon contact to make sulfuric acid. When combination occurs in the gas phase, with inert diluents present like O_2 and N_2, then the acid forms as tiny droplets of mist. Being electrically charged, this mist is exceptionally stable, readily passes through ordinary scrubbing operations, and requires powerful means such as electrostatic precipitation for removal.

Sulfuric acid is miscible in all proportions with water, also with SO_3.

Liquid mixtures of H_2SO_4 and SO_3, which are called oleums, contain no free water, consequently only SO_3 is present in their equilibrium vapors. Similarly, liquid mixtures of H_2SO_4 and water contain no free SO_3, hence only H_2O is found in their equilibrium vapors. It is a curious fact that vapor pressures over 97 to $99\frac{1}{2}\%$ H_2SO_4 are negligible up to at least 100°C, consequently neither H_2O nor SO_3 is found in any equilibrium gas phase over acid in this concentration range.

The presence of acid mists in the off-gases from sulfuric acid plants is undesirable, since these mists represent either a costly recovery problem or a source of yield loss and atmospheric pollution. To prevent mist formation, the air used in acid plants to burn sulfur is first dried by scrubbing with 98% acid, while the absorption of SO_3 from the gases leaving the catalytic converters is also with 97% to 99% acid. Ceramic packed towers rather than plate towers are used for this absorption to avoid corrosion.

Oleum composition is often reported as the per cent by weight of its contained SO_3. Thus 100 lb of a 20% oleum solution contain 20 lb of SO_3 and 80 lb of H_2SO_4. Alternatively an oleum may be characterized in terms of the lb of 100% acid which can be made by addition of water to 100 lb of the oleum. Thus, recognizing that sulfuric acid and SO_3 have molecular weights of 98 and 80, respectively, then 100 lb of a 20% oleum are equivalent to $(80 + 20 \times \frac{98}{80})$ or 104.5 lb of H_2SO_4. Such an oleum is therefore designated as 104.5% acid.

In the plant under consideration, a dry gas at 70°F analyzing 10.5% SO_3, the rest N_2, O_2, and a trace of SO_2 is fed to the bottom of an adiabatic absorption tower, while 98% acid at 70°F is fed to the top. The tower is very efficient, so that it may be viewed as an infinite tower. The hot product acid leaving the bottom of the tower is an oleum, which is in part withdrawn for sale, in part recycled through a heat exchanger to the top of the tower. Before entering this exchanger, water is injected into the recycle stream to reduce its concentration to 98% acid. No SO_3 is ever to appear in the top off-gases. Assuming very high rates of mass transfer:

(a) Can 99.5% acid be produced in this adiabatic tower? Present operating instructions, that is, how many pounds of acid should be fed per mole of gas entering the tower? What is the temperature of the product acid?

(b) Can 106% acid be made in this tower with 98% acid feed? Again present operating instructions.

(c) How would you calculate the maximum concentration that can be obtained from this tower, when no SO_3 is to be lost in the tower off-gases?

(d) Suggest a flow sheet to manufacture 20% oleum from 98% acid and 10.5% SO_3 vapors.

Notes: For this problem, assume molal heat capacities in cal/(g mole)(°C) of gaseous N_2, O_2 and SO_3 are, respectively, 7, 7, and 14. For liquid mixtures of SO_3, H_2O, and H_2SO_4, heat capacities may be assumed constant at 0.33 Btu/(lb)(°F) over the concentration ranges considered. When one pound of SO_3 dissolves isothermally in enough acid of 98 to 100% strength, about 500 Btu of heat are evolved, regardless of the strength of the acid or oleum formed (see Reference 86, page 192, for more accurate enthalpy-concentration data). Vapor pressure information on oleum, presented below, is such that at constant oleum concentration, $\log p_{SO_3}$ plots as a straight line versus reciprocal absolute temperature.

PRESSURES OF SO_3 OVER OLEUMS, MM HG*

H_2SO_4 %	25°C	50°C	70°C	90°C
102	—	1.6	8.5	37.5
103	—	2.1	10.5	44.4
104	0.3	3.0	14.2	56.8
105	0.7	6.0	27.2	103.7
106	2.1	15.7	63.0	217.2
107	6.0	38.1	138.2	434.9
108	—	80.6	249.4	—
112	103.2	409.9	—	—

*Reference (6), Vol. III, 304.

PROBLEM 8-11

In the Wacker process (145), ethylene is oxidized to acetaldehyde below 100°C using an aqueous solution of palladium and copper chlorides as catalyst. In one version of this process, a mixture of ethylene and oxygen is bubbled continuously up through the aqueous catalyst solution, with the acetaldehyde produced being volatilized into the off-gases. The following reactions proceed simultaneously:

$$C_2H_4 + PdCl_2 + H_2O = CH_3CHO + Pd^\circ + 2HCl \qquad \text{(a)}$$

$$2CuCl_2 + Pd^\circ = 2CuCl + PdCl_2 \qquad \text{(b)}$$

$$2CuCl + \tfrac{1}{2}O_2 + 2HCl = 2CuCl_2 + H_2O \qquad \text{(c)}$$

$$C_2H_4(g) + \tfrac{1}{2}O_2 = CH_3CHO(g) \qquad \Delta H^\circ = -50 \text{ kcal}$$

After water-scrubbing to recover acetaldehyde, these off-gases are in part purged, in part recycled. In an alternative version of this process, only ethylene is bubbled through the catalyst solution, whereupon only reactions (a) and (b) occur. The catalyst solution is continuously cycled to a second tower, where it is air-blown for regeneration by reaction (c).

In a preliminary laboratory study (47) to determine the technical feasibility of a single reactor operation using air rather than oxygen, a mixture of air and ethylene was injected at the bottom of a vertical cylinder 2 cm in diameter. This feed gas bubbled up through the catalyst solution contained in the cylinder, with the acetaldehyde formed being swept out of the solution in the off-gases. Liquid water was added as needed to keep the catalyst solution volume constant. The cylinder was jacketed for isothermal operation up to 95°C. Acetaldehyde was recovered from the off-gases by scrubbing with sodium bisulfite solution, which was then analyzed for aldehydes by titration for disappearance of hydrogen ions.

Each data point was obtained by operating at the desired conditions for about a half-hour to approximate steady-state, after which measurements were made. Data recorded were solution composition and depth, temperature, flow rates of ethylene and air, and moles of CH_3CHO produced. "Conversions" were defined as (moles of CH_3CHO formed)/(moles of C_2H_4 fed), and were reported as per cent.

I. For operations at 80°C with a catalyst solution 1.62 molal in $CuCl_2$ and 0.2 molal in $PdCl_2$, using a mixed gas feed of 1.47 cc/sec of O_2 and 0.77 cc/sec of C_2H_4, conversion was roughly proportional to the volume of catalyst liquid present corresponding to the depths studied of from 3–30 cm, with about 27% conversion attained at 30 cm. The gas entered the catalyst solution from below through a single orifice 5 mm in diameter.

II. Conversion was relatively independent of temperature between 75 and 93°C.

III. At a fixed ethylene flow rate, moles of acetaldehyde produced were constant, independent of oxygen flow rate providing the reactor was not "starved" by supplying a stoichiometric deficiency of oxygen.

IV. At higher flow rates of ethylene and air, production of acetaldehyde per unit time for a given depth of immersion and catalyst composition was roughly constant, independent of the flow rates of either gas.

V. A reduction in catalyst concentration to one-quarter of the values mentioned above reduced conversion for any given operating condition by about 50%. In some runs, as discussed below, a white CuCl precipitate formed, causing a change in catalyst concentration and a reduction in conversions.

VI. A fine black precipitate of palladium formed immediately by reaction (a) when air-free ethylene was bubbled through a catalyst solution containing only $PdCl_2$. If $CuCl_2$ was also present in solution, then the black precipitate appeared only after all the Cu^{++} ions have been consumed in reaction (b). No black precipitate appeared at any time with air-oxygen mixtures when using the catalyst solutions studied, in which the mole ratio of $CuCl_2/PdCl_2$ was 8 or greater.

VII. A greenish-white precipitate of CuCl appeared when operating with small catalyst liquid volumes (small immersions) and high conversions. At deep immersions and equal conversions, no precipitate formed.

Smidt (145) reports that reaction (a) is the sum of two reactions: the rapid formation of an ethylene–palladium chloride–complex, which then hydrolyzes to acetaldehyde and the original $PdCl_2$.

(a) An argument develops over interpretation of the above laboratory findings. One man maintains that findings I and II point to mass transfer control. A second man insists that item IV is inconsistent with mass transfer control, and that it is controlled by the rate of hydrolysis of the complex. The first man says that the observed effect of temperature in item II is inconsistent with any chemical reaction rate control. The second man explains the small effect of temperature on rate by saying: "complex formation is exothermal and fast, complex hydrolysis rates increase with temperature but are proportional to complex concentration." Write a memorandum presenting your position.

(b) With a 27% conversion as in part I above but with adiabatic operation and with all feed streams at 25°C what would be the temperature of the reactor and the off-gas stream?

(c) Large plant equipment must be operated outside the explosive range, meaning less than 9% oxygen in ethylene. Discuss the scale-up problems of this reactor, indicating possible design. What further laboratory and pilot studies appear indicated?

Note: Assume the vapor pressure of water over these catalyst solutions to be 75% as great as over pure water at the same temperature. Water vapor pressures are to be found in the usual references. The heat capacities of O_2, N_2, H_2O(g), and C_2H_4 are 7, 7, 8, and 14 Btu/(lb mole)(°F), respectively.

PROBLEM 9-1

Acetylene gas is made by bringing calcium carbide into direct contact with liquid water, whereupon $Ca(OH)_2$ forms along with C_2H_2. This reaction proceeds rapidly with liquid water even at temperatures as low as 10°C, but it is slow with water vapor until well over room temperature. Two types of continuous generators are used, as briefly described below (100). These generators are adiabatic, and for safety are carefully designed to avoid the development of pressures above atmospheric and temperatures above 100°C.

1. *Wet (carbide to water) generators.* Commercial carbide is added to an excess of cold water, with the heat of reaction being carried away as sensible heat of the product streams. Typically, the water and carbide enter

at 25°C, while the acetylene gas and calcium hydroxide slurry leave at 65–80°C. Several alternative designs exist, one involving the introduction of additional carbide whenever the pressure falls, another operating on the Kipp generator principle, etc. The slurry of $Ca(OH)_2$ produced represents a disposal problem, and acetylene is lost in this slurry since acetylene has a noticeable water solubility. The acetylene gas produced leaves these generators saturated with water at the temperature of operation.

2. *Dry* (*water to carbide*) *generators.* Cold liquid water is added to the carbide. In one design carbide is fed continuously to the top of a tray reactor, where it is wetted by spraying with the correct amount of water to keep temperatures below 100°C. The reacting solids attain dryness as the reaction is completed on the last tray. The carbide is fed to the reactor from a double bin system through a screw feeder, and the residue is withdrawn at the bottom through a second bin and screw. These bins are run "full" to minimize gas leakage.

By proper proportioning of water to carbide, the temperature of the reacting solids remains under 100°C, the calcium hydrate is produced dry, and the heat of reaction is removed largely by vaporization of water, rather than as sensible heat of the water slurry as in the wet generators. The advantages of producing a dry hydrate are significant, since acetylene losses by solution are avoided, and disposal of the hydroxide is simplified. Careful control, however, is essential, for if sufficient water is not used, or if the water is not well mixed with the carbide, hot spots may develop in the bed. By reacting with water vapor in the off-gases, incompletely reacted carbide chunks can reach temperatures above 1000°C, causing explosions or the formation of undesired or dangerous materials such as vinylacetylene and peroxides. Given proper operation, these difficulties are avoided.

(a) Acetylene is to be produced continuously in a "wet generator" at atmospheric pressure from commercial carbide, analyzing 85% of CaC_2, the rest are inert substances. How much liquid water must be added per lb of acetylene produced, if the gas and hydroxide slurry leave the reactor at 80°C? The solid carbide and liquid water are both fed at 25°C.

(b) Acetylene is to be produced continuously in a "dry generator" at atmospheric pressure. How much liquid water must be added per lb of acetylene produced, if the maximum system temperature is to be 100°C. The solid carbide and water fed are again at 25°C.

(c) The gases from the wet generator of part (a) of this problem and those from the dry generator of part (b), are both to be cooled to 25°C. How much water is condensed in each case, expressed as moles per mole of C_2H_2 content?

(d) It is proposed to add three moles of liquid water to 1 mole of contained CaC_2 in commercial carbide. Assuming complete reaction, estimate the adiabatic reaction temperature reached in a well-mixed system. The solid carbide and liquid water are fed at 25°C.

Note: The heat capacities of commercial calcium carbide, $Ca(OH)_2$, and CaO are 0.25, 0.2, and 0.19 Btu/(lb)(°F), respectively, while the molal heat capacities for gaseous C_2H_2, gaseous water, and liquid water are 11.5, 9, and 18 Btu/(lb mole)(°F), respectively. The vapor pressure of water over slurries of calcium hydroxide may be assumed the same as over pure water at any temperature (why?). The heat of vaporization of water at 80°C is about 550 cal/g. Values of the enthalpies of formation of CaC_2(c), CaO(c), $Ca(OH)_2$(c), C_2H_2(g), and H_2O(l) are −14.1, −151.8, −235.8, +54.2, and −68.3 kcal/g mole, respectively. Corresponding values of $\Delta F^\circ_{298°K}$ are −15.3, −144.4, −214.3, +50, and −56.7 kcal/g mole.

PROBLEM 9-2

Promoted iron-base catalysts are usually used in the synthesis of ammonia from the elements. Most of these catalysts show a kindling temperature of about 430°C, while the highest temperature to which they can be heated without premature inactivation and particle breakdown is usually around 520–550°C. Since the equilibrium constant for ammonia synthesis is relatively unfavorable between 430–550°C, high pressure is used to drive this gas reaction (which shows a diminution in the number of moles) further towards completion. The reaction is vigorously exothermal, with an enthalpy of formation for NH_3 of about −12.8 kcal/g mole in this temperature range.

Hydrogen at higher pressures and temperatures decarbonizes ordinary steel, weakening it dangerously. To avoid hydrogen damage, the walls of the reactors are cooled internally with incoming cold feed gas, even when using the improved alloys available today. It will be noted that any reactor "internals" such as heat exchangers, etc., which are immersed in the high pressure atmosphere, are not highly stressed, and so represent no serious design problems.

Four general types of reactors have evolved, as shown in Fig. P-12 (54).

1. *Single packed bed.* A single bed of catalyst obviously makes for simple construction. This bed is adiabatic.

2. *Quench converter.* The gas flows through several catalyst beds in series, with interstage cooling by injecting cold, fresh gas between beds. Each bed is adiabatic.

3. *Water cooled converter.* The gas again flows through several catalyst

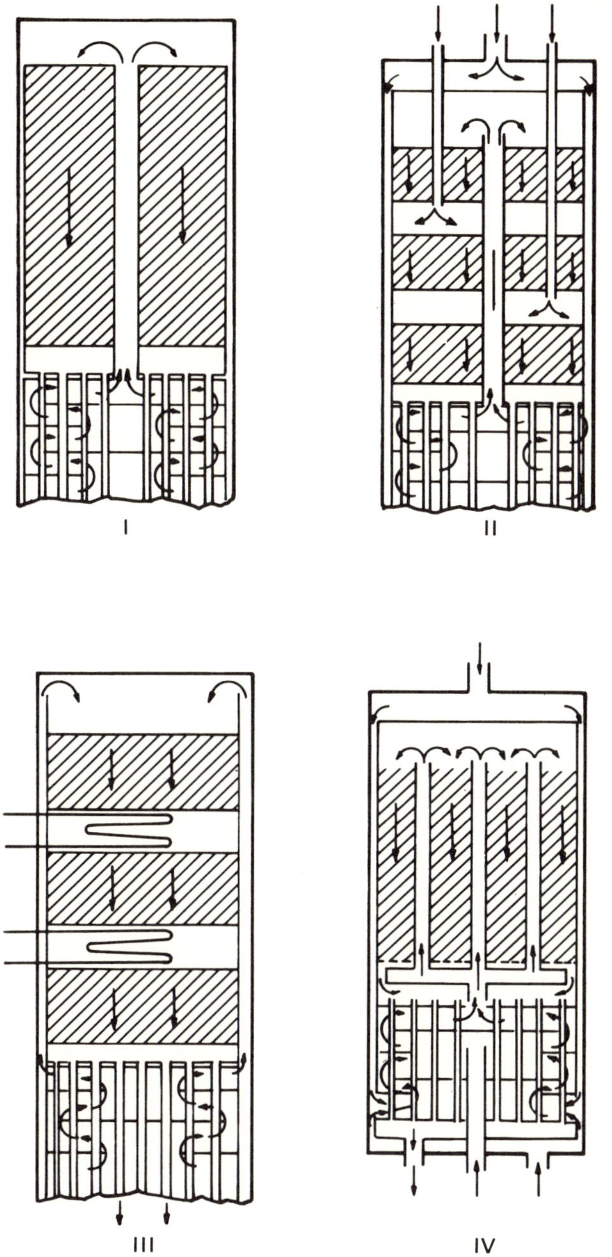

Fig. P-12. Types of fixed-bed converters. I. Adiabatic, single bed. II. Three beds with quench cooling between beds by feed gas. III. Three beds with indirect water cooling between beds. IV. Tubular converter, with countercurrent cooling by feed gas. [Reproduced by permission of H. Hinrichs and J. Niedetsky, *Angew. Chemie, Int. Ed.*, **1** (1962), pp. 206 and 207.]

beds in series, with interstage cooling by coils carrying cooling water. This design is mechanically more difficult because of increased problems of gas leakage and stress concentrations where the water tubes pass through the reactor shell. Each bed is adiabatic.

4. *Tubular converter.* The catalyst is packed in tubes which are cooled externally, either cocurrently or countercurrently, by the feed gases. Construction of the tubular converter is complex, filling the tubes with catalyst is awkward, and radial temperature gradients through the catalyst beds in the tubes are high. Thus, maintenance of any desired temperature history through the catalyst bed is difficult.

Reaction rates at 300 atm and various temperatures for the catalyst under study are presented in Fig. P-13. Note that the ordinate of this figure is defined as (moles of NH_3)/(moles of equivalent gas feed), where "equivalent gas" is that gas obtained after decomposing any ammonia present into its elements. Reactors are to be designed using this catalyst, and to maintain vessel size and investment small, operations close to the optimum conditions are planned. It is consequently decided that the gas leaving each converter stage should not be more than 30°C above the temperature at which gas of this composition attains maximum rates. In no case, however, should the gas temperature exceed 570°C. The gas enters the first bed at 430°C, but enters subsequent beds no cooler than 25°C below the temperature yielding the maximum rate at the composition in question. The data of Fig. P-13 apply over the flow rates considered.

The feed gas entering a converter is to analyze 2% NH_3, the remainder being H_2 and N_2 in a ratio of $\frac{3}{1}$. The heat capacities of H_2, N_2 and NH_3 may be assumed as 7.5, 7.5, and 15 cal/(g mole)(°C), respectively, for this problem. Operations are at 300 atm. The catalyst packing density is 50 lb/cu ft, and is 50% voids. It is proposed to make 100 tons per day of ammonia in these converters.

(a) What is the composition of the gas leaving each converter in the single bed, the three-stage quench converter, and the three-stage water cooled converter? Assume the quench gas to be available in the quench converter at 25°C, and to analyze 2% NH_3.
(b) How were the rate curves in the figure probably obtained? Explain the shape of these curves, and the reason for a maximum in each curve. What is the advantage of reporting the ordinate as moles NH_3 per mole of "equivalent" gas, rather than as mole fraction of ammonia?
(c) How many cu ft of catalyst are required in each of these three converters, for each lb mole of NH_3 produced per hr?
(d) What are the relative advantages of each of these converter types?

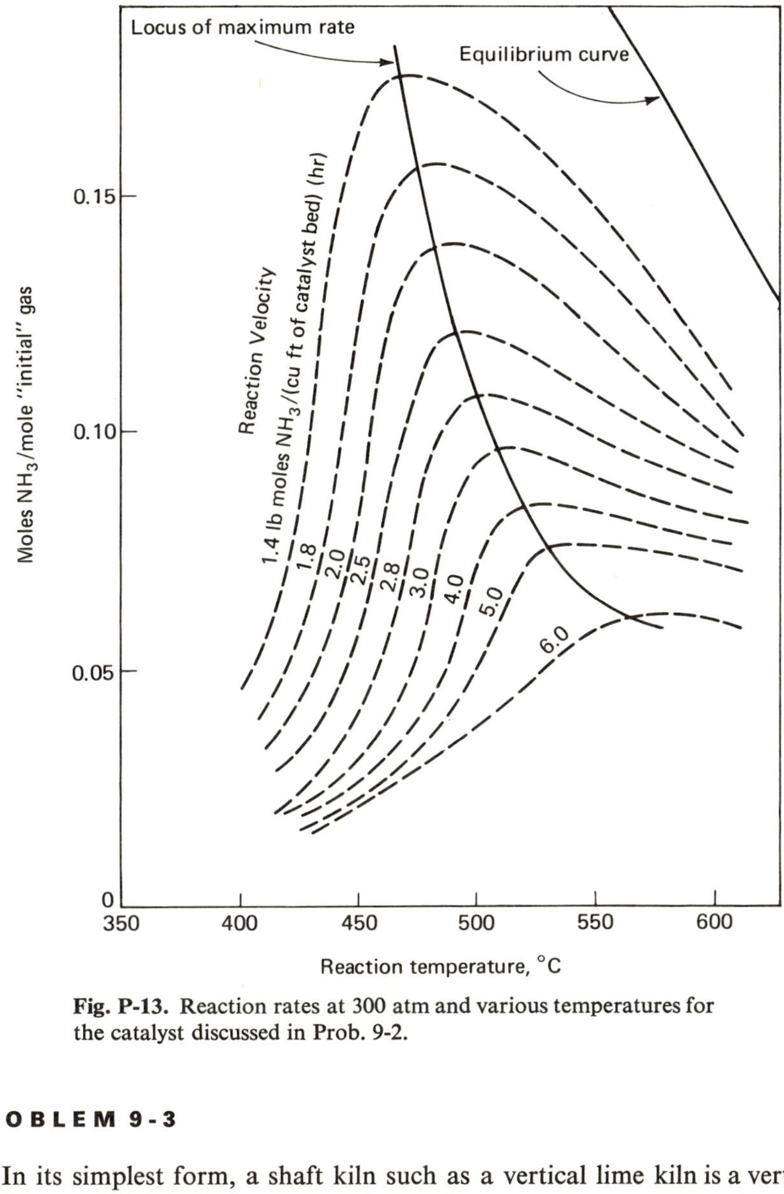

Fig. P-13. Reaction rates at 300 atm and various temperatures for the catalyst discussed in Prob. 9-2.

PROBLEM 9-3

In its simplest form, a shaft kiln such as a vertical lime kiln is a vertical shaft up to 100 ft high, having a cylindrical or square cross section. A device for continuously feeding limestone lumps to the top is provided, and for withdrawing solid product from the bottom. The shaft is filled to a fixed level with a bed of stone moving slowly downward as it decomposes. Fuel gases are injected into the charge through wall ports close to the kiln base, while air for combustion is introduced through a set of lower ports or up through the kiln bottom. Combustion occurs within the kiln bed, with the hot off-gases moving upward, counter to the descending solids.

Gas penetration across the kiln is poor, with the center core of the charge often receiving little gas and consequently little heat while the charge near the walls becomes overheated. Because of such channeling, the product from the lime kilns at Ashtabula (124) contained 10–12% of unburned core material when charging $2 \times 3\frac{1}{2}$ in. limestone lumps and operating at the rated capacity of 90 tons per day. These kilns were 70 ft high, with a square 11×11 ft internal cross section.

By kiln redesign and reconstruction, improved performance has been attained, with output increased to 500 tons per day while reducing unburned core content to under 3%. Heat efficiency remains high. Significant structural changes are reported as follows (see Fig. P-14).

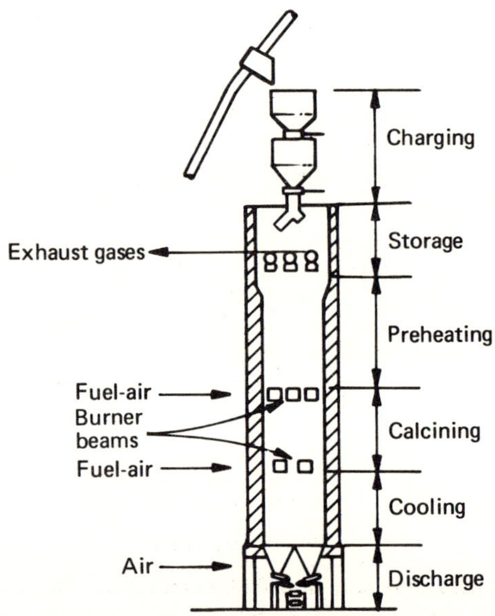

Fig. P-14. Cross section of the vertical lime kiln described in Prob. 9-3. [Reproduced by permission of J. E. Roberts, *Chem. Eng. Progr.*, **59** (1963), No. 10, p. 91.]

(a) Five "burner beams" are installed in each kiln, parallel to one set of opposite walls and extending across the kiln cavity between the other two walls. Two of these beams are in a horizontal plane 11 ft up from the kiln bottom, the other three are in a plane 20 ft up from the kiln bottom. In each plane, the spacing between the neighboring beams and between beam and wall is kept equal.

(b) Each burner beam is a standard 4-in. "I" beam, water-jacketed for cooling to prevent buckling. These beams support and protect pipes carrying air and fuel gas. There are holes in these pipes at every ft or so of length for uniform introduction of gas and air across the kiln space. It is claimed in this way to avoid the maldistribution of heat and fuel which so severely limited kiln capacity and performance with the original design.

(c) There are three take-off pipes installed parallel to the burner beams, again reaching across the kiln, in a plane which is 40 ft above the kiln bottom. These take-off pipes are perforated, and serve to collect uniformly the off-gases across the kiln, thus again avoiding channeling.

The Ashtabula kilns may be viewed as divided into four sections. The first or top section lies above the take-off pipes, and serves to store up 2-hr supply of limestone feed, thus ensuring smooth operation despite any brief interruption in limestone supply. The limestone itself is fed to the top of this section through a double bell system to prevent air leakage at this point. Note that no gas passes through this top section. The second kiln section lies between the take-off pipes and the upper burner beams, and is called the stone preheating section. The third or calcining section lies between the two burner beam levels. The fourth section lies below the bottom burner beams, and is called the lime cooling section. Here heat is exchanged between the product lime and that fraction of the incoming air which enters at the kiln base, rather than at the burner beams.

These kilns operate on CO, and the air fed to the kiln at the base plus that at the burner beams is just enough for complete combustion. The CO, air, and limestone all enter the kiln at 60°F; the calcining zone may be assumed isothermal at 2000°F. The lime is discharged at the kiln bottom at 500°F. If the kiln operates adiabatically, then for each lb mole of product lime:

(a) How many moles of air were introduced at the bottom of the kiln?
(b) How much air and CO were introduced into the bottom burner beams, if neither CO or O_2 exist in the gas leaving the calcining zone?
(c) How much added air and CO were introduced into the top burner beams (if any) to help preheat the $CaCO_3$? It will be assumed that $CaCO_3$ must be preheated in zone 2 to 2000°F before entering this third kiln section, that no limestone decomposition occurs in zone 2, and that no CO or O_2 are present in the kiln top gases.
(d) What will be the top off-gas temperature?

Note: For this problem, assume molal heat capacities for $CaCO_3$, CaO, CO_2, O_2, CO, and N_2 to be 22.5, 12, 10.5, 7, 7, and 7 Btu/(lb mole)(°F), respectively.

$$CO + \tfrac{1}{2}O_2 = CO_2 \qquad \Delta H^\circ = -66 \text{ kcal/g mole}$$
$$CaCO_3 = CaO + CO_2 \qquad \Delta H^\circ = +42.6 \text{ kcal/g mole}$$

PROBLEM 9-4

Acetaldehyde (CH_3CHO, boils at 21°C, miscible with water) dimerizes in the presence of alkali to form aldol ($CH_3CHOH \cdot CH_3CHO$, again water soluble), evolving 26.5 kcal/g mole of aldol formed. The German I.G. Farben Co. operated a large aldol plant during World War II, (8) in which the reactor consisted of 16 loops of pipe externally cooled with brine at 10°C. These loops were connected in parallel, with the liquid reaction mixture circulating through each pipe isothermally at 20°C and at a velocity of 11.5 cm/sec. The time for circulation in each loop was 6 min. Total feed to this reactor system of 16 loops plus recirculating pump was 4 cu m of pure liquid acetaldehyde per hour (0.78 g/cc., density), introduced just after the discharge side of the recirculating pump, while 75 liter/hr of a 10% KOH solution were introduced on the suction side of this pump. Such a separation of feed points avoided resin formation resulting when these two feed streams were introduced closer together. A fraction of the recirculating liquors was withdrawn as product just before the point where fresh KOH solution was introduced. Operations were continuous and in the liquid phase throughout, "residence time" in the reactor (namely the volume of the reactor divided by the volume of feed/hr) was reported at $1\frac{1}{2}$ hr, and the product stream showed 40% conversion.

The reaction to aldol is assumed second order in acetaldehyde at a given level of caustic concentration, or

$$r = \frac{dn_A}{V d\theta} = -k\left(\frac{n_A}{V}\right)^2$$

where r is the rate of acetaldehyde disappearance in g moles/(l)(sec), n_A is g moles of acetaldehyde present, and V is volume in l. It may be assumed for purposes of this problem, that the density of the liquid phase does not change as reaction proceeds, and that volume % equals weight % of acetaldehyde and aldol.

(a) How long was each loop of pipe? What was the diameter of each pipe?

(b) What was the average concentration of CH_3CHO in volume % in the recirculating liquor immediately downstream from the point at which fresh CH_3CHO is added? Evaluate k_2 in the rate expression.

(c) What is the average overall coefficient of heat transfer between this pipe and the brine coolant (whose temperature is relatively constant at 10°C).

(d) It is proposed that a well-stirred tank could be substituted for this pipe reactor. Acetaldehyde would enter continuously as product is withdrawn. Estimate the size of this tank. If the tank were cylindrical with the diameter equal to the liquid depth, would the side wall area of the tank be sufficient for cooling purposes? The overall coefficient of heat transfer in the tank is to be assumed the same as in the pipe loops.

(e) At least one serious reactor explosion is reported to have occurred in this plant. Comment on possible causes and desirable precautions.

PROBLEM 9-5

The granular catalyst from a certain fluid bed operation must be regenerated after brief exposure to the reactions of the primary process. Up to the present, this regeneration has been accomplished in the pilot plant by batchwise fluidization in reducing gases (a mixture of H_2 and CO) for 9–11 min at 600°C. Exposure for periods shorter than 9 min results in incomplete regeneration. Exposure for more than 11 min yields an inactivated material, which, however, fortunately can be reactivated after being brought into contact with the reactants of the primary process. The change in catalyst weight during regeneration is negligible.

Batchwise regeneration of the catalyst is considered unsuitable for the large scale plant, consequently regeneration is to be carried out in a continuous fluid bed operation. Catalyst will be fed and withdrawn at a rate of F lb/min and the fluid bed is to contain W lb of catalyst at all times. It will be assumed that catalyst will be withdrawn from the bed by use of a simple overflow pipe, and that at any time, each catalyst particle in the bed has an equal chance of leaving via the overflow pipe. It is proposed that design should be such that W/F is 10 min. The quantity F will be 100 lb/min.

(a) Regenerating in this fashion, what fraction of the material being withdrawn will have a residence time lying between 9 and 11 min?

(b) It is now proposed to operate two beds in series with the new beds of weight W_{I} and W_{II} so chosen that

$$W_{\text{I}} = W_{\text{II}} = \frac{W}{2}$$

Feed is as before to bed I, overflows from here to bed II, and then out of the system. What fraction of the product now had a total residence time in the two beds of from 9–11 min?

PROBLEM 9-6

The mineral apatite is a complex calcium fluorophosphate of varying composition sometimes represented as $Ca_5(F \cdot Cl \cdot OH)(PO_4)_3$. Ore of commerical value contains from 30–36% of P_2O_5. Fluorine content is usually 3% or so, which is much too high for a cattle feed, and should be reduced to 0.1% or less to be safe. Moreover, to be assimilated by an animal, phosphorus should be present as the orthophosphate. A common test for orthophosphate in suitable form is its solubility in ammonium citrate.

A well-known method for reducing the fluorine content of apatite is to mix it with phosphoric acid or sulfuric acid, followed by calcining at elevated temperature (17). In test work on the possible use of sodium acid phosphate to replace these acids, a batch of 1000 lb of a certain phosphate rock (100-mesh) was mixed with a strong aqueous solution containing 160 lb of NaH_2PO_4, dried, and the resulting solid was reground. Defluorination was accomplished by fluidizing the acidulated material in hot gas generated by burning methane with an excess of air (55). For rock from a foreign deposit heated batchwise at constant temperature, the logarithm of the fluorine content per 100 lb of the bed material formed a straight line when plotted against time. In a batch run at 2300°F, the fluorine content of the solid fell from an initial 3.3 lb/100 lb of fluoride-free charge to 0.2 lb/100 lb of fluoride-free charge in 120 min. The rate of defluorination was not significantly affected by minor changes in flue gas composition.

(a) A continuous operation in the same equipment is proposed using the same bed weight and operating temperature as in the batch run. Assuming that the bed temperature is fixed at 2300°F, at what rate should solid feed be introduced into the bed, if the product is to contain 0.2 lb of fluorine per 100 lb of fluorine-free charge? An associate states that Eq. (9-3) cannot be used without modification here because x_0, the initial fluorine concentration, is not zero. Comment.

(b) The product from the first fluid bed is run continuously through a second fluid bed of equal size, again at 2300°F. What is the average fluorine content of the product from this second bed?

(c) Optional. How could part (a) be solved, were the defluorination rate to be proportional to the second power of the residual fluorine in the solid?

PROBLEM 9-7

The fluid bed can be used as a reactor in which particles are made to grow in size. Thus, when liquid heavy petroleum residues are sprayed directly into a hot bed of fluidized coke particles, the carbon formed in cracking is found to be uniformly distributed over the surfaces of the particles present (16). Again, when spraying an aqueous solution of an inorganic salt directly into a bed of particles of this salt fluidized by a hot gas, the solid salt formed in evaporation is found uniformly distributed over the surfaces of the particles present. New particles may not appear unless conditions are such that particle breakage occurs. When very fine particles are present, then the tackiness of the freshly formed surface (this tackiness usually lasts for only a very short while) may result in agglomeration. So long as the bed is well fluidized and the liquid spray is fine, such agglomeration is usually insignificant, if the fines present are not below 100 μ in size.

(a) The increase in particle diameter over any time period in all of these processes is always the same for all particles independent of their size. What does this finding mean with regard to the mechanism of growth?
(b) A mixture of spherical particles of sodium bromide (density, 3.2 g/cc) contains an equal amount by weight of particles 0.2 mm and 0.4 mm in diameter. One pound of this mix is fluidized with hot air, whereupon a solution of NaBr is sprayed directly into the bed through a submerged nozzle. Operations are smooth, hence the new solid formed is uniformly distributed over the particle surfaces. What are the diameters of the particles present after the bed weight has increased to 2 lb? What is now the weight fraction of the larger particles?
(c) In a continuous operation, spherical seeds of NaBr all 0.2 mm in diameter are added at a rate of 0.01 lb/min to a bed of this salt, well fluidized with hot air. A NaBr solution is sprayed into the bed at such a rate that 0.1 lb/min of NaBr is so introduced. All particles in the bed have an equal chance of being removed, regardless of age in the bed. It may also be assumed that all particles larger than 2.0 mm are immediately removed by a screening procedure, and that the number of particles in the bed is equal to 20 times the number fed per min. Estimate the bed weight. Estimate the percentage by weight of the overflow which consists of particles larger than 1 mm in diameter.

PROBLEM 9-8

Silica can exist in at least three crystalline forms, namely crystabolite, tridymite, and quartz. Phase relations are complex (126), with quartz being thermodynamically stable at normal temperatures. Quartz is converted to crystabolite at elevated temperature, and once so formed, crystabolite reverts to quartz only very slowly at lower temperatures. Unlike other forms of silica, crystabolite (melts at 1625°C) undergoes little expansion with temperature above 430°F, and consequently is valuable in making molds for metal casting and for high temperature furnace linings.

A batch of a New York State quartz sand was heated in a fluid bed.

Conversion to crystobalite appears to be first-order in W, the weight fraction of the sand present as quartz (99)

$$-\frac{dW}{d\theta} = kW$$

where θ is hr and k is 2.0 reciprocal hr at 1500°C. This quartz sand is to be fed continuously at a rate of 3 tons/hr to a single fluid bed of 30 tons of sand, in which the mixing of solids is complete. The bed is at 1500°C.

(a) Product withdrawal is through an overflow pipe, and each particle in the bed has an equal chance of entering this overflow stream, regardless of residence time. What is the crystabolite content of the product stream?

(b) It is considered necessary to cool the product stream rapidly to 1000°C to prevent reversion to quartz. These product solids are therefore dropped into a second fluid bed which serves to preheat the air for combustion. Natural gas burned in the reactor is injected directly into the reactor bed without preheating. The off-gases from the reactor proceed to a third fluid bed, which serves to preheat the quartz sand feed. The air, natural gas, and silica sand all enter this three-bed system at 25°C. What are the temperatures of beds 2 and 3, when operating with the theoretical ratio of fuel to air?

Assume the hot combustion products will have a superficial velocity through the fluid bed of 3 ft/sec, and that under these conditions, the "expanded" fluid bed will have a density of 80 lb/cu ft, regardless of the per cent conversion of quartz to crystobalite. Standard heats of formation of CH_4, CO, CO_2, and H_2O are -18, -26.5, -94, and -58 kcal/g mole, respectively, at 250°C. Values of MC_p for carbon, H_2, O_2, N_2, CO, CO_2, H_2O(g), and CH_4 are 3.5, 7, 7, 7, 7, 10.5, 10.5, and 17.5 cal/(g mole)(°C), respectively. The average values of MC_p for silica and crystobalite may be taken as 18.5 cal/(g mole)(°C). Assume the enthalpy of transition of quartz to crystobalite to be +1000 cal/g mole (at 1100°C).

PROBLEM 9-9

Phthalic anhydride is produced by the air oxidation of naphthalene in a suitable reactor. The product phthalic anhydride is recovered by passing the gases leaving the reactor through a partial condenser where 40–60% of the total anhydride is removed as liquid. The gases from the partial condenser now flow through *switch condensers*, where most of the remaining anhydride is recovered as a solid, crystalline deposit on the cooling surfaces. These condensers are typically tubular heat exchangers, with the anhydride-containing gases outside the tubes. The name *switch condensers* derives from the fact that as one condenser is filling up with condensing solids, a second has been "switched" off stream for harvesting. Harvesting involves passing hot liquid through the condenser tubes to melt the contained solids. Several factors are considered in condenser design (125).

(1) Spare condenser capacity must be installed to allow for periodic cleaning; (2) the capacity of each condenser for condensed naphthalene should be as large as possible as compared to condenser weight, otherwise the heat needs of the condenser itself per cycle become excessive, (3) it is stated that the use of hot and cold oil as a heat transfer medium rather than a water-steam combination is advantageous, since this reduces condenser metal weight, (4) product gas must contact cooling surfaces adequately and with reasonable residence times, otherwise gases leaving the switch condensers will be supersaturated in anhydride, and (5) fire and explosion hazards are reduced by using finned tubes spaced as close together as possible. If the tubes nearest the gas entrance ports are closely spaced, however, then condensed naphthalene seals the spaces between them and so stops operation before the downstream tubes can function.

(a) The feed to the reactor of the plant under study is reported as 14 lb of air/lb of naphthalene. For 100 lb of naphthalene fed, 95 lb of phthalic anhydride are produced along with CO_2 and H_2O as the only other products of the oxidation with no naphthalene surviving in the tail gases. The condenser system is at 35 psia. What fraction of this phthalic anhydride is theoretically recoverable by condensation as liquid which can be immediately drained from the condenser, rather than as solid?

(b) The switch condensers operate with oil which is at 140°F while condensing anhydride, and at 350°F during melting. During condensation, what in theory is the lowest attainable loss of anhydride in the tail gas, expressed as lb of anhydride/lb of naphthalene fed?

(c) What reason is there for the statement that condenser weight can be reduced by use of oil rather than water as a heat transfer medium?

(d) Propose a design for a commercial-sized switch condenser which would have a relatively large capacity for solid condensate and so could be kept onstream for a longer period of time. It would, of course, be desirable to recover as much anhydride as possible in liquid form.

Data:

VAPOR PRESSURE OF PHTHALIC ANHYDRIDE

Temperature, °C	Condition	Pressure, mm Hg
96.5	solid	1
121.3	solid	5
131	melts	9.2
202.3	liquid	100
284.5	liquid	760

PROBLEM 9-10

Aromatic hydrocarbons are not only more difficult to oxidize than olefins and paraffins, but also offer several resistance points to further attack as oxidation proceeds. Thus, 1 mole of naphthalene reacts with 1.5 moles of oxygen to form α-naphthoquinone, but under more severe oxidizing conditions it reacts with 4.5 moles of oxygen to form phthalic anhydride (PA):

$$C_{10}H_8(g) + 4\tfrac{1}{2}O_2 = C_8H_4O_3(g) + 2CO_2 + 2H_2O \qquad \Delta H = -428 \text{ kcal}$$

with 9 moles of oxygen to form maleic anhydride:

$$C_{10}H_8(g) + 9O_2 = C_4H_2O_3(g) + 6CO_2 + 3H_2O \qquad \Delta H = -874 \text{ kcal}$$

or with 12 moles of oxygen for complete oxidation:

$$C_{10}H_8(g) + 12O_2 = 10CO_2 + 4H_2O \qquad \Delta H = -1232 \text{ kcal}$$

Commercial production of PA is by oxidation of naphthalene (or ortho-xylene) using excess air, a vanadium pentoxide catalyst, and careful temperature control. Both fixed bed and fluid bed reactors are used for this highly exothermal oxidation, and each reactor type has its proponents. In modern fixed bed reactors, the catalyst is packed into a multiplicity of tubes cooled externally by a circulating molten salt bath. In fluidized bed reactors, cooling is usually by heat transfer surfaces within the bed. Substantially all of the naphthalene in the feed is oxidized in a single pass through either type of reactor, with relative advantages as follows (81, 125).

1. Temperature control. Temperature is constant to within 1–2°F even in a large fluidized bed containing up to 30 tons of catalyst. Temperature spreads are unavoidably greater in fixed beds, though claims are made that temperature can be kept within ±9°F of the optimum when operating with externally cooled, catalyst-packed tubes of smaller diameters. Depending on the catalyst, optimum PA yields are attained within the range of 650–725°F with both fixed bed and fluidized reactors.

2. Air-naphthalene ratios. Excess air is invariably used in naphthalene oxidation to avoid by-product formation. Large quantities of excess air are undesirable, however, since product recovery is thereby made more difficult and pumping costs are increased. To be outside the explosive range, the feed should contain 20 lb of air/lb of naphthalene, or higher. For safety, fixed beds are usually operated at feed ratios of $\frac{25}{1}$ or higher. Fluidized beds, on the other hand, are considered safe to operate in the explosive range because the high particle concentration prevents development of an ignition wave, consequently feed ratios of around $\frac{12}{1}$–$\frac{15}{1}$ are used. Thus, liquid naphthalene is pumped directly into the fluid bed, where it promptly vaporizes as it mixes into the bed. With fixed beds, the hot air stream carrying naphthalene must be prepared externally.

3. Power needs. Power savings due to the smaller gas volumes passing through the fluid bed are offset in part by its greater pressure drop.

4. Dusting. Catalyst attrition and dusting are not insignificant in fluid beds. Proponents of fluid bed reactors claim that attrition is minimized by use of abrasion-resistant particles, and that dusting is minimized by use of cyclones. A final filter is installed in the off-gas pipe from fluid beds to keep dust out of the product.

5. After-burning. A catalyst is required by this reaction over the temperature range operated, and so with fixed beds, reaction zone is limited to the catalyst bed itself. In fluidized operation, because of bubble formation, a little naphthalene occasionally breaks through the bed and finds some catalyst dust particles in the vapor space, on which reaction occurs. Being outside the bed, this dust particle is now unable to surrender the heat of reaction to the bed itself, consequently its temperature increases to a point where it catalyzes complete combustion of the surrounding phthalic anhydride to CO_2 and H_2O. Temperatures as high as 1000°C can result, with loss of product and fire hazard. This undesirable development is called *afterburning*.
To avoid after-burning, which can be catastrophic, the gases leaving a fluidized bed are rapidly cooled to below the kindling temperature for this reaction (about 500°F) by introducing either cold, inert gas or cold catalyst, or by heat transfer surface. Operating with low air-naphthalene ratios also helps reduce after-burning tendencies.

6. Yields. It is claimed that yields are slightly higher in fixed beds, apparently because of reduced backmixing and avoidance of any after-burning.

7. Raw materials. Ortho-xylene and naphthalene are alternative raw materials, with slightly different catalyst formulations being preferred for optimum performance. In changing from one raw material to the other, catalyst replacement with fixed bed operation is far more difficult than in fluid bed operations. New plants installed in recent years have, in most cases, operated fluid bed reactors.

(a) Estimate the off-gas composition on a dry basis for a feed consisting of air and naphthalene in a weight ratio of $\frac{15}{1}$, where 95 lb of phthalic anhydride are produced per 100 lb of naphthalene feed.
(b) Plan the "start-up" of a fluidized bed reactor for naphthalene.
(c) What problems are involved in operating such a reactor at less than design capacity? At somewhat above design capacity?
(d) How should the operation of this fluid bed reactor system be controlled?

PROBLEM 9-11

Industrial formaldehyde production from methanol is by either (1) passing air containing more than 30 mole % of methanol through a silver catalyst at atmospheric pressure, or (2) passing a mixture of less than 6.7% of methanol in air through an iron-activated molybdenum oxide catalyst at atmospheric pressure. The composition range of 6.7 to about 30% of methanol in air represents the explosive limits of this system. Only about two-thirds of the methanol feed is converted to formaldehyde in the silver process, consequently methanol must be separated from the formaldehyde product by distillation. Moreover, since operations are at the relatively high temperature of around 600°C, small but undesired amounts of formic acid are produced. In the oxide process, on the other hand, operating temperatures are relatively low, being below 400°C, hence formic acid production is minimal. High conversions of methanol to formaldehyde are attained because of the large excess of oxygen present. This high dilution results in low formaldehyde concentrations in the product gas, thus requiring larger equipment (converters, blowers, scrubbers, etc.) than in the silver process.

To avoid the low formaldehyde concentrations just described, it is proposed (36) to cause the reacting gases to flow continuously in series through a number of adiabatic catalyst beds. A small amount of fresh liquid methanol is fed to the gas stream at the entrance to each bed. The reacting gas is cooled between beds in part by this injected methanol and in part by heat transfer to cooling coils. The formaldehyde concentration of the gas stream obviously increases from bed to bed as the incremental methanol is oxidized, while the oxygen content falls off correspondingly. Because of the high formaldehyde concentrations finally attained, important economies in recovery equipment size are claimed. Formaldehyde passes through the successive catalyst beds without serious decomposition. The methanol injection per stage is small, thus keeping the temperature rise across each bed small and "securing prolonged catalyst life." To protect the catalyst further, the partial pressure of the oxygen in the feed stream may be reduced by recycling part of the waste gas from the final formaldehyde scrubber back to the feed into the first catalyst bed.

In one application of this idea, formaldehyde is to be made by the air oxidation of methanol over an oxide catalyst in a reactor containing 11 sequential adiabatic catalyst beds. The gases pass through these 11 beds in series, being augmented by injection of liquid methanol at 25°C between each pair of beds. The gases from the last bed pass into a cooler-condenser for removal of substantially all the water, alcohol, and formaldehyde, with the condenser off-gases in part purged, in part recycled. Thus, the gases enter each of the 11 beds at 270°C leave at the same temperature T_x and, in each case, are cooled between beds back to 270°C by the injected cold methanol and then by water-cooled coils. The feed gas to the first bed is an equimolal mixture of air and recycle gas to which methanol has been added, and this mixture enters the first bed at 270°C. The same quantity of methanol per unit time is fed to each bed. The formaldehyde concentration of the gas stream obviously increases from bed to bed as the incremental methanol is oxidized, while the oxygen concentration falls correspondingly.

Apart from a little oxidation to CO and H_2O, the only reaction occurring to any significant degree is

$$CH_3OH(g) + \tfrac{1}{2}O_2 = CH_2O(g) + H_2O(g)$$

The equilibrium constant for this reaction is very favorable. It may be assumed that the mole ratio of CH_2O/CO in the gas from each bed is the same as in the gas from the last bed.

(a) The purge gas (dry) analyzes 0.72% CO, 0.12% O_2, the rest N_2. What is the composition of the condensate obtained, in mole %? The feed of methanol is such that the mole ratio of methanol/nitrogen in the off-gases from the last bed is 0.015.

(b) What is T_x?

(c) The catalyst particles are $\frac{1}{16}$ in. granules. The following simplified rate expression may be assumed to apply over the small temperature interval studied:

$$r = \frac{dn_F}{W d\theta} = \alpha T^3$$

T is in °K, α is a constant, and r is in terms of lb moles of formaldehyde formed per lb catalyst/hr. If the first bed is 6 in. deep, what is the approximate depth of the sixth bed?

Notes: Heat of vaporization of methanol is roughly 10 kcal/g mole. For this problem, assume MC_p for O_2, N_2, CO, H_2O, CH_3OH, and CH_2O all gaseous are, respectively, 7, 7, 7, 10.5, 21, and 14 cal/(g mole) (°C), while $\Delta H°$ for the formation reactions are 0, 0, −28, −58, −48, and −27.7 cal/g mole.

PROBLEM 9-12

A possible new use for metallic arsenic is under study by a development group. If the work is successful, then the necessary metal is to be obtained by hydrogen reduction of granular arsenic trioxide, of which a large inventory is on hand:

$$As_2O_3 + 3H_2 = \tfrac{1}{2}As_4 + 3H_2O(g)$$

It is suspected that rates of this reaction will become significant over about 650°K.

You are asked to propose a flow sheet for an operation to produce 10 tons per day of the metal, indicating preferred operating conditions and probable fuel needs. Some detail on the reactor and product recovery system should be provided, especially since a colleague remarks "a fluid bed operation may lead to undesirably high oxide volatilizations into the product gas." This statement deserves brief discussion. In designing the system, it is to be remembered that the toxicity hazards of arsenic are great, making it important to minimize contamination and to monitor all effluent streams from the plant.

Arsenic melts at 1090°K, develops sublimation pressures of 10 and 760 mm Hg at 710 and 883°K, respectively, and exists primarily as As_4 in the gas phase below 1000°K.

Arsenic trioxide melts at 586°K, has vapor pressures of 10, 40, 100, and 760 mm Hg at 533, 572, 605.5, and 730.3°K. Thermodynamic data are presented below (70, 162) with the elements and compounds existing in their normal states of aggregation (gaseous, liquid, or solid) at the indicated temperature and 1 atm total pressure.

Compound	$\Delta H^\circ_{298°K}$ Formation	ΔF° Formation, kcal/g mole 800°K	900°K	1000°K
As_2O_3	−157	−115.0	−107.85	−102.7
H_2O	− 57.8	− 48.64	− 47.35	− 46.03

Substance	$(H_T - H_{298°K})$, kcal/g mole 800°K	900°K	1000°K
H_2	3.52	4.23	4.94
H_2O	4.30	5.25	6.21
As_2O_3(c)	33	37.5	42.4
As_4	9.7	11.67	13.64

Arsenic oxide exists as As_4O_6 in the gas-phase, has a heat of vaporization of 14.4 kcal/g mole of As_4O_6 at 733°C.

PROBLEM 10-1

The Shell process for ethanol production (14, 105) involves the direct combination of ethylene and water vapor over 85% phosphoric acid as catalyst at 570°F and 66 atm. This phosphoric acid catalyst is carried on $\frac{1}{4} \times \frac{1}{4}$ in. pellets of cellite, a calcined diatomaceous earth consisting primarily of porous aluminum silicate. Acid stronger than 85% favors ether production while weaker acid is a less effective catalyst. Temperatures over 570°F benefit rates but also cause acid decomposition, so that P_2O_5 as well as water vapor appears in the vapor phase. At 570°F, 85% phosphoric acid exerts a water vapor pressure of 35 atm, with no acid or P_2O_5 in the vapor phase (18). An increase in water vapor pressure over 35 atm at 570°F therefore causes the acid to become more dilute, while a decrease causes it to become more concentrated.

Assume the hydration of ethylene to be the only reaction occurring:

$$C_2H_4 + H_2O(g) = C_2H_5OH(g)$$

The equilibrium constant for this reaction is as follows (25):

$$\log_{10} K = \frac{2132}{T} - 6.241$$

where K is in reciprocal atm and T is in °K. A simplified flow sheet of a plant operating with an ethylene feed containing 3% N_2 is presented in Fig. P-15. The off-gases leaving the reactor are to be in equilibrium with the foregoing reaction and with 85% phosphoric acid. The gases leaving the condenser are at 25°C, and the overall alcohol yield on the ethylene fed is 97%. The solubility of ethylene and nitrogen in the liquid condensate may be assumed negligible. The normal vapor pressures of pure water and pure ethyl alcohol at 25°C are 23.8 and 59 mm Hg, respectively.

Assuming the perfect gas laws:

(a) The mole fractions of alcohol and water vapors in the recycle gases at

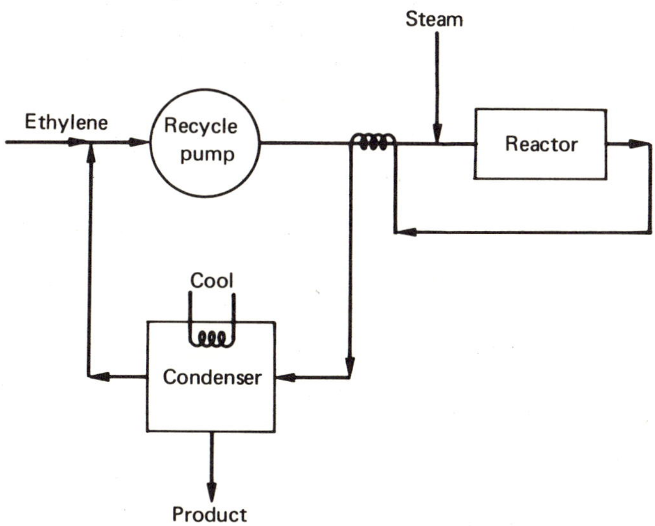

Fig. P-15. Ethyl alcohol production for Prob. 10-1.

25°C leaving the condenser are considered to be negligible. Is this a justifiable assumption?

(b) What is the composition of the reactor off-gases?

(c) How many moles of gas leave the reactor per mole of fresh 97% ethylene fed to the system?

(d) What is the weight % alcohol in the water-alcohol mix leaving the condenser?

Optional. Repeat the foregoing problem, recognizing that the real gas laws apply. Values of fugacities, the critical constants, etc., can be obtained from the usual references.

PROBLEM 10-2

The two important barium ores are witherite ($BaCO_3$), found in Europe, and barytes, found in the United States. Both materials are extremely insoluble, and are converted to more reactive forms (the oxide) or a water-soluble form (the sulfide) for use. Like other carbonates, barium carbonate decomposes thermally to the oxide and CO_2:

$$BaCO_3 = BaO + CO_2$$

Unfortunately, impractically high temperatures are required for this reaction because of relatively low dissociation pressures: 0.1 atm at 1482°K, 0.25 atm at 1558°K, 0.5 atm at 1622°K, 1.0 atm at 1694°K, and 10 atm at 1994°K (68). To reduce these decomposition temperatures to more manageable levels, a pure carbon such as carbon black is mixed with the witherite, whereupon the mixture is heated indirectly in a retort. The following reaction occurs along with the barium carbonate decomposition (22):

$$C + CO_2 = 2CO$$

$$\Delta F^\circ = 40{,}740 - 41.7T \text{ kcal/g mole of } CO_2$$

(a) What are the equilibrium partial pressures of CO and CO_2 over a mixture of $BaCO_3$ and solid carbon at 1482°K? At what temperature is the total dissociation pressure over this mixture equal to 1 atm?

(b) It has been suggested that BaO might be made in a fluid bed operation burning methane and air. Comment.

The heats of formation at 25°C of CO, CO_2, H_2O, and CH_4 are −26.4, −94, −57.8, and −17.9 kcal/g mole, respectively. Barium hydroxide, which melts at 690°K, has a dissociation pressure of 0.072 atm at 1022°K and of 1.0 atm at 1261°K [Reference (61), Vol. III, 299]. Dissociation of barium oxide to the elements is negligible over the temperatures considered here. Values of K_p for the "water-gas" reaction are presented in Prob. 1-1.

PROBLEM 10-3

Butadiene is produced today largely by the catalytic dehydrogenation of butanes and the butenes. Operations are not simple, since (a) temperatures around 1100°F are required, leading to undesired by-product formation due to cracking, (b) large amounts of relatively high temperature heat must be supplied to the endothermal dehydrogenation reactions, and (c) partial pressures of the butanes and butenes in the reaction mixture must be kept low to attain reasonably attractive equilibrium yields. Alternative reaction schemes having fewer disadvantages are being sought, with promise being shown by the so-called iodative dehydrogenations (74). These involve the gas-phase reaction of the hydrocarbons with elemental iodine to form the iodides, followed by splitting HI out of these iodides to form olefins. Iodine is unique among the halogens in that rates of both halogenation and dehydrohalogenation are large around 1000°F. The overall reactions may be written:

$$C_4H_{10} + I_2 = C_4H_8 + 2HI \qquad \log K_{800^\circ K} = 0.346 \qquad \Delta H^\circ_{800^\circ K} = 22 \text{ kcal}$$

$$C_4H_8 + I_2 = C_4H_6 + 2HI \qquad \log K_{800^\circ K} = 0.27 \qquad \Delta H^\circ_{800^\circ K} = 25 \text{ kcal}$$

To force these reactions to completion as well as to reduce iodine requirements per pass through the reactor, oxygen or air is mixed with the reaction gases, whereupon elemental gaseous iodine is regenerated:

$$2HI + \tfrac{1}{2}O_2 = H_2O + I_2 \qquad \log K_{800^\circ K} = 11.63$$

$$\Delta H^\circ_{800^\circ K} = -55.7 \text{ kcal}$$

The foregoing reaction is reportedly faster than the hydrocarbon combustion reactions at 1000°F, consequently little hydrocarbon oxidation occurs.

Iodine is costly, and as such must be retained in the plant system. A liquid scrubbing system for iodine recovery from a dilute hydrocarbon mixture would require a large and costly battery of columns. An attractive alternative is to remove iodine quantitatively from the hot gases by reaction with a nonvolatile solid "acceptor" like nickel oxide:

$$\text{NiO(c)} + 2\text{HI(g)} = \text{NiI}_2\text{(c)} + \text{H}_2\text{O(g)} \qquad \Delta H^\circ_{800^\circ K} = -35.2 \text{ kcal}$$

$$\log K_{800^\circ K} = 3.25$$

Little vapor pressure information is available for nickel iodide, but it appears that these pressures become significant only above about 500°K. Nickel iodide reacts with oxygen at a reasonable rate even below 1000°F to form elemental iodine and NiO:

$$NiI_2 + \tfrac{1}{2}O_2 = NiO + I_2 \qquad \Delta H^\circ_{800^\circ K} = -20 \text{ kcal} \qquad \log K_{800^\circ K} = 8.2$$

Recovery operations of butadiene from the reactor off-gases in the dehydrohalogenation plant just described would be similar to those of orthodox butane dehydrogenation plants. That is, these gases contain hydrogen, methane, some C_2's and C_3's, butanes, butenes, butadiene, etc., from which the C_4's and higher are separated by absorption and stripping, followed by butadiene recovery by extractive distillation with a material such as furfural or acetonitrile.

Propose a flow sheet for a plant for iodative dehydrogenation of butanes to butadiene, using NiO as an "acceptor." This flow sheet need not be extended beyond the point where the dilute reactor off-gases enter the butadiene recovery section of the plant. Indicate heat requirements at all points in the process, and how these are to be met. To the extent possible, show relative stream magnitudes, and discuss operating problems.

PROBLEM 10-4

It is suggested by a clever engineer that the Solvay reaction (10-33) might somehow be carried out using the ion-exchange resins described in Chap. 3. You are asked to discuss this suggestion, presenting possible flow sheets. Estimate the energy requirements of your best process, in Btu/lb of sodium carbonate product.

PROBLEM 10-5

In a Solvay plant with a flow sheet as in Fig. 10-2, it is found that the gases leaving the cleanup towers analyze 10% CO_2, 2% CO, the rest N_2. The purge gas analyzes 1.2% CO, 48.8% CO_2, the rest N_2. All the gases from the decomposer enter the precipitating tower in this operation, along with recycled kiln top gases, and no gases other than the purge leave the system until they leave the cleanup towers.

As in so many industries, a special terminology has developed in the Solvay industry. Thus any ammonia which can be volatilized from filtered aqueous solutions by simple boiling is called *free*, and such free ammonia is present as NH_3, NH_4OH, NH_4HCO_3, and/or $(NH_4)_2CO_3$. "Fixed ammonia" is present as the salt of a strong acid, such as NH_4Cl, and cannot be volatilized from an aqueous solution of the salt by simple boiling. "Total ammonia" is the sum of the free and fixed ammonia.

The liquor from the bottom of the ammonia still containa no ammonia and is approximately neutral. This ammonia still functions correctly, so that lime is used here to free only the fixed ammonia. The clear liquors from the bicarbonate filter entering the ammonia still analyze (for each g mole Cl^- ion) 0.3 g moles Na^+ ions, 0.9 total g moles NH_3, and 0.15 g moles CO_2. The Na_2CO_3 produced contains no chloride.

(a) How could the liquors from the bicarbonate filter have been analyzed?
(b) For each mole of chloride ion fed to the NH_3 still, how many g moles of Na_2CO_3 are produced?
(c) The ammonia still is operating correctly. What is the ratio of NH_3 to CO_2 in the off-gases from this still?
(d) How many tons of $CaCO_3$ are consumed per ton of Na_2CO_3 produced?
(e) How many tons of coke (10% of ash) are consumed per ton of Na_2CO_3 produced?

PROBLEM 10-6

Acetylene is a rather unstable material, with standard free energies and enthalpies of formation from the elements at 25°C of +54.2 kcal and +50 kcal/g mole, respectively. Acetylene at room temperature and pressure will decompose into its elements when in contact with a hot surface, but this decomposition is limited to the vicinity of the hot surface and is not violent. As the gas temperature or pressure is increased, conditions are quickly reached where spontaneous detonations can occur. Additions of diluent gases like water vapor and hydrocarbons increase threshold temperatures and pressures below which detonations do not occur. The probability of explosions also decreases as the size of the container is decreased. Thus, acetylene is safe in the pressure cylinders of commerce (about 230 psig at 70°C) because these are completely filled with a porous mass of cellular material, consequently no pockets of appreciable size exist containing "free" gaseous acetylene. To increase capacity these cylinders contain acetone, in which acetylene is highly soluble. Acetylene cylinders are filled using slow-running piston compressors, built with heavy walls capable of withstanding large pressure shocks (100).

Acetylene in cylinders represents less than 10% of the shipping weight, and these cylinders must, of course, be returned empty for reuse. Calcium carbide may be viewed as an inexpensive throw-away container for acetylene, in that over $\frac{1}{3}$ lb of acetylene can be generated per lb of carbide. Commercially, about 3000 kw-hr of electricity are consumed per ton of carbide (80% of CaC_2, the rest is CaO) produced.

Acetylene is made industrially not only from calcium carbide, but also by the high temperature cracking of hydrocarbons. One method of generating the high temperature heat needed for these endothermal cracking reactions is by passing the hydrocarbons through an electric arc. Exposure to elevated temperatures must be for a controlled and brief (0.01 sec) time period, since the acetylene so formed in turn rapidly decomposes into its elements at these temperatures. Quenching is therefore essential, and is by direct injection of a water spray into the moving gas stream at a suitable distance downstream from the electric arc.

The off-gases from a German arc plant operating on methane are reported to analyze 13% C_2H_2, 52% H_2, 1% C_2H_4, 25% CH_4, 2% CO, and 7% N_2. These gases leaving the arc at 1600°C are water-quenched to below 700°C at which temperature no further reactions occur.

Recovery of acetylene from the product gas mixture is by scrubbing with a solvent such as acetone or dimethylformamide. A young engineer suggests that this acetylene might be more advantageously recovered by direct conversion to calcium carbide, to be made by passing the gases before or after quenching through a bed of lime. If necessary, this lime could be mixed with carbon to improve conversion:

	kcals (39)		
	$\Delta F^\circ_{1000^\circ K}$	$\Delta F^\circ_{1920^\circ K}$	$\Delta H^\circ_{1950^\circ K}$
$C_2H_2 + CaO = CaC_2 + H_2O$	+20.3	+14.6	−24.8
$H_2O + C = CO + H_2$	− 1.9	−32.8	+31.9

Comment on this proposal. If you think it deserves study, what laboratory or pilot investigations should be made?

PROBLEM 10-7

(a) For esterification reactions, the numerical values of ΔF° are close to zero, making the values of the associated equilibrium constants close to unity. Thus for the liquid phase reaction of ethanol with acetic acid, the "equilibrium constant" K is reported to vary with C, the initial mole ratio of acetic acid to acetic acid plus ethanol as follows Reference (75), Vol. 8, 321:

$$K = 1.31 + 4.86C$$

The amount of mineral acid added to the system as catalyst also shifts the equilibrium composition somewhat. Explain how and why this equilibrium constant is not independent of composition at constant temperature. In the liquid phase, esterifications of equimolal quantities of ethanol with acetic acid, 65.6% of the reagents have reacted at equilibrium at 100°C to form ethyl acetate, and 67.3% at 200°C.

Estimate the enthalpy change per mole of ester formed. What assumptions are involved in this estimate?

(b) Esterification reactions can be driven to completion despite being "equilibrium reactions" by removing one or both of the products of reaction, namely, ester and water. Complete conversion of ethanol and acetic acid to ethyl acetate is reportedly attained in the following continuous process, which is incompletely described to you as follows: an equimolal mixture of acetic acid and ethanol, plus a little sulfuric acid to serve as catalyst, is fed to the middle plate of a fractionating column called the reactor. The distillate contains alcohol, water, and ester, and is fed to the middle plate of a second column called the separator, from which the ternary azeotrope leaves overhead. The bottoms from the separator, containing alcohol and water, are returned to the esterification column at a plate below the feed. Water is added to the separator distillate, whereupon two immiscible liquid layers form because of the limited mutual solubility of water and the ester. The ester-rich top layer, containing perhaps 10% of water and alcohol in solution, is fed to a third fractionating column called the drier, from which the ester product stream is obtained.

Draw a flow sheet of this operation. What is the composition of the bottoms from the reactor, and what is probably done with it? Can dry ester be produced in the drier? Does this dry ester leave as bottoms or as distillate? What is done with the water-alcohol stream from this column? What happens to the bottoms of the separating tower? It is maintained that a large excess of alcohol can be maintained in the reactor, despite equimolal feed of alcohol and acid to the system, thus benefiting kinetics. Do you agree?

(c) It is pointed out that both rate and equilibrium conversion would be benefited by having acetic acid rather than ethanol in excess within the reactor tower. Draw and justify a flow sheet whereby this could be attainable with equimolal feed of ethanol and acetic acid to the plant. Is there any basis for preferring the scheme of part (b) above?

Notes: Atmospheric boiling points and compositions of various reagents are as follows: 77.2°C for ethyl acetate, 78.3°C for ethanol, 118.8°C for acetic acid, and 100°C for water. For the azeotropes (57):

	Weight %			
	Ethanol	Ester	Water	Boiling Point, °C
ethanol-water	95.6	—	4.4	78.15
ethanol-ethyl acetate	30.8	69.2	—	71.8
ethyl acetate-water	—	91.8	8.2	70.4
alcohcol-ester-water	9.0	83.2	7.8	70.3

PROBLEM 10-8

The use of selective catalysts sometimes makes possible the operation of a sequence of reactions, each at a different temperature, whereby an overall reaction having an unfavorable constant can nevertheless be made to proceed far toward completion. Thus the "reforming" of CH_4 with steam to hydrogen and CO_2 has a favorable constant only at elevated temperatures, where, however, the "water-gas shift reaction" also always occurs:

$$CH_4 + 2H_2O = CO_2 + 4H_2 \qquad K_{p,\,1100°F} = 1.07 \qquad K_{p,\,800°F} = 0.0024 \tag{a}$$

$$H_2O + CO = CO_2 + H_2 \qquad K_{p,\,1100°F} = 2.65 \qquad K_{p,\,800°F} = 9.03 \tag{b}$$

The desired high conversion of methane to hydrogen obtainable by reaction (a) is diminished by the unavoidable formation of CO and H_2O by reaction (b). A selective catalyst for the shift reaction (b) exists, however, in the form of iron and zinc oxides. Industrially, the high temperature gases from the reformer are cooled, more steam is added, and the mixture passed over a shift catalyst which accelerates reaction (b), but has little effect on reaction (a). The adverse effect of reaction (b) on hydrogen production is therefore reversed in the shift reactor.

(a) Gas suitable for ammonia synthesis was made in the older TVA facilities (16) by reacting steam with natural gas (96% CH_4, 4% C_2H_6) at 21 psia to yield a gas analyzing (dry basis) 74.3% H_2, 0.8% N_2, 11.7% CO_2, 10.5% CO, and 2.7% CH_4. The nitrogen in this gas somehow resulted from leak of air into the system. How many moles of steam were fed per mole of methane entering the reformer? Do these analytical data appear to be internally consistent? The actual off-gas temperature is reported as 1100°F. If the off-gases were at equilibrium with respect to reaction (a), what would the temperature have been? Repeat for reaction (b). Was the catalyst in this reformer "selective"?

(b) The reformer product gas is now mixed with more steam and with flue gases from burning more of this natural gas with air. This mixture passes through a shift converter, from which it emerges at 20 psia and 800°F, analyzing (dry basis) 59.9% H_2, 21.3% N_2, 16.8% CO_2, 1.9% CO, and 0.1% N_2. What probably happened to the methane left in the gas leaving the reformer? The nitrogen content here is obviously below the levels at which it can be accurately determined. This gas may be assumed at equilibrium with respect to reaction (b). Is it at equilibrium with respect to reaction (a)? Is the shift catalyst used "selective"?

(c) Approximate the gas analysis which would have been attained, had a nonselective catalyst been used in the shift reactor, allowing both reactions to reach equilibrium.

Note: In modern plants, improved materials of construction have made possible the use of higher temperatures and pressures in the reformer, with economies attained as a consequence (75).

PROBLEM 10-9

Calcium cyanamide ($CaCN_2$), an important fertilizer material, is made in a two-step process. First, a mixture of lime and coke are heated to a high temperature in an electric furnace to form calcium carbide:

$$CaO + 3C = CaC_2 + CO \qquad \Delta F°_{298°K} = 95.4 \text{ kcal} \qquad \Delta H°_{298°K} = 111 \text{ kcals}$$

The carbide produced is inevitably contaminated with the ash carried in the coke. This impure carbide is cooled, ground to fine particle size, and then reheated to about 1000°K. Upon passing nitrogen gas through this hot carbide, cyanamide forms:

$$CaC_2 + N_2 = CaCN_2 + C \qquad \Delta F°_{298°K} = -56 \text{ kcals}$$

It is evident that, in this two-step process, the free energy released by the second reaction is not utilized in the first. Moreover, the cyanamide produced is contaminated not only by impurities in the carbide used, but also by the carbon formed in the reaction.

It has been proposed that an alternative route to cyanamide might involve reacting $CaCO_3$ with ammonia at around 1600°F (83):

$$CaCO_3 + 2NH_3 = CaCN_2 + 3H_2O(g) \qquad K_{p,\,1600°F} = 0.027 \qquad \text{(a)}$$

This reaction utilizes cheap, abundant raw materials, avoids high temperature electrothermal processing, and would produce a pure "water white" pure cyanamide of interest for chemical synthesis.

A skeptical associate doubts the feasibility of reaction (a). "Ammonia is very unstable at high temperature, wants to decompose to the elements, and will no doubt do so long before reaction (a) occurs. The following reaction sequence is then indicated:

$$2NH_3 = N_2 + 3H_2 \qquad K_{p,\,1600°F} = 1.65 \times 10^7 \qquad \text{(b)}$$

$$N_2 + 3H_2 + CaCO_3 = CaCN_2 + 3H_2O \qquad \text{(c)}$$

The equilibrium constant for reaction (c) is obviously most unfavorable, so that no $CaCN_2$ will form. It is therefore a waste of time and money to explore these reactions."

A laboratory study was nevertheless undertaken, in which a gas at atmospheric pressure analyzing 25% CO_2 and 75% NH_3 was passed through a heated fixed bed of $CaCO_3$ at about 1600°F (33). At a relatively early stage in the analysis, a "grab" sample of the off-gas was found to analyze (wet basis) 19% NH_3, 12% H_2O, 66% CO_2, 1% N_2, 0.6% H_2, 1.0% CO, and 0.8% HCN. (There was some concern over the accuracy of analysis for components present at concentrations less than 1%.) Calcium cyanamide was present in the solid product at the end of the run. It was stated that the following reactions also occurred:

$$CaCO_3 = CaO + CO_2 \tag{d}$$

$$H_2 + CO_2 = H_2O + CO \tag{e}$$

$$CO + NH_3 = HCN + H_2O \tag{f}$$

Equilibrium constants and approximate values of $\Delta H°$ for all of these reactions except (c) are

Reaction	$\Delta H°$, kcal	K_p	
	1600°F	1600°F	1700°F
(a)	+60	0.027	0.074
(b)	+27	1.65×10^7	2.83×10^7
(d)	+39.6	0.70	1.55
(e)	+8.0	1.238	1.457
(f)	+11.5	0.0123	0.0154

(a) With regard to the pilot run just described, were any of reactions (a), (b), (c), (d), (e), and (f) near equilibrium? Briefly discuss the remarks of the critical associate in the light of these findings.

(b) At the time of the gas analysis, how many moles of solid $CaCN_2$ were being produced/mole of gas feed? Of CaO? Estimate the fraction of the ammonia in the feed which was decomposed to the elements.

(c) Is it thermodynamically possible for the CaO produced in this operation to be converted to $CaCN_2$? It was suggested that production of CaO would be avoided by adding enough CO_2 to the feed to "stall" reaction (d). Do you agree?

(d) Propose a flow sheet, including a suitable reactor, indicating how the heat of these endothermal reactions should be supplied, how materials efficiency should be maximized, etc. Ammonia, pure limestone, natural gas, and coke are available.

PROBLEM 11-1

In a Japanese plant for making potassium bromate electrolytically (110) a current of 500 amp is passed through 24 cells in series, with 90 V overall. Electrolyte is fed continuously to the system at a constant rate of 32 l/hr. This flow is split equally between two banks of 12 cells with series flow through the cells in each bank. The electrolyte feed (pH around 9) contains 215 g/l of KBr, 22 g/l of $KBrO_3$, and 2 g/l of $K_2Cr_2O_7$, which acts as a corrosion inhibitor. The product liquors from the two cell banks are mixed, and analyze 90 g/l of KBr and 270 g/l of $KBrO_3$. This liquor leaves at the cell temperature of 70°C, and is cooled to precipitate $KBrO_3$. After adding water and fresh KBr, this liquor is recycled to the cell.

The cells themselves are concrete boxes, with stainless steel cathodes and lead dioxide-coated iron anodes for corrosion resistance. No diaphragm is used in these cells, so that the caustic potash produced combines with the bromine evolved to form the desired bromate ions. Reaction is encouraged by electrolyte mixing caused by gases evolved at the electrodes. To minimize the hazard of hydrogen explosions, a large volume of air is blown continuously through the cells over the electrolyte surface, diluting the off-gases to far below the explosive limits.

(a) Chromate is not decomposed under the cell operating conditions. What is the chromate concentration in the mixed product liquor from the cell banks?

(b) The reactions producing bromate are stated to be as follows:

anode: $Br^- = \frac{1}{2}Br_2 + e$

cathode: $e + K^+ + H_2O = K^+ + OH^- + \frac{1}{2}H_2$

solution: $6OH^- + 3Br_2 = BrO_3^- + 5Br^- + 3H_2O$

Are there any other half-cell reactions which deserve mention? These reactions are analogous to those presented earlier in the text of this chapter for caustic-chlorine cells. What is the current efficiency of bromate production?

(c) How many standard liters per hr of oxygen are probably produced at the anode?

PROBLEM 11-2

A continuously operating caustic soda chlorine cell of the diaphragm type is not performing satisfactorily, in that the product caustic soda liquor is contaminated with chlorate and carbonate, while the anode gas contains relatively large amounts of impurities. The current consumption is 3700 amp at 4.6 V, while the cell is fed with 70 l/hr of the usual saturated brine (sp. gr. 1.18) analyzing 24% NaCl, 0.1% HCl by weight. The product liquor from the cathode compartment has a density of 1.14 g/cc and analyzes 16.3% NaCl, 5.6% NaOH, 0.12% Na_2CO_3, and 0.14% $NaClO_3$. The anode gas analyzes (dry basis) 72% Cl_2, 5.3% H_2, 0.8% CO_2, 7.1% O_2, and the rest is N_2. The cathode gas is all hydrogen.

(a) How many g equivalents of chlorine appear per hour as molecular chlorine in the off-gas? What is the current efficiency of this chlorine production?

(b) Present explanations for each of the following in the anode gas: N_2, O_2, CO_2, and H_2. Explain the presence of both carbonate and chlorate in the cell liquors. How can the high cell voltage be explained? What recommendations for improved operations can be made?

(c) How many equivalents of current per hour are represented at the anode by the following: the Na_2CO_3 and $NaClO_3$ is the product liquor, the Cl_2, O_2, and CO_2 in the anode gas? How nearly do these account for the total current flowing?

(d) How many equivalents of current per hour at the cathode are represented by the following: OH^- to neutralize the H^+ in the feed; the OH^-, $CO_3^=$, and ClO_3^- in the product liquor; the OH^- made (and simultaneously destroyed at the cathode) while producing O_2 and CO_2 at the anode? How nearly do these account for the total current flowing?

PROBLEM 11-3

In caustic soda-chlorine cells of the diaphragm type, saturated brine is fed continuously to the anode compartment, passes through the porous diaphragm into the cathode compartment, and then on out of the cell. Because of the vigorous stirring resulting from the gas bubbles rising through the electrolyte on both sides of the diaphragm, both anolyte and catholyte may be considered well mixed, and so concentration gradients in the idealized cell exist only across the diaphragm. The brine flow velocity through the diaphragm must exceed the back-migration velocity of the OH^- ions from the catholyte to the anolyte side of the diaphragm. If brine flow through the diaphragm falls below this velocity, then OH^- ions find their way back into the anolyte, react with the chlorine gas bubbles and generally reduce cell efficiency.

The diaphragm cell under discussion is designed to operate at 3.6 V and 175°F with a current density of 70 amp/sq ft. This current density is equal for anode, cathode, and diaphragm, since all three have equal areas in this cell. The diaphragm is of matted asbestos fiber $\frac{1}{8}$ in. thick, with 50% voids, and is located between anode and cathode, which are spaced about 1.5 in. apart. The hot product, cell liquor analyzes 8.7% NaOH, 15.3% NaCl by weight, with a density of 1.12 g/cc. The pure NaCl brine feed to the anode compartment has a density of about 1.18 g/cc.

Ion migration velocities in an electrolyte are proportional to the voltage gradient. Hydroxide ions have a mobility in aqueous systems of 0.0021/cm/sec at 25°C for a voltage gradient of 1 V/cm, and about twice this mobility at 175°F. The solution leaves the anode compartment by flowing through the diaphragm pores and has a specific resistivity of 0.2 ohms/cc at 70°C.

(a) For each sq ft of electrode area, how many lb/hr of brine are fed to the cell?
(b) Assuming piston flow, what is the average linear velocity of the brine moving through the pores of the diaphragm?
(c) The voltage drop through the solution between the electrodes of this cell is obviously (3.6–2.7) or 0.9 V and this voltage drop serves to move the ions. Recognizing that the diaphragm is 50% voids, estimate the voltage gradient in V/cm through the diaphragm.
(d) What is the average migration velocity of OH^- ions through diaphragm pores? Compare this with the answer to part (b).
(e) Could a greater percentage conversion of NaCl to NaOH be attained in this cell? By altered cell design? Discuss.

PROBLEM 11-4

In some localities, it is economically attractive to electrolyze by-product aqueous HCl to the elements. The deNora type cell assembly developed for continuous operation (2, 52) contains 50 cells, each approximately 6 ft × 6 ft × 3 in. These cells are stacked together face to face as the leaves in a filter press with the large faces vertical. The separator between each pair of neighboring cells is an impermeable graphite sheet which acts as a "bipolar electrode," in that one side is the anode of one cell and the other side is the cathode of the neighboring cell. The two end plates of the cell stack act as terminal anode and terminal cathode for the assembly. Such a battery of 50 cells in series typically operates at 115 V overall, with current densities of 100 amp/sq ft.

A porous membrane of polyvinyl chloride, which is relatively impermeable to gas but permeable to ions, is inserted between every anode and cathode, separating each cell into an anode compartment and a cathode compartment. Hydrochloric acid 30% by weight enters the bottom of each cell through appropriate feed holes, rises on both sides of each membrane, and leaves through appropriate spill holes at the top. Electrolyte mixing in the anode and in the cathode compartments is vigorous due to gas evolution, consequently the product liquor leaves each cell at the concentration of the electrolyte within the cell. This concentration is adjusted to that for maximum electrical conductivity, namely, 18% by weight HCl. This spent acid is again brought up to 30% strength by cycling through an HCl absorption tower, and then back to the cells.

(a) Sketch your impressions of the details of the deNora type cell-stack construction.
(b) A typical installation of 50 deNora cells operates at 3500 amp and 115 V overall, produces 5.65 short tons of chlorine per day, with a cell temperature of 80°C. What is the average voltage efficiency of these cells? The current efficiency? What are the possible reasons why the current efficiency is below 100%?
(c) Considering the cell battery to be completely adiabatic, at what temperature will the feed liquor enter the cell, if the product liquor leaves at 80°C?

The standard free energy and standard enthalpy of formation of HCl in aqueous solution analyzing 18% by weight of HCl are about —26 and —38 kcal/g mole, respectively, at 80°C. The enthalpy change of condensing 1 mole of water vapor into a sufficiently large quantity of this solution so that no significant concentration change occurs is about 10.5 kcal/g mole. The enthalpy change of dissolving 1 mole of gaseous HCl in a sufficiently large amount of 18% solution so that no significant concentration change recurs is about 16 kcal/g mole (exothermal). The partial vapor pressures of H_2O and HCl over this solution analyzing 18% by weight of HCl are 240 and 12 mm Hg, respectively, at 80°C. The heat capacities of the acid liquors may be taken as 0.75 cal/(°C)(g).

PROBLEM 11-5

In the electrolysis of an aqueous HCl solution, the H^+ ions are found to have five times the migration velocity of the Cl^- ions. In electrolyzing an HCl solution, therefore, the liquor concentrations at the anodes tend to diminish more rapidly than at the cathodes. Wherever the anolyte concentrations fall to much under 10%, electrochemically produced oxygen begins to appear in the anode off-gas, which reduces current efficiencies and complicates chlorine recovery. Desirably, all liquor concentrations are kept at 18% by weight HCl, since this is the concentration of maximum electrical conductivity (52). An advantage of operating cells on a continuous flow basis, rather than batchwise, is that these desired electrolyte concentrations can be readily maintained constant. It should be remembered that the vigorous evolution of gas from the electrodes on both sides of the cell diaphragm helps keep the anolyte and the catholyte well mixed in each cell.

(a) A 30% by weight HCl solution is fed separately to all the anode compartments and to all the cathode compartments of the deNora cells of the preceding problem. The product liquors from each analyze 18% by weight of HCl. Can this be accomplished with an equal feed rate to each compartment?

(b) This battery of 50 cells produces 5.65 tons of chlorine daily. To maintain an 18% strength in both anolyte and catholyte products, estimate the feed rate needed to each anode compartment in lb of a 30% HCl solution per hr. What is the proper feed to each cathode compartment? Assume no liquor flow through the diaphragm. Vapor pressure information is presented in the preceding problem.

PROBLEM 11-6

Severe electrode corrosion problems have been encountered in cells for direct electrolysis of HCl to chlorine and hydrogen. It has been suggested (125) that these difficulties might be avoided in the electrolysis of a polyvalent metal chloride, reducing the metal from a higher to a lower valence state at the cathode while generating chlorine at the anode. The lower chloride in solution is withdrawn from the cell, converted back to the higher chloride by reactions with HCl and air, and recycled to the cell. As illustrated below for the copper system, the sum of these two reactions is the Deacon reaction:

$$CuCl_2 \longrightarrow CuCl + \tfrac{1}{2}Cl_2 \qquad \text{(a)}$$

$$CuCl + HCl + \tfrac{1}{4}O_2 \longrightarrow CuCl_2 + \tfrac{1}{2}H_2O \qquad \text{(b)}$$

$$\overline{HCl + \tfrac{1}{4}O_2 \longrightarrow \tfrac{1}{2}Cl_2 + \tfrac{1}{2}H_2O}$$

While iron or chromium chlorides can also be used in reactions of type (a) and (b), hydrogen production during the electrolysis can be totally avoided at the cathode only with the copper salt, hence copper chloride is the preferred reagent.

Oxidation of the spent cuprous cell liquors to the cupric state by reaction (b) involves acidifying with HCl and then bubbling air through this liquor at 90°C and normal pressures. Bubble towers (see Chap. 10 for description) and packed towers are stated to be not very satisfactory reactors because of limited gas–liquid interfacial areas. Preferred operation involves using turbo mixer to disperse air thoroughly into the cuprous liquor and then recycling the oxidized liquor to the cells.

The advantage claimed for this electrolytic process over the classic gas phase Deacon process (see Prob. 1-4) is that the chlorine generated is pure, not mixed with nitrogen and unreacted oxygen, thus avoiding the need for costly recovery equipment. Advantages over direct HCl electrolysis are (1) less energy consumption per unit of chlorine produced because of lower voltages, (2) no serious corrosion at the cell anode, (3) simpler cell construction because no cathode gas is generated, and (4) smaller cell construction because higher current densities are practical.

An experimental cell (125) contains flat, parallel anodes and cathodes but no diaphragm. The anodes are solid sheets of dense graphite, located on either side of, and parallel to, the flat cathode made of porous graphite. Regenerated cupric chloride solution is fed directly into the narrow space (apparently less than 1 in. wide) between anode and cathode. The product liquor is withdrawn from the cell through the porous cathode, which has a hollow center for easier liquor flow. By such removal, the cuprous chloride leaves the cell almost as rapidly as it is formed, and recombination by contact with the chlorine gas bubbling up through the electrolyte is minimized. Current efficiencies are reported to be lower than when a diaphragm is inserted between electrodes to prevent contact of the chlorine gas bubbles with the cathode, but the increased voltage drop through the diaphragm reportedly more than offsets this gain in current efficiency.

An experimental cell having cathode face area of 10 sq ft operates at 80°C at 1.8 V and 4000 amp. When feeding 70 gal/hr of an aqueous solution weighing 10.4 lb/cu ft and containing 15% $CuCl_2$, 20% HCl by weight, the product liquors contain 3.7% CuCl, 10.2% $CuCl_2$, and 20.2% HCl. The cell off-gases are reported to analyze 86% Cl_2, 2% HCl, and 12% H_2O on a weight basis.

(a) What was the current efficiency of this operation? Explain why the current efficiency was not higher.

(b) It was found that at constant current with increasing electrolyte flow, the current efficiency increased to a maximum and then decreased again. Propose an explanation for this finding.

PROBLEM 11-7

Industrial aluminum cells operate open to the atmosphere at about 1000°C. The cell container is a steel box lined with carbon blocks. The cathode is a layer of molten aluminum covering the bottom of the box, on which floats an electrolyte layer of molten cryolite (Na_3AlF_6) containing a little dissolved alumina. Carbon anodes dip down into this electrolyte from above, with the flat end faces of these electrodes parallel to and kept a short distance above the aluminum cathode surface. The cell feed is alumina (Al_2O_3), which from time to time is added directly to the electrolyte surface as a fine powder of carefully controlled particle size.

The alumina feed dissolves in the cryolite. Molten aluminum metal forms at the cathode, while oxygen discharged at the anode reacts with the carbon electrode to form a mixture of CO and CO_2. Fresh electrode carbon must be continuously supplied to replace that lost by oxidation, and this is done by slowly moving the electrodes downward as they are consumed.

Molten metal and the anode off-gases are the two products of the cell. The gas escapes continuously, but the metal is harvested only at intervals by aspirating from the molten aluminum cathode through a suitable pipe.

(a) The anode reactions are of particular interest, and three different explanations for the appearance of the CO — CO_2 mixture are presented:

I. The primary anode reaction is as follows:

$$O^{=} + C = CO + 2e$$

The CO formed then decomposes in part:

$$2CO = C + CO_2$$

II. The primary reaction is as follows:

$$O^{=} + \tfrac{1}{2}C = \tfrac{1}{2}CO_2$$

Part of the CO_2 formed then attacks the electrode:

$$CO_2 + C = 2CO$$

III. The primary reaction is as in II, but part of the CO_2 formed is reduced to CO by back reaction with the aluminum metal, some of which is always present as tiny droplets in the electrolyte.

To determine which of these three explanations is correct, a test of rather short duration was made in a small experimental cell (114). The carbon anode, carefully protected from air oxidation, was weighed before and after the run, and was found to have lost 0.312 lb of carbon. During this period, 1271 amp-hr of current were consumed at 5.7 V, while cell temperature was 1000°C. No carbon fines were detected anywhere within the electrolyte system during the operation. The free energies of formation of Al_2O_3, CO, and CO_2 at 1000°C are −307, −94.6, and −53.5 kcal/g mole, respectively.

(a) It is maintained that these findings show that explanation III is the correct one. Do you agree?

(b) Current efficiencies in commercial cells are found by comparing the weight of metal made over a prolonged time period with the current consumed. Such a time period will inevitably include variations in electrolyte compositions, temperatures, electrode geometry, etc. Current efficiencies determined in this way are obviously unsatisfactory either for control or for study of the effect of operating variables on cell performance.

It has been proposed (101) that, if mechanism III above is correct, current efficiencies at any moment can be calculated from the gas analysis alone. The equation proposed is as follows:

$$\% \text{ current efficiency} = \frac{\% CO_2}{2} + 50$$

where $\% \ CO_2$ is the per cent of CO_2 in the anode off-gas. Derive this equation, and comment.

(c) Large cell operators have objected to this equation for cell efficiency calculation, pointing out that air leaks into a cell, and that gas analysis varies somewhat across a cell.

Recognizing these difficulties, what was the efficiency of a large cell whose off-gases were found to analyze 50% of N_2, 8% of O_2, 3.5% of CO, and the rest is CO_2? Are the objections of the cell operators valid?

PROBLEM 11-8

Steel is iron containing usually from 0.10–0.40% by weight of carbon, manganese, cobalt, etc. These alloying ingredients, though present in small concentration, have a very large effect upon steel's physical properties such as strength, ductility, and hardness.

In industry today, steel is produced in the form of very large ingots, which are reduced to size by squeezing in enormous rolling mills. To make the metal sheet 5.5 mils thick suitable for tin plate requires a total of about 300 kw-hr/ton in the various manufacturing and rolling operations.

It is proposed that a sheet of iron 5.5 mils thick might be made continuously by electrolytic means. The composition of this metal sheet might then be adjusted in some subsequent heat treatment. It is claimed by one engineer that the cell could be fed directly with the strong electrolyte solution made by dissolving iron ore in the acid cell liquors. It is maintained that electrolytically produced sheet might compete successfully with that made today by orthodox routes.

The feasibility of being able to design a cell which will produce a sheet of iron continuously by electrolysis from an aqueous iron solution is to be discussed. The ultimate goal is to make a sheet $\frac{1}{12}$ in. thick and at least 6 ft wide at a rate of at least 1 ft/sec. You are asked to prepare a memorandum which prepares the way for a discussion within the research and development department of your company. In this memorandum you are to raise and comment on the critical technical questions at issue, and to indicate what, if any, initial steps should be undertaken in the laboratory for preliminary screening of this proposal. Consideration is to be limited to the preparation of the iron sheet, with subsequent adjustment of its composition to be considered in a later meeting.

PROBLEM 11-9

In the electrolysis of water, as pointed out in the text, hydrogen is evolved at the cathode and oxygen at the anode by the half-cell reactions (11-3a) and (11-3b). These reactions occur in the reverse direction in a hydrogen-oxygen fuel cell, forming water while generating electrical energy. Extensive progress has been made in developing the so-called porous gas diffusion type electrodes for these gas-consuming type half-cell reactions. For operations below 100°C in an aqueous electrolyte, a conducting electrode material has been developed comprising a corrosion-resistant wire cloth coated with a porous layer of carbon black and platinum black particles bonded with Teflon. This porous cloth, having a thickness of perhaps 1 mm, is usually positioned vertically in a cell, separating the electrolyte on one side from the reagent gas on the other. The electrolyte penetrates only a small way into the pores, despite a hydrostatic head of 2 ft and more, because of the unfavorable contact angle of aqueous solutions against Teflon. On the other hand, the reagent gas (hydrogen or oxygen, depending on whether the anode or cathode is discussed), easily passed through the pores to the meniscus of the liquid. This gas dissolves in the electrolyte in the pores and diffuses to the submerged platinum black surfaces where the half-cell reactions occur. Typically, the oxygen electrode may show a polarization of 0.3 V at 200 amp/sq ft, while the hydrogen electrode is perhaps one-tenth as great.

Before such gas diffusion type electrodes become available, the following claim appears in a patent (64): "A process for producing oxygen at a rate upwards of 340 g/kw-hr of electrical energy which comprises electrolyzing between an anode and an air depolarized cathode an aqueous solution essentially comprising potassium hydroxide, mingling catholyte . . . with the anolyte, and collecting the oxygen thereby produced." The cathode used here was a porous carbon cylinder, into which oxygen or air was pumped.

(a) An associate says that this cell is really an oxygen concentration cell, with oxygen supplied at the cathode and produced at the anode. He claims that it does not differ in principle from a copper refining cell. Do you agree? Your associate says further that such a cell could be made to serve as an oxygen compressor, would be isothermal and have no moving parts. Do you agree?

(b) What is the theoretical energy requirement of this cell, in grams of oxygen produced per kw-hr? Compare with the 340 g/kw-hr mentioned in the claim. Compare theoretical and actual voltages.

(c) Propose a system design, including the cell design and associated equipment of all types for producing pure oxygen at atmospheric pressure from air by this operation. Use as electrode material the newer porous, gas diffusion cloth material described earlier.

PROBLEM 11-10

The following excerpts are from a U.S. patent (15):

"Various methods have heretofore been proposed to reduce the cost of operation of electrolytic cells, especially those for the production of alkali metal hydroxides, such as caustic soda and chlorine, by reducing the amount of electrical energy necessary to operation of such cells. The energy requirements are, in general, the sum of the potential necessary to discharge chlorine at the anode, overcome polarization of the anode, overcome resistance of the electrolyte, overcome polarization at the cathode, and discharge hydrogen at the cathode. Especially since hydrogen is a by-product of the electrochemical reactions involved, and is considered to be of less value than the basic products of chlorine and, in the case of sodium chloride electrolysis, caustic soda, it has been appreciated for some time that, if the main products of the reactions involved could be obtained without production of hydrogen, a considerable saving could be effected."

Claim number 1 of this patent reads:

"An electrolyte cell for the electrolysis of aqueous solutions of electrolytes including a container for the electrolyte, one of the vertical walls thereof formed by a permeable diaphragm, an anode within the container, a cathode outside the container in juxtaposition to the diaphragm, and means to pass electric current between said anode and said cathode, said cathode including a porous, oxygen-activating, water-repellent conductor of electricity, and a discontinuous nonelectrically active solution-wettable spacer between said cathode and said diaphragm, said spacer comprising vertically running, laterally spaced, asbestos cords which extend below the lower edge of both the diaphragm and the porous member."

(a) Propose a sketch of this cell which could be the basis for pilot studies. What are the electrode reactions indicated, and the overall cell reaction?

(b) What advantages and disadvantages, if any, are in prospect relative to the usual type of diaphragm cell used in caustic soda production?

PROBLEM 11-11

Baizer's success in electrolytic hydrodimerization of acrylonitrile (AN) to adiponitrile (ADN) encouraged further investigations along similar lines. Arad et al. (3) studied the nonelectrolytic synthesis of ADN from AN, using as reducing agent an amalgam containing 0.009 g moles of sodium metal per 100 g of mercury. This amalgam was dropped into 15 cc of an acidified aqueous 5% AN solution contained in a suitable vessel kept at 5°C by external cooling. When adding enough amalgam to react with 30% of the AN present, the sodium was found to react completely, with all three of the following reactions occurring simultaneously:

$$2Na(Hg) + 2CH_2{:}CHCN + 2HCl = 2NaCl + CN(CH_2)_4CN \quad \text{(a)}$$

$$2Na(Hg) + CH_2{:}CHCN + 2HCl = CH_3CH_2CN + 2NaCl \quad \text{(b)}$$

$$2Na(Hg) + 2HCl = H_2 + 2\,NaCl \quad \text{(c)}$$

Reaction (b) making propionitrile (PN) predominated at low acidities. As acid strength increased, PN production diminished while both ADN and H_2 production increased, with maximum conversion of 25% of the sodium to ADN attained at 18% acid, another 70% of the sodium forming H_2 by reaction (c), and the remainder forming PN by reaction (b). Acid conditions appeared essential, since no ADN formed in neutral solution.

New perspectives on the amlagam process were afforded by Matsudo (90) who reported preparing a solution of 35.3 g of acrylonitrile plus a trace of *p*-methoxyphenol, 9.0 g of water, and 70.3 g of 1,3-dimethylurea. This solution was placed in a 500-ml four-necked flask containing 50 ml of mercury. An amalgam containing 0.511% by weight of sodium was now added dropwise over a 15-min period while maintaining the temperature at 25°C, and bubbling CO_2 into the systems to keep the pH at 8 to 9. The solution was then diluted with water and the nitriles extracted with dichloromethane, from which the nitriles were recovered by evaporation and drying. Chromatographic analysis showed a yield of 3.08 g ADN and 0.12 g PN. It was found that dimethylurea could be successfully replaced by other polar solvents such as methylformamide, dimethyl sulfoxide, and tetraethylammonium *p*-toluenesulfonate.

(a) Matsuda reports that the yield of ADN was 96.2% based on acrylonitrile, and 84.3% based on the sodium used. How much amalgam was added, if all the sodium reacted?
(b) What appear to be the critical differences between Matsuda's work and that of Arad?
(c) Assume that large quantities of amalgam containing 0.5% by weight

sodium could be made available from a mercury caustic soda-chlorine plant. Propose a suitable flow sheet for making ADN from such an amalgam, and discuss. Acrylonitrile boils at 77.6°C, has a water solubility of 8.0% by weight at 24°C (water solubility in AN at 24°C is about 3% by weight), is extremely reactive and must be stabilized for storage with hydroquinone, etc., to avoid polymerization.

Bibliography

1. Andrussov, L., *Angew. Chem.*, **48** (1955), 593.

2. Anonymous, *Chem. Eng.*, **74** (1964), 172.

3. Arad, Y., M. Levy, I. R. Miller, and D. Vofsi, *J. Electrochem. Soc.*, **114** (1967), 899.

4. Asselin, G. F., "Advances in Petroleum Technology and Refining," Vol. 9, New York: Wiley-Interscience Publishers, 1964.

5. Atroschenko, V. L., and I. P. Kushnarenko, *Intern. Chem. Eng.*, **4** (1964), 581.

6. Baddour, R. F., "Process Ion Exchange," to be published.

7. Baizer, M. M., *J. Electrochem. Soc.*, **111** (1964), 215, 223.

8. BIOS Report No. 1650.

9. Bodenstein, M., *Trans. Electrochem. Soc.*, **51** (1927), 365.

10. ———, *Z. Physik. Chem.* (*Leipzig*), **100** (1922), 68.

11. Boldt, J. R., and P. Quineau, "The Winning of Nickel," Van Nostrand Co., 1967.

12. Boynton, D. E., W. B. Nichols, and H. M. Spurlin, *Ind. Eng. Chem.*, **51** (1959), 489.

13. Brooks, A., *J. Am. Chem. Soc.*, **715** (1953), 2464.

14. Brown, E. H., and C. D. Whitt, *Ind. Eng. Chem.*, **44** (1952), 615.

15. Butler, Clarence A., "Electrolytic Cell," U.S. Patent 2,681,887 (June 22, 1954).

16. Burt, R. B., *Ind. Eng. Chem.*, **46** (1954), 2479.

17. Butte, Charles A., "Manufacture of Defluorinated Dicalcium Phosphate," U.S. Patent 2,442,969 (May 16, 1946).

18. Carle, T. C., and D. M. Stewart, *Chem. Ind.* (*London*), 1962, 830.

19. Carberry, J. J., *Chem. Eng. Sci.*, **9** (1959), 189.

20. Chilton, T. H., "The Manufacture of Nitric Acid," *Chem. Eng. Progr. Monograph Series*, **56**(3) (1960).

21. Cooper, E. D., D. F. Black, and C. B. Anderson, *Chem. Eng. Progr.*, **61** (1965), 89.

22. Coughlin, J. P., "Contributions to the Data on Theoretical Metallurgy," *U.S. Bur. Mines Bull.*, **542** (1954).

23. Daroux, W. G., *J. Appl. Chem.* (*London*), **3** (1953), 241.

24. De Donder, T., and G. Van Lerberghe, *Bull. Sci. Acad. Roy. Belg.*, **12**, Ser. 5 (1926), 151.

25. Dodge, B., "Chemical Engineering Thermodynamics," New York: McGraw-Hill, Inc., 1944, 491.

26. Dunlop, D. D., L. I. Griffin, and J. F. Moser, *Chem. Eng. Progr.*, **54** (1958), 39.

27. Dupasquier, J., *Ann. Fac. Sci. Univ. Marseille*, **21** (1951), 155.

28. Eklund, R. B., "The Rate of Oxidation of Sulfur Dioxide with a Commercial Vanadium Catalyst," Uppsala: Almqvist & Wiksell, 1956.

29. Elliott, John F., and Molly Gleiser, "Thermochemistry for Steelmaking," Reading, Mass.: Addison-Wesley Publishing Co., 1960.

30. Eucken, A., and M. Jakob, "Der Chemie-Ingenieur," Leipzig: Akademische Verlagsgesellschaft MBH, 1937.

31. Fair, J. R., *Chem. Eng.*, **74** (1967), 67.

31a. Fariss, R. H., "Catalysis in Practice," A symposium, London: Institution of Chemical Engineers, June 1963, p. 51.

32. Fedotieff, P. P., *Z. Angew. Chem.*, **44** (1904), 1644.

33. Feinman, S., *Ind. Eng. Chem., Process Design Develop.*, **3** (1964), 241.

34. Fergussen, R. R., and T. Gabor, *J. Electrochem. Soc.*, **111** (1964), 585.

35. Ferm, R. J., *Chem. Rev.*, **57** (1957), 621.

36. Field, E., "Formaldelyde Synthesis," U.S. Patent 2,504,402 (April 18, 1950).

37. Field, J. H., H. E., Benson, G. E. Johnson, J. S. Tosh, and A. J. Forney, *U.S. Bur. Mines Bull.*, **597** (1962).

38. Findlay, A., A. N. Campbell, and N. D., Smith, "The Phase Rule," New York: Dover Publications, Inc., 1951.

39. Gamson, B. W., and R. H. Elkins, *Chem. Eng. Progr.*, **49** (1953), 203.

40. Glenn, R. A., E. E. Donath, and R. J. Grace, *Am. Chem. Soc. Div. Fuel Chem., Preprints*, **10**(4) (1966), 97.

41. Gilliland, E. R., *Chem. Can.*, **2** (1950), 1.

42. Gilliland, E. R., E. A. Mason, and R. C. Oliver, *Ind. Eng. Chem.*, **45** (1953), 117.

43. Glass, W. B., "Solid Catalyzed Reaction in a Fluidized Bed," Ph.D. Thesis, Cambridge: M.I.T., 1956.

44. Glassner, Alvin, "Thermochemical Properties of the Oxides, Fluorides and Chlorides to 2500°K," *Argonne Natl. Lab., U.S. At. Energy Comm.*, **ANL 5750**, 1957.

45. Glasstone, Samuel, "An Introduction to Electrochemistry," 10th ed., Princeton, N. J.: D. Van Nostrand Co., 1942.

46. "Gmelins Handbuch der Anorganischen Chemie," 8th ed., Weinheim, Verlag Chemie, G.m.b.H., 1961.

47. Goldsmith, R. L., "Liquid Phase Catalytic Oxidation of Ethylene to Acetaldehyde," S.B. Thesis, Cambridge: M.I.T., 1962.

48. Goldstein, R. F., and A. L. Waddams, "The Petroleum Chemicals Industry," 3rd ed., London: E. & F. N. Spon, Ltd., 1967.

49. Graham, J. J., and P. F. Way, and S. Chase, *Chem. Eng. Progr.*, **58** (1962), 96.

50. Groll, H. P. A., and G. Hearne, *Ind. Eng. Chem.*, **31** (1939), 1530.

51. Gross, P., Aluminum Congr., Paris 1954, **1**, 167.

52. Hampel, C. A., "Encyclopaedia of Electrochemistry," New York: Reinhold Co., 1964.

53. Harriott, Peter, *Chem. Eng.*, **68** (1961), 81, 165.

54. Hinricks, H., and J. Niedetzky, *Angew. Chem. Int. Ed.*, **1** (1962), 206.

55. Hollingsworth, C. A., and J. C. Williams, U.S. Patent 2,839,377, "Defluorination of Phosphate Rock" (Feb. 28, 1951).

56. Homer, H. W., *J. Soc. Chem. Ind.* (*London*) *Trans.*, **60** (1941), 213.

57. Horsley, L. H., "Azeotropic Data," *Advan. Chem. Ser.*, **6** (1952).

58. Hottel, H. C., *Symp. Combustion 4th, Cambridge, Mass., 1952*, Baltimore: Williams & Wilkins Co., 1953, Chap. 9.

59. Hottel, H. C., and W. R. Hawthorne, *3rd Symp. Combustion, Madison. Wis.*, Baltimore: Williams & Wilkins Co., 1949.

60. Howard, W. B., B. H. Wood, and E. J. Kaltenbacker, *Chem. Eng. Progr.*, **57** (1961), 70.

61. "International Critical Tables," 1st ed., New York: McGraw-Hill, Inc., 1928.

62. "JANAF, Thermochemical Data," Midland, Mich.: Dow Chemical Co., 1960.

63. Janecke, Ernst, *Anorg. Chem.*, **51** (1906), 132; **52** (1907), 358; **53** (1907), 319.

64. Janes, M., "Electrolytic Method for Producing Oxygen," U.S. Patent 2,390,591 (Dec. 5, 1945).

65. Johnstone, H. F., *Chem. Eng. Progr.*, **44** (1948), 657.

66. Johnstone, H. S., L. Foering, T. Yu-Sheng, and G. H. Messerly, *J. Am. Chem. Soc.*, **73** (1951), 2319.

67. Kauter, C. T., and W. Leitenberger, *Chem. Eng. Tech.*, **25** (1953), 697.

68. Kelley, K. K., and C. T. Anderson, "Contributions to the Data on Theoretical Metallurgy: IV Metal Carbonates," *U.S. Bur. Mines Bull.*, **348** (1935).

69. Kelley, K. K., "Contributions to the Data on Theoretical Metallurgy: 7, The Thermodynamic Properties of Sulfur," *U.S. Bur. Mines Bull.*, **406** (1937).

70. ———, "Contributions to the Data on Theoretical Metallurgy-High Temperature Heat, Content Heat Capacity and Entropy Data," *U.S. Bur. Mines Bull.*, **584** (1960).

71. ———, "Energy Requirements and Equilibria of Magnesium Chloride," *U.S. Bur. Mines Tech. Paper*, **676** (1945).

72. Kellogg, H. H., *Trans. AIME*, **236** (1966), 602.

73. Keyes, F. G., and H. Horo, *J. Am. Chem. Soc.*, **44** (1922), 479.

74. King, R. W., *Hydrocarbon Process. Petrol. Refiner*, **45** (11). (1966), 189.

75. Kirk, R. E., and D. F. Othmer, "Encyclopedia of Chemical Technology," 1st and 2nd eds., New York: John Wiley & Sons, Inc., 1965.

76. Kitchener, J. A., "Ion Exchange Resins," London: Methuen & Co. Ltd., 1957.

77. Knibbs, N. V. S., "Lime and Magnesia," London: Ernest Benn, Ltd., 1924.

78. Kobe, K. A., and H. R. Crawford, *Petrol. Refiner*, **37** (1958), 7, 125.

79. Kolbel, H., H. Langemann, and J. Platz, *Dechema Monograph*, **41** (1962), 225–244.

80. Kubachewski, O., and E. L. Evans, "Metallurgical Thermochemistry," London: Butterworth-Springer Ltd., 1951.

80a. Kusik, L. C., private communications.

81. Landau, R., and H. Harper, *Chem. Ind. (London)*, **1961**, 1143.

82. Lebedev, B. G., and V. A. Levitskii, *Russ. J. Phys. Chem.*, **35** (1961), 1380; and **36** (1962), 329.

83. Letts, R. W. M., "Process Evaluation by Chemical Equilibrium Computation," Buffalo Meeting, A.I.Ch.E., 1962.

84. Levenspiel, O., "Chemical Reactor Engineering," New York: John Wiley & Sons, Inc., 1962.

85. Lewis, W. K., E. R. Gilliland, and T. K. Roy, *Trans. Indian Inst. Chem. Engrs.*, **7** (1954–1955), 33.

86. Lewis, W. K., A. H. Radasch, and H. C. Lewis, "Industrial Stoichiometry," 2nd ed., New York: McGraw-Hill, Inc., 1954.

87. Locker, L. D., "Reaction of Silicon with Hydrogen Chloride in a Fluidized Bed," S.M. Thesis, Cambridge: M.I.T., 1963.

88. Mantell, C. L., "Electrochemical Engineering," 4th ed., New York: McGraw-Hill, Inc., 1960.

89. Margolin, S. V., and L. V. Gallagher, "The Production of Crystobalite from Quartz in a Fluid Bed," *Chem. Eng. Progr. Symp., Ser.* **62** (62), (1966), 56–62.

90. Matsudo, Fujo, *Tetrahedron Letters*, **49** (1966), 6193.

91. McAdams, W. H., "Heat Transmission," 3rd ed., New York: McGraw-Hill, Inc., 1954.

92. McGannon, Harold E., "Making Shaping and Treating of Steel," 8 ed., Pittsburgh, Pa.: U.S. Steel Corp., 1964.

93. McKee, R. H., and J. R. Heard, Jr., *Trans. Electrochem. Soc.*, **65** (1934), 301, 327.

94. McKewan, W. M., *Trans AIME*, **218** (1960), 2.

95. Meissner, F., E. Schwiedessen, and D. P. Othmer, *Ind. Eng. Chem.*, **46** (1954), 727.

96. Meissner, H. P., C. L. Kusik, and W. H. Dalzell, *Ind. Eng. Chem., Fundamentals*, **8** (1969), 659.

97. Meissner, H. P., and E. M. Thode, *Ind. Eng. Chem.*, **43** (1951), 129.

98. Meissner, H. P., and F. C. Schora, "Process for Producing Lead," U.S. Patent 3,281,237 (Oct. 25, 1966).

99. ———, *Trans. AIME*, **224** (1962), 460.

100. Miller, S. A., "Acetylene, Its Manufacture, Properties and Uses," New York: Academic Press, Inc., 1965.

101. Minn, J. C., *J. Metals*, **18** (1966), 308.

102. Mittasch, A., *J. Phys. Chem.*, **40** (1902), 1.

103. Moe, J. M., *Chem. Eng. Progr.*, **58** (1962), 33.

104. Motoo, W., *Sci. Rept. Tohoku Univ., Ser. 1*, **22** (1933), 1229.

105. Nelson, C. R., and K. L. Courter, *Chem. Eng. Progr.*, **50** (1954), 529.

106. Newton, R. H., and B. F. Dodge, *J. Am. Chem. Soc.*, **55** (1933), 4747.

107. Nicholls, P., "Underfeed Combustion, Effect of Preheat and Distribution of Ash in Fuel Beds," *U.S. Bur. Mines Bull.*, **378** (1934).

108. Oliver, G. D., and S. E. J. Johnson, *J. Am. Chem. Soc.*, **76** (1954), 4721.

109. Olsen, J. C., and O. Direnga, *Ind. Eng. Chem.*, **33** (1941), 204.

110. Osuga, T., and K. Sugino, *J. Electrochem. Soc.*, **104** (1957), 448.

111. Othmer, R. F., K. Kon, and T. Igarishi, *Ind. Eng. Chem.*, **48** (1956), 1258.

112. Park, J. D., A. F. Benning, F. B. Downing, J. F. Laucius, and R. C. McHarness, *Ind. Eng. Chem.*, **39** (1947), 354.

113. Pearson, T. G., "Chemical Background of the Aluminum Industry," *Roy. Inst. Chem. (London), Lectures, Monographs, Rept.* No. 3, 1955.

114. Pearson, T. G., and J. Waddington, *Discussions Faraday Soc.*, No. 1 (1947), 307.

115. Pidgeon, L. M., and J. A. King, *Trans. Faraday Soc.*, **96** (4) (1948), 197.

116. Pings, C. J., *Chem. Eng. Progr.*, **59**(12) (1963), 90; *J. Am. Inst. Chem. Engrs.*, **10** (1964), 934; *Ind. Eng. Chem., Fundamentals*, **4** (1965), 260.

117. Pohlenz, J. B., and L. O. Stine, *Hydrocarbon Process. Petrol. Refiner*, **41** (1962), 191; also **43** (1964), 232.

118. Potter, Edmund C., "Electrochemistry, Principles and Applications," London: Cleaver-Hume Press Ltd., 1956.

119. Potter, C., and S. Baron, *Chem. Eng. Progr.*, **47** (1951), 473.

120. Preuner, G., and W. Schupp, *Z. Physik. Chem.* (Leipzig), **68** (1909), 129.

121. Purdon, F. F., and V. W. Slater, "Aqueous Solutions and the Phase Diagram," London: Edward Arnold & Co., 1946.

121a. Rehani, D. N., T. K. Marajanan, and L. K. Doraiswamy, *Ind. Eng. Chem., Process Design Development*, **4** (1965), 403.

122. Reid, R. C., and T. K. Sherwood, "Properties of Gases and Liquids," New York: McGraw-Hill, Inc., 1958.

123. Reinders, W., *Z. Anorg. Chem.*, **93** (1915), 202.

124. Reinders, W., and F. Goudriaan, *Z. Anorg. Allgem. Chem.*, **126** (1923) 85.

125. Riley, W. L., *Hydrocarbon Process. Petrol. Refiner*, **42** (1963), 167; *Trans. Inst. Chem. Engrs.*, **37** (1959), 305.

126. Roberts, A. L., "Kinetics of High Temperature Process," ed. W. D. Kingery. New York: John Wiley & Sons, Inc., 1959.

127. Roberts, J. E., *Chem. Eng. Progr.*, **59** (1963), 88.

128. Roberts, C. P., *Chem. Eng. Progr.*, **46** (1950), 456.

129. Robinson, C. S., and E. R. Gilliland, "Elements of Fractional Distillation," 4th ed., New York: McGraw-Hill, Inc., 1950.

130. Rossini, F. D., *Am. Petrol. Inst. Res. Project 44*, "Selected Values of Properties of Hydrocarbons and Related Compounds, Tables w and x," Texas A & M Univ., College Station, Tex., 1960.

131. Rossini, F. D., "Selected Values of Chemical Thermodynamic Properties," *U.S. Natl. Bur. Std. Circ.*, **500** (1952).

132. Satterfield, C. N., and T. K. Sherwood, "The Role of Diffusion in Catalysis," Reading, Mass.: Addison-Wesley Publishing Co., 1963.

133. Schenck, R., and W. Hamplemann, *Metal Erz*, **1** (1913), 283.

134. Schönbeck, R., H. König, K. Krzemicki, and L. Kalhofer, *Angew. Chem. Intern. Ed.*, **5**(7) (1966), 642.

135. Schreinemakers, F. A. H., *Z. Physik. Chem.* (Leipzig), **11** (1893), 76.

136. Schutt, H. C., and S. B. Zdonik, "Making Ethylene I," *Oil Gas J.*, **54**(41) (1956), 98.

137. Sconce, J. S., "Chlorine, Its Manufacture, Properties and Uses," New York: Reinhold Co., 1962.

138. Seidell, A., and W. F. Linke, "Solubilities of Inorganic and Metal Organic Compounds," Princeton, N. J.: D. Van Nostrand Co., 1958.

139. Sherwood, E. M., and J. M., Jr., Blocher, *J. Metals*, **17** (1965), 595.

140. Sherwood, Peter W., *Petrol. Eng.*, **28** (1949), 129.

141. ——— *Petrol. Refiner*, **34** (3) (1955), 201.

142. ——— *Ind. Chemist*, **39** (1963), 242.

143. Sherwood, T. K., and R. L. Pigford, "Absorption and Extraction," 2nd ed., New York: McGraw-Hill, Inc., 1950.

144. Silsby, R. I., and E. W. Sawyer, *J. Appl. Chem.* (*London*), **6** (1956), 347.

145. Smidt, J., W. Hafner, R. Jera, R. Sieber, J. Sedlmeier, and A. Sabel, *Angew. Chem. Intern. Ed.*, **1** (1962), 81.

146. Sorgenti, H. A., and G. F. Sachsel, *Ind. Eng. Chem.*, **52** (1960) 101–104.

147. Stull, D. R., *Ind. Eng. Chem.*, **41** (1949), 1968.

148. Thiele, E. W., *Ind. Eng. Chem.*, **31** (1939), 916.

149. Tosh, J. S., J. H. Field, H. E. Benson, and W. P. Haynes, "Study of the System," *U.S. Bur. Mines Rept. Invest.* No. 5484 (1959).

150. Truesdale, E. C., and R. K. Waring, *J. Am. Chem. Soc.*, **63** (1941), 1610.

151. Tu, C. M., H. Davis, and H. C. Hottel, *Ind. Eng. Chem.*, **26** (1934), 749.

152. Ullmanns "Enzyklslopädie der technischen Chemie," 3rd ed., Munich-Berlin: Urban and Schwarzenberg, 1960.

153. Uspenskaya, L. N., and A. G. Bergman, *Zh. Obshch. Khim.*, **25** (1955), 1277, 2028.

154. Van Wormer, K., "The Reaction of Methane with Iron and Iron Oxides in a Fluidized Bed," Sc.D. Thesis, Cambridge: M.I.T., 1961.

155. Veatch, F., J. L. Caldahan, J. D. Idol, and E. C. Milleger, *Chem. Eng. Progr.*, **56** (1960), 65.

156. Vetter, Klaus J., "Electrochemical Kinetics, Theoretical Aspects," New York: Academic Press, Inc., 1967.

157. Wagman, D. D., "Selected Values of Chemical Thermodynamic Properties," *U.S. Natl. Bur. Std. Tech. Note*, **270** (1968).

158. Walker, J. F., "Formaldelyde," 3rd ed., New York: Reinhold Publishing Corp., 1964.

159. Wenner, R. R., and E. C. Dybal, *Chem. Eng. Progr.*, **44** (1948), 275.

160. Wesselhoft, R. D., J. M. Woods, and J. M. Smith, *J. Am. Inst. Chem. Engrs.*, **5** (1959), 361.

161. White, W. B., S. M. Johnson, and G. B. Dantzig, *J. Chem. Phys.*, **28** (1958), 751.

162. Wicks, C. E., and F. E. Block, "Thermodynamic Properties of 65 Elements," *U.S. Bur. Mines Bull.*, **605** (1963).

163. Wood, R. K., and W. F. Stevens, *J. Appl. Chem.* (*London*), **14** (1964), 325.

164. Zdonick, S. B., E. J. Green, and L. R. Hallee, *Oil Gas J.*, **64** and **65** (Nov. 1966), 15 articles.

165. Zeleznick, Frank J., Gordon, Sanford, *Ind. Eng. Chem.*, **60** (1968), 6, 27.

166. Zimmerman, C. C., and R. York, *Ind. Eng. Chem.*, *Process Design Develop.*, **3** (1964), 254.

167. Zwolinski, Bruno J., "Selected Values of Properties of Chemical Compounds," Thermodynamics Res. Center, Texas A & M Univ., College Station, Tex., Dec. 31, 1967.

Subject Index

Problem Index

* Problems occur on pages 237–374. Number refers to the number of the problem in which the topic is treated.